Cryptosporidiosis of Man and Animals

Editors

J. P. Dubey
Zoonotic Diseases Laboratory
Livestock and Poultry Sciences Institute
Agricultural Research Service
U.S. Department of Agriculture
BARC-East
Beltsville, Maryland

C. A. Speer
Veterinary Molecular Biology
Montana State University
Bozeman, Montana

R. Fayer
Zoonotic Diseases Laboratory
Livestock and Poultry Sciences Institute
Agricultural Research Service
U.S. Department of Agriculture
BARC-East
Beltsville, Maryland

CRC Press
Taylor & Francis Group
Boca Raton London New York

CRC Press is an imprint of the
Taylor & Francis Group, an **informa** business

First published 1990 by CRC Press
Taylor & Francis Group
6000 Broken Sound Parkway NW, Suite 300
Boca Raton, FL 33487-2742

Reissued 2018 by CRC Press

Library of Congress Cataloging-in-Publication Data

Cryptosporidiosis of man and animals/editors, J. P. Dubey, C. A. Speer, R. Fayer
p. cm.
Includes bibliographical references.
ISBN 0-8493-6401-9
1. Cryptosporidiosis. I. Dubey, J. P. II. Speer, C. A. III. Fayer, R.
RC136.5.C79 1990
616.9'36—dc20 90-1611

A Library of Congress record exists under LC control number: 90001611

Publisher's Note
The publisher has gone to great lengths to ensure the quality of this reprint but points out that some imperfections in the original copies may be apparent.

Disclaimer
The publisher has made every effort to trace copyright holders and welcomes correspondence from those they have been unable to contact.

ISBN 13: 978-1-315-89216-0 (hbk)
ISBN 13: 978-1-351-07126-0 (ebk)

Visit the Taylor & Francis Web site at http://www.taylorandfrancis.com and the
CRC Press Web site at http://www.crcpress.com

PREFACE

Cryptosporidium has recently emerged as the most prevalent and widespread intestinal parasite. At least one species infecting man and virtually all other mammals, *Cryptopsporidium parvum,* is ubiquitous, zoonotic, and pathogenic. Other species infect birds, fish, and snakes. All known species and their host relationships are discussed.

Although the genus *Cryptosporidium* was named in 1910, it was not considered a pathogen until 1971, when it was found associated with diarrhea of calves. The first human infections were reported in 1976, but not until the severity of infections in patients with acquired immunodeficiency syndrome (AIDS) dramatized the need for information on this parasite was there a widespread stimulation of research efforts. In the 1980s, over 800 papers have been published on diagnosis, the life cycle, transmission, epidemiology, immunity, pathogenesis, chemotherapy, disinfection, cultivation, and other areas.

In attempting to provide a current and comprehensive review of cryptosporidiosis in man and animals, this book is divided into 10 chapters. Chapter 1 is introductory and provides a general review of history, taxonomy, structure, life cycle, pathogenesis, pathology, clinical signs and symptoms, immunity, epidemiology, treatment, prevention, and control. Chapter 2 emphasizes techniques including diagnostic methods for identifying the tissue and fecal stages, the maintenance of the parasite in the laboratory, collections, purification and preservation, preparation of antigens, and serological techniques. Chapter 3 deals with *Cryptosporidium* in surface waters and in drinking water. Successive chapters address cryptosporidiosis in various hosts including man, cattle and other ruminants, rodents, dogs, cats and other mammals, pigs, horses, fish and reptiles, and birds. There is an exhaustively complete bibliography and cross-referenced index. Individual chapters are written by specialists with first-hand knowledge of the subject area.

This book attempts to provide a broad coverage of current information needed by public health workers, physicians, veterinarians, parasitologists, technicians, and various biologists who encounter or work with this parasitic disease.

J. P. Dubey
C. A. Speer
R. Fayer

THE EDITORS

J. P. Dubey, B.V.Sc. and A.H., M.V.Sc., Ph.D., received his veterinary degree in 1960 and Masters in Veterinary Parasitology in 1963 from India. He obtained a Ph.D. in Medical Microbiology in 1966 from the University of Sheffield, England. He obtained postdoctoral training with Dr. J. K. Frenkel, Department of Pathology and Oncology, University of Kansas Medical Center, Kansas City, from 1968 to 1973. From 1973 to 1978, he was Associate Professor of Veterinary Parasitology, Department of Pathobiology, Ohio State University, Columbus. He was then Professor of Veterinary Parasitology, Department of Veterinary Science, Montana State University, Bozeman, from 1978 to 1982. He is presently Project Leader, Zoonotic Diseases Laboratory, Livestock and Poultry Sciences Institute, Agricultural Research Service, U.S. Department of Agriculture, Beltsville, MD.

Dr. Dubey has spent over 25 years researching *Toxoplasma* and related coccidian parasites of man and animals. He has published over 350 refereed research papers in international journals. In 1985 he was chosen to be the first recipient of the Distinguished Veterinary Parasitologist Award by the American Association of Veterinary Parasitologists. In 1990, Dr. Dubey received the Ralston Purina Small Animal Research Award in parasitology and the Beltsville Agricultural Research Center Scientist of the Year Award.

C. A. Speer, Ph.D., received a B.S. degree in zoology from Colorado State University in 1967, and M.S. and Ph.D. degrees in zoology from Utah State University in 1969 and 1972. During 1972 to 1973, Dr. Speer was Assistant Professor of Histology at the University of Texas Medical Center, Houston. From 1973 to 1975, he was a research associate of the U.S.-A.I.D. Malaria Immunity and Vaccination Project at the University of New Mexico, Albuquerque. From 1975 to 1983, Dr. Speer was Assistant, and then Associate Professor of Microbiology at the University of Montana, Missoula. During 1983 to 1986, he was Associate Professor and Director of the electron microscope facility in the Department of Veterinary Science/Veterinary Research Laboratory at Montana State University, Bozeman. In 1986, Speer was appointed Professor and Head of the Department of Veterinary Science and Director of the Veterinary Research Laboratory, and in 1989, Head of Veterinary Molecular Biology at Montana State University.

Dr. Speer has published more than 130 refereed research papers on the *in vitro* cultivation, immunity, ultrastructure, pathogenesis, and experimental infections of coccidian, trypanosomal, and malarial parasites. His current research is aimed at identifying, characterizing, and cloning coccidian antigens that induce protective immune responses. Dr. Speer received the Mershon Award for Excellence in Research in 1986 from the Montana Academy of Sciences, and Honorary Alumnus Award in 1987 from the College of Natural Sciences, Colorado State University, Fort Collins.

R. Fayer, Ph.D., received his B.S. degree in 1962 from the University of Alaska, and his M.S. and Ph.D. degrees in zoology from Utah State University in 1964 and 1968. From 1968 to 1972, he was Zoologist with the Coccidiosis Project, Beltsville Parasitology Laboratory, Agricultural Research Service (ARS), U.S. Department of Agriculture (USDA), Beltsville, MD. He was Project Leader for the *Sarcocystis* Project from 1972 to 1978 at the Animal Parasitology Institute (API), ARS, USDA, and then Laboratory Chief for Ruminant Parasites Laboratory, API, ARS, USDA, from 1978 to 1984. He was Director of the API, ARS, USDA, from 1984 to 1988 and now serves as Research Leader, Zoonotic Diseases Laboratory, Livestock and Poultry Sciences Institute, ARS, USDA, Beltsville, MD.

Dr. Fayer has spent over 25 years researching *Cryptosporidium* and related coccidia of animals and man. He has published over 175 papers in scientific journals. He received the USDA Superior Service Award and Medal as well as the American Society of Parasitologists' H. B. Ward Medal. He is an internationally recognized scientist specializing in research on protozoan diseases.

ACKNOWLEDGMENTS

We wish to thank Joan Haynes, Judy Sirk, and Eva Kovacs for secretarial assistance, and Robert B. Ewing and Andy Blixt for technical assistance with illustrations. This book was supported in part by funds from the Montana State University Agricultural Experiment Station (MSUAES), Bozeman, MT. MSUAES Journal Series No. J-2425.

CONTRIBUTORS

Kenneth W. Angus, B.V.M.S., F.R.C.V.S., D.V.M.
Moredun Research Institute
Edinburgh, Scotland

Byron L. Blagburn, Ph.D.
Associate Professor
Department of Pathology
College of Veterinary Medicine
Auburn University
Auburn, Alabama

William L. Current, Ph.D.
Senior Research Scientist
Infectious Disease Research
Lilly Research Laboratories
Lilly Corporate Center
Indianapolis, Indiana
and Adjunct Professor
Department of Pathology
Indiana University School of Medicine
Indianapolis, Indiana

J. P. Dubey, B.V.Sc. and A.H., M.V.Sc., Ph.D.
Project Leader
Zoonotic Diseases Laboratory
Agricultural Research Service
United States Department of Agriculture
Beltsville, Maryland

Ronald Fayer, Ph.D.
Research Leader
Zoonotic Diseases Laboratory
Agricultural Research Service
United States Department of Agriculture
Beltsville, Maryland

Charles W. Kim, Ph.D.
Professor
Departments of Microbiology and Medicine
State University of New York
Stony Brook, New York

David S. Lindsay, Ph.D.
Research Fellow
Department of Pathobiology
College of Veterinary Medicine
Auburn University
Auburn, Alabama

Lance E. Perryman, D.V.M., Ph.D.
Professor
Department of Veterinary Microbiology and Pathology
Washington State University
Pullman, Washington

Michael W. Riggs, D.V.M., Ph.D.
Assistant Professor
Department of Comparative and Experimental Pathology
College of Veterinary Medicine
University of Florida
Gainesville, Florida

Clarence A. Speer, Ph.D.
Professor and Head
Veterinary Molecular Biology
Montana State University
Bozeman, Montana

Charles R. Sterling, Ph.D.
Professor
Department of Veterinary Science
University of Arizona
Tucson, Arizona

Beth L. P. Ungar, M.D.
Associate Professor
Department of Medicine and Preventive Medicine
Uniformed Services University of the Health Sciences
Bethesda, Maryland

Steve J. Upton, Ph.D.
Assistant Professor
Division of Biology
Kansas State University
Manhattan, Kansas

TABLE OF CONTENTS

Chapter 1

GENERAL BIOLOGY OF *CRYPTOSPORIDIUM*

R. Fayer, C. A. Speer, and J. P. Dubey

TABLE OF CONTENTS

I. INTRODUCTION AND HISTORY

In 1907 Ernest Edward Tyzzer (Figure 1) clearly described a protozoan parasite he frequently found in the gastric glands of laboratory mice.[806] He recognized asexual and sexual modes of reproduction followed by spore (oocyst) formation, noting the presence of an attachment organelle at all stages. Furthermore, he described fecal transmission of spores which contaminated food. Noting that the taxonomic position was uncertain, Tyzzer nevertheless offered the name *Cryptosporidium muris*. In 1910 Tyzzer described the same organism in much greater detail, proposing *Cryptosporidium* as a new genus and *C. muris* as a new species.[807] He extended the host range from *Mus musculus* to Japanese waltzing mice and English mice and speculated that sporozoites liberated from oocysts that matured in the gastric glands might be a source of autoinfection. Although Tyzzer credited J. Jackson Clarke with the first published description of a parasite resembling *Cryptosporidium*, neither the dimensions of the organism nor the figures presented by Clarke support Tyzzer's comment.[161] Tyzzer's original description of the life cycle remains remarkably accurate based on confirmation by recent investigators using electron microscopy, except that Tyzzer interpreted the location of the developmental stages of the parasite as extracellular.

In 1912 Tyzzer identified and named a new species — *Cryptosporidium parvum*, once again providing great morphologic detail.[808] He distinguished the new species from *C. muris* by experimentally infecting mice and showing that *C. parvum* was smaller and developed only in the small intestine. He concluded that *C. parvum* was not strictly extracellular, but could not be considered intracellular because it appeared to protrude from the host cell surface. He noted that organisms of similar appearance to *C. parvum* and having the same relation to the host cells were found in the small intestine of the rabbit.

Tyzzer was first to report on avian cryptosporidiosis.[809] He found all developmental stages in the chicken cecal epithelium. They were illustrated but were not well described, probably because he thought they were *C. parvum*, which he had described previously in mice.

For nearly half a century after Tyzzer's work, *Cryptosporidium* was not regarded as economically or medically important and was therefore of interest primarily to taxonomists and serendipidists. A total of 21 species of *Cryptosporidium* found in fish, reptiles, birds, and mammals were named in large part on the assumption that each host species supported a separate species of *Cryptosporidium* (Table 1).

In 1955 the first report linking cryptosporidiosis with morbidity and mortality also named a new species — *Cryptosporidium meleagridis*.[740] Diarrhea and a low death rate were found in 10- to 14-d-old turkey poults. However, it was not until 1971 when *Cryptosporidium* was found to be associated with bovine diarrhea[607] that veterinary interest was stimulated.

In 1976 Nime and colleagues[591] as well as Meisel and colleagues[540] independently reported the first cases of human cryptosporidiosis. After that, relatively few cases were subsequently reported until 1982, when physicians in Boston, Los Angeles, Newark, New York, Philadelphia, and San Francisco reported to the Centers for Disease Control that 21 males had severe protracted diarrhea caused by *Cryptosporidium* in association with Aquired Immune Deficiency Syndrome (AIDS).[42] Medical and veterinary interest in the epidemiology, diagnosis, treatment, and prevention increased substantially thereafter throughout the world.

From 1981 through 1985 the development of several methods to clearly identify the oocyst stage in fecal samples provided health care workers and scientists with the reliable techniques for noninvasive diagnosis. These techniques enabled large-scale epidemiologic surveys, facilitated testing efficacy of chemotherapeutics and disinfectants, and permitted experimental studies on the biology of the organism.

FIGURE 1. Ernest Edward Tyzzer. (Courtesy of National Library of Medicine, Negative No. 85-205.)

II. TAXONOMY

Cryptosporidium is one of several genera of protozoans in the phylum Apicomplexa that are referred to as coccidia (Table 2).[236] Generally, coccidia develop in the gastrointestinal tract of vertebrates through all or part of their life cycle. Some require or are capable of extraintestinal asexual development and are referred to as tissue cyst-forming coccidia. These genera include *Besnoitia, Caryospora, Frenkelia, Hammondia, Neospora, Sarcocystis,* and *Toxoplasma.* The genera of coccidia that develop only in gastrointestinal or respiratory epthelium without formation of tissue cysts include *Eimeria, Isospora,* and *Cryptosporidium.* Species of *Cryptosporidium* have been named according to the host in which the parasite was found (Table 1). Recent reviews and pertinent cross transmission studies place the validity of many of these species in doubt. For example, *C. ameivae* is a nomen nudem.[458] *C. crotali, C. ctenosauris, C. lampropeltis,* and *C. vulpis* were believed to belong to the genus *Cryptosporidium* based on oocyst morphology, but recent data indicate that the stage reported as the oocyst is actually the sporocyst stage of *Sarcocystis.*[236] Another doubtful species is *C. curyi,* the oocysts of which from cats are five to six times larger than those of *C. parvum.* In fish, *C. nasorum* is the only named species and therefore remains valid pending cross-transmission studies and comparative studies

TABLE 1
Species of *Cryptosporidium*

Species	Author	Host
C. agni	Barker and Carbonnel, 1974	*Ovis aries* (domestic sheep)
C. ameivae	Arcay de Peraza and Bastardo de San Jose, 1969	*Ameiva ameiva* (lizard)
C. anserinum	Proctor and Kemp, 1974	*Anser anser* (domestic goose)
C. baileyi[a]	Currentet al., 1986	*Gallus gallus* (domestic chicken)
C. bovis	Barker and Carbonell, 1974	*Bos taurus* (ox)
C. crotali	Triffit, 1925	*Crotalus confluentus* (rattlesnake)
C. ctenosauris	Duszynski, 1969	*Ctenosaura similis* (lizard)
C. cuniculus	Inman and Takeuchi, 1979	*Oryctolagus cuniculus* (domestic rabbit)
C. curyi	Oggassawara et al., 1986	*Felis catis* (domestic cat)
C. felis	Iseki, 1979	*Felis catis* (domestic cat)
C. garnhami	Bird, 1981	*Homo sapiens* (human)
C. lampropeltis	Anderson et al., 1968	*Lampropeltis calligaster* (king snake)
C. meleagridis[a]	Slavin, 1955	*Meleagris gallopayo* (turkey)
C. muris[a]	Tyzzer, 1910	*Mus musculus* (domestic mouse)
C. nasorum[a]	Hoover et al., 1981	*Naso literatus* (fish)
C. parvum[a]	Tyzzer, 1912	*Mus musculus* (domestic mouse)
C. rhesi	Levine, 1980	*Macaca mulatta* (rhesus monkey)
C. serpentis[a]	Levine, 1980	*Elaphe guttata* (corn snake), *E. subocularis* (rat snake), *Crotalus horridis* (timber rattlesnake), *Sanzinia madagascarensus* (Madagascar boa)
C. tyzzeri	Levine, 1961	*Gallus gallus* (domestic chicken)
C. vulpis	Wetzel, 1938	*Vulpes vulpes* (European common fox)
C. wrairi	Vetterling et al., 1971	*Cavia porcellus* (guinea pig)

[a] Accepted as valid species

TABLE 2
Taxonomic Classification of *Cryptosporidium*

Classification	Name	Biological characteristics
Phylum	Apicomplexa	Apical commplex with polar rings, rhoptries, micronemes, conoid, and subpellicular microtubules
Class	Sporozoasida	Locomotion of mature organisms by body flexion, glia, or undulation
Subclass	Coccidiasina	Life cycle with merogony, gametogony, and sporogony
Order	Eucoccidiorida	Merogony present; in vertebrates
Suborder	Eimeriorina	Male and female gametes develop independently
Family	Cryptosporidiidae	Homoxenous with development just under surface membranes of host cell; oocyst without sporocysts and with four sporozoites; microgametes without flagella

From Fayer R. and Ungar, B. L. P., *Microbiol. Rev.*, 50, 458, 1986. With permission.

with similar genera such as *Epieimeria*. *C. serpentis*[457] remains as the only valid species in reptiles.

In birds, *C. meleagridis*[740] and *C. baileyi*[194] and in mammals, *C. muris*[807] and *C. parvum*[808] are accepted as distinct and valid species; *C. parvum* appears to be infectious for all species of mammals including humans.

Neither subspecies nor strains are recognized or used to designate most coccidia. However, there is a clear recognition of biological differences within some species of coccidia and some studies have noted that *C. parvum* exhibits such variations. Three isolates in experimentally infected mice showed no morphological or life cycle variation.[195] Chromosomal DNA from five isolates could not be differentiated by field inversion gel electrophoresis.[539] In contrast, two isolates produced markedly different clinical responses in experimentally infected calves[236] and five isolates were differentiated from one another by two-dimensional gel electrophoresis sporozoite protein patterns.[538] No traits of great enough variation have been reported to indicate that more than one species might exist within the taxon *C. parvum*.

III. HOST SPECIFICITY

Transmission studies have been conducted in which oocysts of *Cryptosporidium* were obtained from an animal of one species and inoculated into animals of other species to determine if the parasite was infective for the recipients. Such studies have yielded valuable information regarding the epidemiology and taxonomy of the genus. Generally, isolates of *Cryptosporidium* spp. from mammals have been infective for other mammals (Table 3).[236] It is most probable that *C. parvum* was the species transmitted in all such studies. Attempts to infect avians with oocysts from mammals has yielded mixed results. Chickens inoculated intratracheally with oocysts of *C. parvum* had light infections in the trachea, larynx, and primary bronchi, and small numbers of oocysts were found in feces of a few birds.[474] Chickens inoculated orally with oocysts of *C. parvum* had no developmental stages in mucosal smears, tissue sections, or feces.[474] Transmission from mammals to other vertebrate classes has not been reported.

Oocysts of *Cryptosporidium* have been transmitted from one avian species to another. Oocysts from quail and pheasants were infective for chickens.[597] Oocysts of *C. baileyi* were mildly infective for turkeys when inoculated orally[473,194] and developed extensively throughout the respiratory tract of turkeys when inoculated intratracheally.[473] Oocysts from chickens,[468,194] quail, and pheasant[596] were not infective for mammals.

Transmission between other vertebrate classes has not been reported. There are several reported failures in the ability to transmit *Cryptosporidium* from one host to another[194,474,597] etc. Successful or unsuccessful transmission can be affected not only by species specificity, but also by the route of inoculation of oocysts, the age and condition of the oocysts, and the age and immune status of the recipient. Therefore, negative transmission studies are not always wholly indicative of host specificity.

IV. EPIDEMIOLOGY

Transmission is by fecal-oral spread of the oocyst stage. Cross transmission studies previously discussed in this chapter demonstrate a lack of host specificity for *C. parvum*. Isolates from humans are infectious for a variety of mammals, and isolates from one mammalian species are infectious for others (Table 3). Zoonotic infection has been reported also.[17,186,512,644,648,664] Many such reports implicate cattle or other livestock as sources of human infection. The high prevalence of cryptosporidiosis in neonatal calves, the close human contact during their care, and the human contact during the milking of their dams provide opportunities for transmission to farmers, handlers, and veterinarians. Similar opportunities for zoonotic transmission exist when persons are closely associated with infected wildlife, zoo animals, and companion animals (Table 5).

Water for drinking or swimming can serve as a vehicle for transmission of the oocyst stage (See Chapter 3). Contaminated surface water has been implicated as the source for several cases of human cryptosporidiosis in New Mexico.[46,258] The high prevalence of oocysts in untreated surface water is thought to come from runoff from nearby fields, pastures, or areas of wildlife or livestock activity.[368,602] In this regard it has been suggested that the agricultural use of sewage sludge could pose a health risk.[37] Fecal contamination of drinking water in rivers or wells in countries regularly beset by severe seasonal rains, such as Bangladesh, or in other countries

TABLE 3
Transmission of *Cryptosporidium* Infection among Host Species via Oocysts

Oocyst source	Recipient species	Infection of recipients[a]	Ref.
Cat	Cat	+	370
	Human	(+)	419, 464
	Mouse	–	370
	Guinea pig	–	370
Cattle	Cat	+	61
	Cattle	+	235, 556, 811, 825
	Chicken	–	404, 474, 596, 619, 647, 811
	Dog	+	61
	Goat	+	326
	Guinea pig	+	811
	Human	(+)	17, 186, 664
	Mouse	+	326, 622, 647, 674, 811
	Pig	+	325, 556, 813, 821
	Rabbit	+	326
	Rat	+	326, 674, 811, 816
	Sheep	+	33, 326, 811, 816, 817
Chicken	Calf	+	474
	Chicken	+	375, 468
	Cotton rat	–	468
	Duck	+	194, 468
	Gerbil	–	468
	Goat	–	194
	Guinea pig	–	468, 854
	Hamster	–	468
	Mouse	–	194, 468
	Pig	–	468
	Quail	–	194, 468
	Rat	–	468
	Turkey	+	194, 468, 473
Deer	Mouse	+	814
Human	Cat	+	186
	Cattle	+	186, 811
	Chicken	–	597
	Dog	+	186
	Goat	+	186
	Human	+	70, 102, 420
	Mouse	+	186, 597, 674
	Pig	+	558, 597
	Rat	+	674
Guinea pig	Sheep	+	597, 811
	Chicken	–	854
	Cattle	+	239a
	Guinea pig	+	36, 854
	Mouse	–; +	36, 854
	Rabbit	–	854
	Sheep	+	36
	Turkey	–	854
Mouse	Cattle	+	418, 647, 825
	Mouse	+	418
	Pig	+	813
	Rat	–	807
	Goat	+	326
Pig	Cattle	+	825
	Pig	+	403
	Human	(+)	319

TABLE 3 (continued)
Transmission of *Cryptosporidium* Infection among Host Species via Oocysts

Oocyst source	Recipient species	Infection of recipients[a]	Ref.
Pheasant	Chicken	+	597
	Mouse	–	597
Quail	Chicken	+	597
	Mouse	–	597
	Cat[b]	+	372
	Dog[b]	+	372
	Guinea pig[b]	+	372
	Mouse[b]	+	372
	Rabbit[b]	+	372
Rat	Sheep	+	811
Sheep	Mouse	+	815
	Pig	+	813
	Rat	+	811, 815
	Sheep	+	15, 815

[a] Parentheses indicate circumstantial evidence.
[b] *C. muris.*

Adapted from Reference 236.

during especially rainy seasons could explain increases in human infections reported during such times.[180] Large-scale outbreaks of cryptosporidiosis in the industrialized countries have been associated with contamination of community drinking water. Outbreaks have been reported in San Antonio, TX;[200] Carrollton, GA;[322] Sheffield and Oxfordshire, England;[704] and Ayrshire, Scotland[743] (see Chapter 3 for details). Simultaneous acquisition of cryptosporidiosis and giardiasis by travelers in localities where waterborne *Giardia* has long been a problem[389] further emphasizes the importance of water as a source of infection.

The possibility of airborne transmission of the oocyst stage, although difficult to document in humans or other mammals, is recognized as the source of respiratory cryptosporidiosis in poultry (see Chapter 9 for details). *Cryptosporidium* also has been identified in the respiratory tree of at least 14 persons.[356] One case is reported in which a veterinary scientist wearing protective clothing and gloves was caring for a calf with cryptosporidiosis. She smelled the air from the open end of an inserted stomach tube and later developed clinical intestinal cryptosporidiosis.[355] A technician employed in the laboratory of one of the authors (RF) became infected and was hospitalized with severe cryptosporidiosis less than a week after cleaning oocysts from bovine feces. Although an aerosol from centrifugation is suspected, hand to mouth transmission cannot be ruled out.

Transmission by arthropods associated with feces in areas lacking good sanitation might be another source of infection. In Peru, *C. parvum* oocyst wall material was detected by immunofluorescence microscopy on flies in two homes using specific monoclonal antibody.[774]

Fomites might also serve as vehicles for transmission of oocysts. This possibility was suggested after studying an outbreak of cryptosporidiosis in a day care center where separate groups of children having no personal contact with one another became infected.[11] During this time adult staff members remained asymptomatic and seronegative for antibody to *Cryptosporidium* and food was provided by each child's respective family. The only apparent link among the groups of children were the play rooms and the toilet facilities which were used by them all.

Transmission via contaminated food seems probable. However, there has been no clear documentation of such a source of infection.

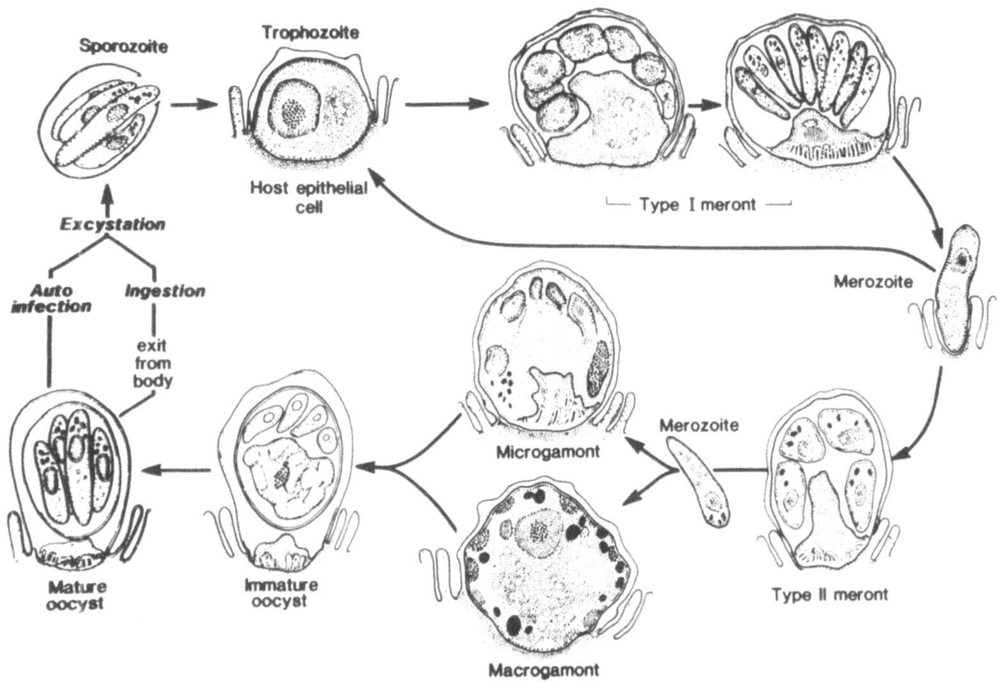

FIGURE 2. Life cycle of *Cryptosporidium*.

Close personal contact appears to be an important source of transmission. Studies of day care centers, households, health care personnel, hospital patients, and sexual partners implies that most transmission in these cases are person-to-person rather than common source.[180] Spread is probably hand to mouth through bathing or cleaning an infected person, or possibly by contact with contaminated clothing, bed linens, potty chairs, bed pans, or other soiled articles. Some spread might be through anal-oral sexual practices.

The significance of asymptomatic carriers of infectious *Cryptosporidium* is unknown. Reports of prevalence in asymptomatic persons range from 6% in India and 10% in Liberia to 27% in New York City.[179,180,219] See Chapter 4 for details.

V. LIFE CYCLE

The morphologic stages of *Cryptosporidium* and their sequence of development resemble other coccidia and are depicted in Figure 2 and Plate 1*. When oocysts in food, water, or the general environment are ingested or inhaled by a suitable host, sporozoites excyst and parasitize epithelial cells of the gastrointestinal or respiratory tract. The process of excystation of sporozoites from oocysts usually requires exposure to reducing conditions, pancreatic enzymes, and bile salts for most coccidia. Exposure of oocysts of *Cryptosporidium* to these stimuli usually results in a high percentage of excystation. However, some sporozoites of *Cryptosporidium* can excyst in warm aqueous solutions without any of the aforementioned special stimuli.[234,673,782,824] The ability of these sporozoites to excyst spontaneously explains in part why *Cryptosporidium* can infect extraintestinal sites such as the conjuctiva of the eye[325,472,669] and the respiratory tree[7,119,217,279,286,314,325,349,356,669,789,795,882] and why autoinfection by endogenous oocysts is possible. For details of *in vitro* excystation, see Chapter 2.

* Plate 1 appears following page 24.

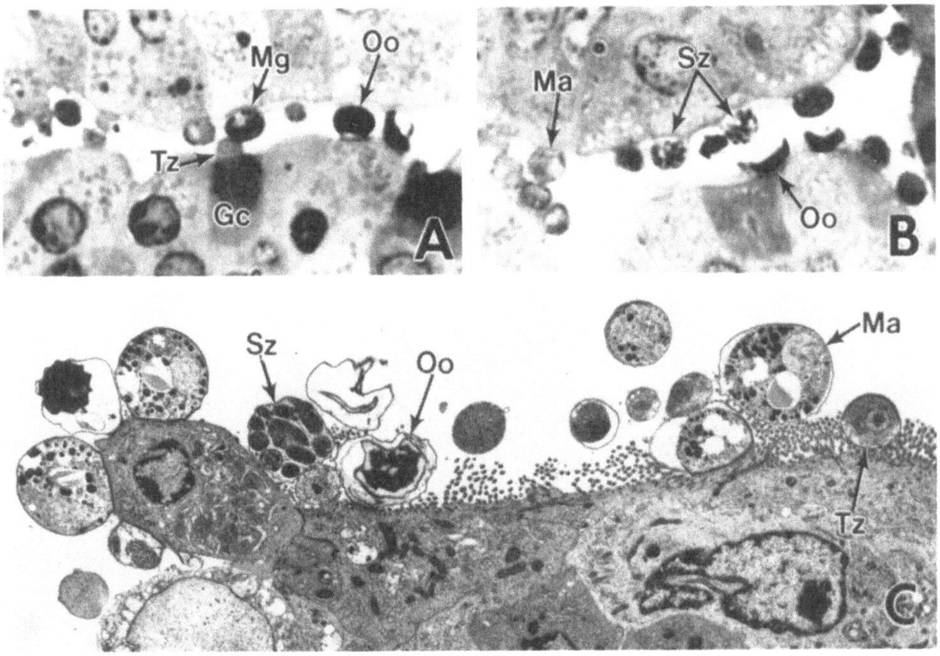

FIGURE 3. Light (A and B) and transmission electron (C) micrographs of *C. parvum* in mouse small intestine. Several parasite stages are shown at the luminal surface of the intestinal epithelium; Gc, Goblet cell; Ma, macrogamont; Mg, microgamont; Oo, oocyst; Sz, schizont; Tz, trophozoite. (Magnification [A and B] × 1200; [C] × 3400.) ([C] Courtesy of Dr. Harley Moon, National Animal Disease Center, ARS, USDA, Ames, IA.)

Sporozoites and subsequent developmental stages are all found at the luminal surface of the epithelium (Figures 3 and 4). They often appear to be attached superficially to the host cell, but are actually intracellular, beneath the host cell membrane, and extracytoplasmic. At the interface of the parasite and the host cell, each stage has a "feeder" or "attachment" organelle which is unique to the genus *Cryptosporidium*. The function of the organelle has not been determined, but appears substantially to increase the surface area between the parasite and host cell, facilitating exchange of materials.

The first change observed after the banana-shaped sporozoite becomes intracellular is its differentiation into a spherical trophozoite with a prominent nucleus. Asexual multiplication referred to as merogony or schizogony results when this nucleus divides. Two morphologic types of schizonts have been observed. Type I schizonts contain six to eight nuclei. As the schizont matures, each nucleus becomes incorporated into a merozoite, a form closely resembling the sporozoite. Each merozoite can potentially invade another host cell where it can develop into another type I schizont or a type II schizont which ultimately produces four merozoites. It is thought that merozoites from type II schizonts initiate sexual multiplication (gametogony) when they leave the schizont and parasitize new host cells. In the new cells, merozoites differentiate into either microgamonts or macrogamonts, the male and female stages, respectively. Early in development, microgamonts become multinucleate and upon maturation, each nucleus becomes incorporated into the equivalent of a sperm, the microgamete, which will fertilize the equivalent of an ovum, the macrogamont. After fertilization, the macrogamont develops into an oocyst, which sporulates *in situ*. When sporogony is complete, each oocyst contains four sporozoites. Most oocysts leave the body. Those in the gastrointestinal tract are excreted with the feces and those in the respiratory tract are carried out with respiratory or nasal secretions. It is thought that some oocysts do not leave the body, but release sporozoites that can repeat the cycle of schizogony, gametogony, and sporogony.[197] Some observers have reported

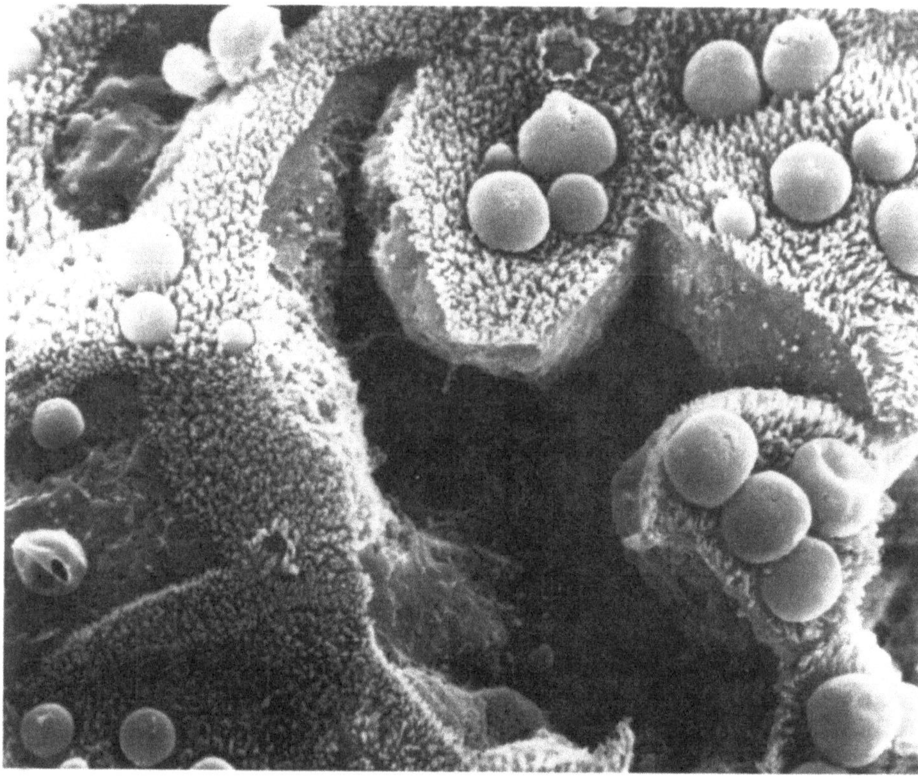

FIGURE 4. Scanning electron micrograph (SEM) of numerous cryptosporidia on the surface of sheep intestinal epithelial cells; note that some of the epithelial cells have sloughed (arrow). (Magnification × 3900.)

that oocysts with relatively thicker walls leave the body to transmit infection to other hosts, whereas oocysts with relatively thinner walls release sporozoites that autoinfect the host.[190,197]

Autoinfection is uncommon among the other coccidia during development in the gastrointestinal tract. Genera such as *Besnoitia, Eimeria, Hammondia, Isospora,* and *Toxoplasma* are not infectious until they sporulate outside the body. Genera such as *Frenkelia* and *Sarcocystis* sporulate endogenously, but are infective only for another host species. The life cycle of *Neospora* is not known. Other than *Cryptosporidium,* only *Caryospora* is known to sporulate *in situ* and to autoinfect.

The time between acquisition of *Cryptosporidium* infection and the shedding of oocysts, referred to as the prepatent period, varies with the host and species of cryptosporidia. Experimentally determined prepatent periods range from 2 to 7 d in cattle,[825] 5 to 10 d in cats,[370] 2 to 14 d in dogs,[61] 3 to 6 d in pigs[556,813] and 2 to 5 d in lambs.[816] Based on known accidental ingestion, the prepatent period ranges from 5 to 28 d in humans[17,102,390,674] with a mean of 7.2 d.[390] The patent period or duration of oocyst shedding ranges from 1 to 12 d in cattle,[12,825] 3 to 33 d in dogs,[61] and 5 to 14 d in pigs.[556,813,821] In eight immunocompetent humans for whom numerous and frequent and appropriately timed specimens were available, the patent period was 18 to 31 d, but three humans who had negative specimens were later found to shed oocysts on days 50, 75, and 89.[390] The timing of development of specific stages in the host is not known with certainty, but mature type I schizonts, mature type II schizonts, gamonts, and oocysts have been found *in vitro* at 12, 24, 48, and 72 h, respectively, after inoculation of cell cultures with sporozoites.[189]

Ultrastructurally, the oocysts of *Cryptosporidium* spp. appear similar to those of other

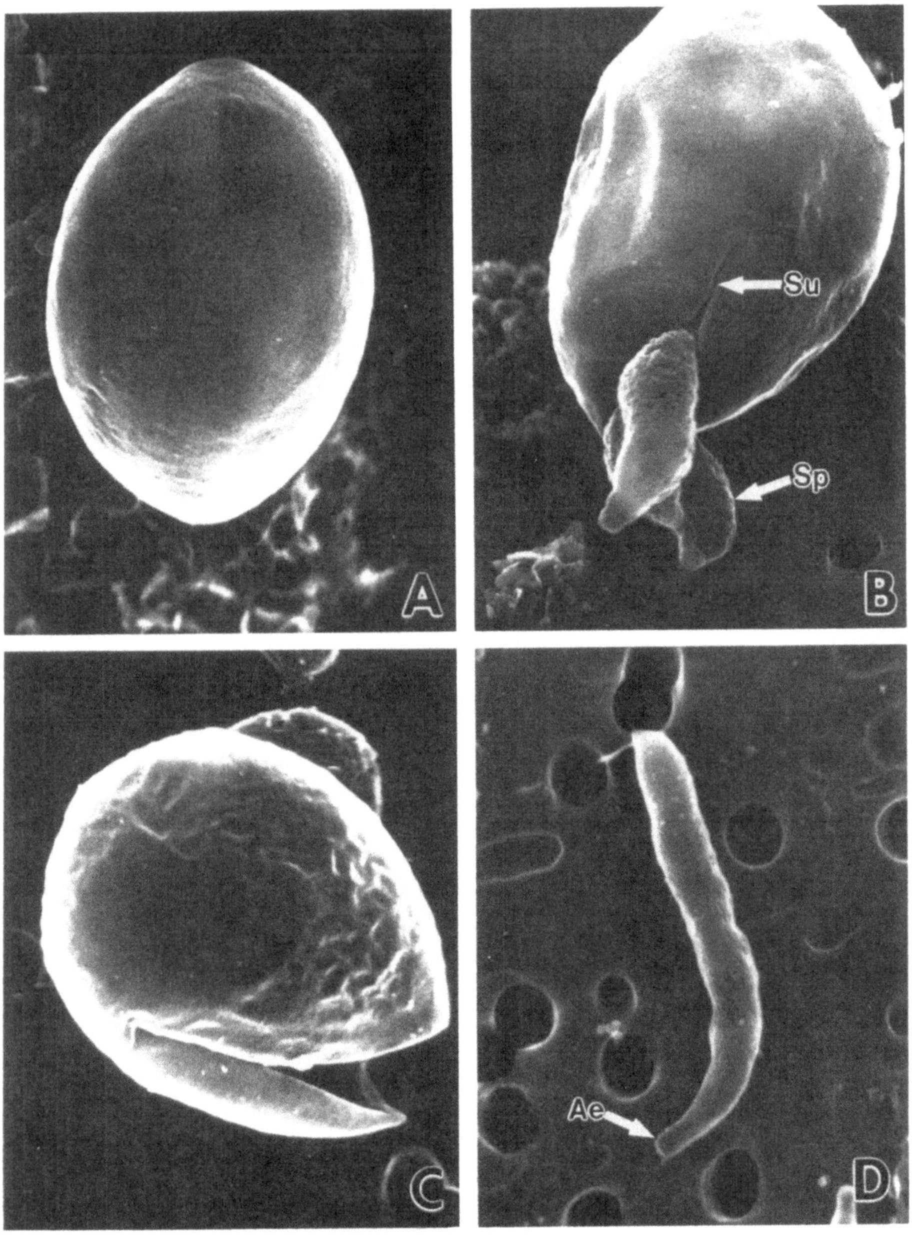

FIGURE 5. SEM of *C. parvum* oocysts and excysting sporozoites. (A) Intact oocyst prior to excystation. (Magnification × 16,000.) (B) Three sporozoites (Sp) excysting from oocyst simultaneously via the cleaved suture (Su). (Magnification × 16,000.) (C) Empty oocyst. (Magnification × 16,000.) (D) Excysted sporozoite; Ae, apical end. (Magnification × 14,000.) ([A, B, and C] from Reduker, D. W., Speer, C. A., and Blixt, J. A., *J. Protozool.*, 32, 708, 1985. With permission.)

coccidia with the oocyst wall consisting of distinct inner and outer layers. However, *Cryptosporidium* oocysts also appear to be unique among the coccidia by having a suture at one end of the oocyst wall. Upon exposure to excysting fluid (see Chapter 2), the suture undergoes dissolution to create an opening in the wall through which the sporozoites escape (Figure 5). Oocysts of other coccidia, such as certain *Eimeria* spp., have a thin area at one pole called a micropyle which is resistant to trypsin and bile. Trypsin-bile sensitive sutures similar to that of oocysts of *C. parvum* have been found in the sporocyst walls (not oocyst walls) of other coccidians such as *Sarcocystis, Toxoplasma, Isospora,* and *Goussia.* The presence of a suture may reflect a close phylogenetic relationship between *Cryptosporidium* spp. and the sarcocystids or perhaps the calyptosporids (i.e., *Goussia, Calyptospora*). However, the suture in *Crytosporidium* is located in what is being referred to as the "oocyst wall". If it is a true oocyst wall, then *Cryptosporidium* is phylogenetically unique in that an oocyst suture has yet to be described for any other coccidian taxon.

The sporozoites and merozoites of *Cryptosporidium* spp. are ultrastructurally similar to those of other coccidia and possess most of the organelles typical of the phylum Apicomplexa such as pellicle (plasmalemma plus a double inner membrane complex), micronemes, electron-dense bodies, nucleus, ribosomes, apical rings, and subpellicular microtubules. They lack certain organelles that are characteristically found in most coccidians such as conoid, rhoptries, mitochondria, micropores, and typical polar rings. Sporozoites and merozoites have a cylindrical-shaped, electron-dense collar which may be analogous to a polar ring; it is located at the anterior tip closely applied to the inner membrane complex (Figure 6).

Sporozoites and merozoites attach to epithelial cells via their anterior ends and become enveloped by host cell microvilli until they eventually reside within a parasitophorous vacuole (Figure 6). Following attachment to an epithelial cell, ultrastructural changes can be seen in the apex of the host cell as well as in the parasite, resulting in the formation of the feeder or attachment organelle (Figure 6). At the site of contact, the host cell develops an electron-dense layer immediately above its terminal web of microfilaments and the parasite forms an apical invagination of that portion of its plasmalemma immediately above the apical rings, which disappear. The apical invagination enlarges as the electron-dense collar moves laterally, eventually occupying a site close to the lateral margin of the electron-dense layer. Simultaneously, a fibrous layer that spans the area between the electron-dense collar is formed between the invaginated parasite plasmalemma and the electron-dense layer of the host cell, and the parasite plasmalemma in this region becomes highly convoluted, increasing its absorptive area (Figure 6). The feeder organelle remains attached to the residual bodies of schizonts, but eventually separates from fully formed merozoites (Figure 7). The residual bodies of mature macrogamonts and microgamonts eventually separate from the feeder organelle (Figures 8 and 10).

Trophozoites contain an undifferentiated cytoplasm, a well-developed feeder organelle, and a nucleus with a prominent nucleolus (Figure 6). Trophozoites give rise to schizonts or gamonts. Schizonts and microgamonts undergo two to four nuclear divisions preceding merozoite or microgamete formation. Macrogamonts do not undergo nuclear division.

During schizogony, the nucleus divides two or three times depending on whether four or eight merozoites are formed. There are no apparent division spindles, nuclear plaques, nor centrioles. After the formation of four or eight nuclei, merozoites form simultaneously at the surface of the schizont. At various sites around the margin of the schizont, anlagen of developing merozoites appear immediately beneath the parasite plasmalemma. At each site of merozoite formation there is a double inner membrane complex, an electron-dense collar, and electron-dense granules immediately above a single nucleus. As merozoites bud at the schizont surface, the pellicle extends inward, incorporating within the merozoite bud a nucleus, electron-dense granules, a few micronemes, ribosomes, and cytoplasm. As they near maturity, merozoites pinch off from the residual body, then become more elongate and form more micronemes and ribosomes (Figure 7). The host cell membranes surrounding mature schizonts lyse, releasing the merozoites (Figure 9) which can infect other host cells.

Microgamonts are relatively rare compared to other stages of *Cryptosporidium* spp. Immature microgamonts are similar to schizonts, but can usually be distinguished from schizonts by their more compact and smaller nuclei. Early microgamonts are surrounded by a single surface membrane. In later stages a double membrane appears at various sites around the margin of the gamont where microgametes have started to form (Figure 10). The body of each microgamete

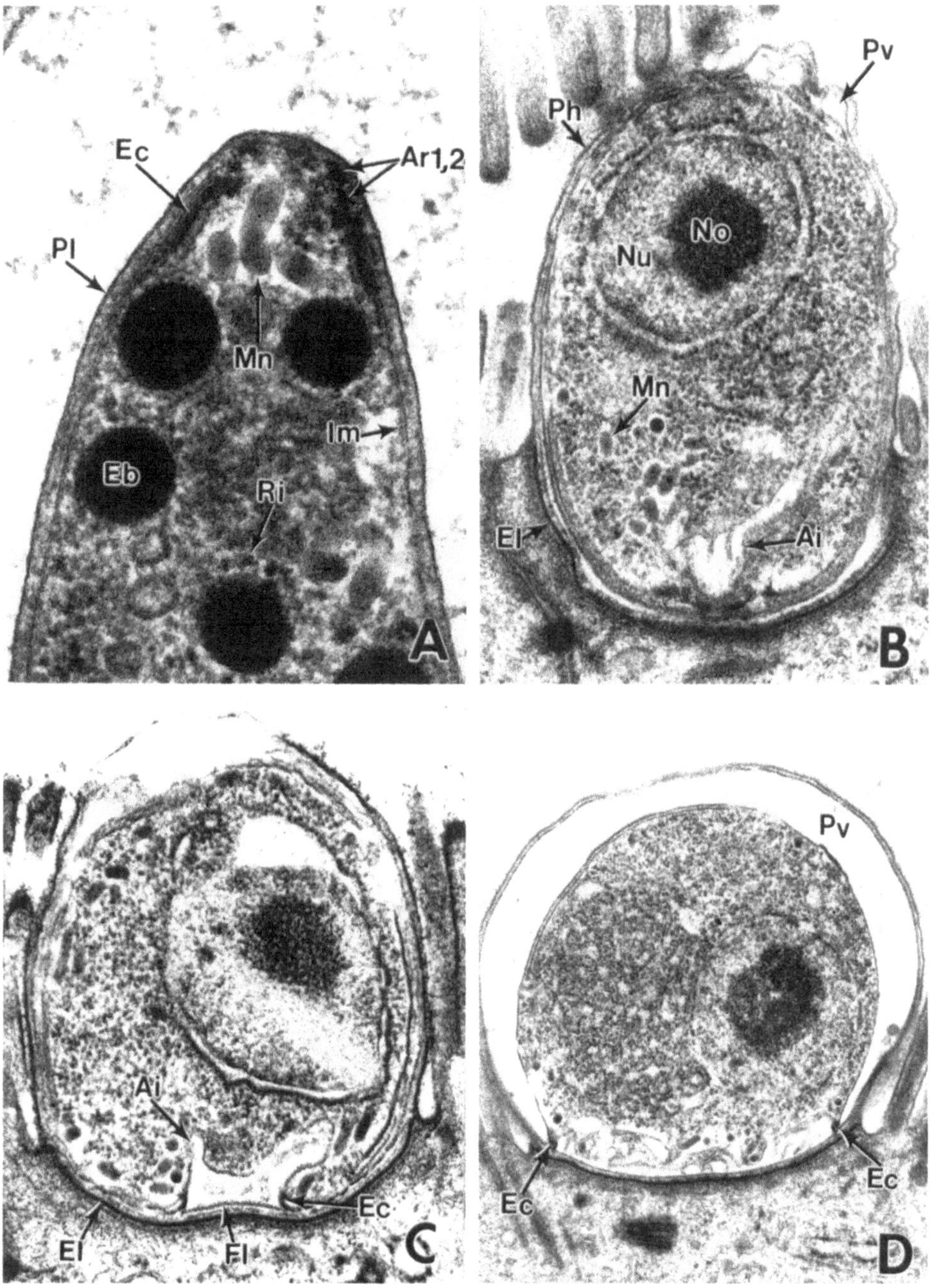

FIGURE 6. Transmission electron micrographs (TEMs). (A) Apical complex of merozoite of *C. parvum* showing anterior rings (Ar1,2), electron-dense bodies (Eb), electron-dense collar (Ec), inner membrane of pellicle (Im), micronemes (Mn), plasmalemmma of pellicle (Pl), and ribosomes (Ri). (Magnification ×77,300.) (B) Merozoite in early stage of attachment to an intestinal epithelial cell; Ai, anterior invagination; El, electron-dense layer of attachment zone; No, nucleolus; Nu, nucleus; Ph, plasmalemma of host cell; Pv, parasitophorous vacuole. (Magnification ×32,500.) (C) Slightly more advanced stage of attachment; note the enlarged anterior invagination (Ai), electron-dense collar (Ec), electron-dense layer (El) and fibrous layer (Fl) of attachment organelle. (Magnification ×34,000.) (D) Trophozoite with electron-dense collar (Ec) located at margins of attachment organelle. (Magnification ×18,000.)

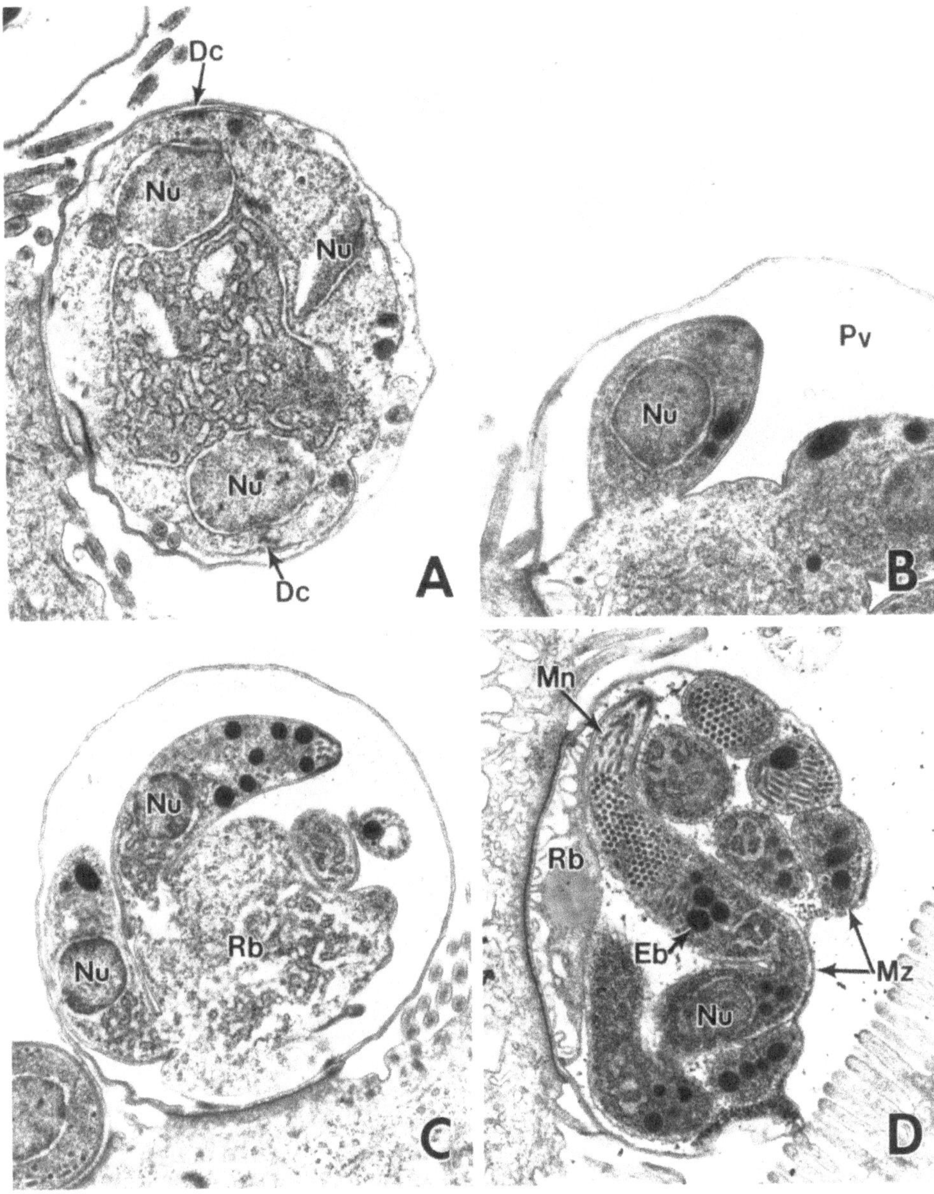

FIGURE 7. TEMs of schizogony. (A) Schizont showing early stage in the development of the apical complexes (Dc) of merozoites immediately above schizont nuclei (Nu). (Magnification × 14,000.) (B) Merozoite budding at surface of schizont; Nu, nucleus of budding merozoite; Pv, parasitophorous vacuole. (Magnification × 18,000.) (C) Schizont with merozoites in an advanced stage of budding from a residual body (Rb). (Magnification × 14,000.) (D) Mature schizont with merozoites (Mz). Residual body (Rb) includes the attachment organelle. Eb, electron-dense body; Mn, microneme; Nu, nucleus of merozoite. (Magnification × 14,000.)

is formed by protrusion of a nucleus from the gamont surface. In intermediate stages of microgamete formation, the nucleus incorporated into the microgamete bud contains equal amounts of electron-lucent and electron-dense chromatin. At this time, an apical cap, which consists of two closely applied membranes, forms over the anterior end of the microgamete (Figure 10). As microgametes mature, they eventually pinch off the gamont surface. The residual body is surrounded by a single membrane and contains numerous ribosomes, endoplasmic reticulum, and a few micronemes (Figure 10).

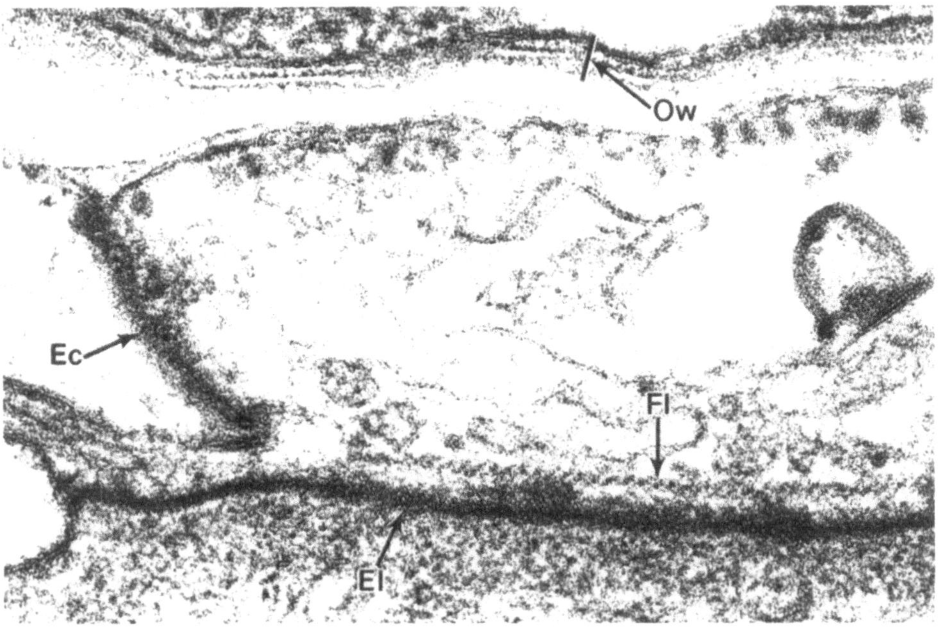

FIGURE 8. High-magnification TEM showing portion of an attachment organelle associated with a developing oocyst. Note that the oocyst wall (Ow) has separated from the attachment organelle. Ec, electron-dense collar; El, electron-dense layer; Fl, fibrous layer. (Magnification × 99,000.)

Microgametes are bullet shaped and small (approximately 1.4 × 0.5 µm in *C. parvum*) and lack flagella and mitochondria, which are characteristically found in microgametes of other coccidia (Figure 11). They contain a condensed nucleus which occupies most of the microgamete. They are surrounded completely by a plasmalemma; a single inner membrane extends approximately 2/3 the length of the microgamete. Eight microtubules originate from a conical structure at the anterior end and progress posteriorly in close association with the outer surface of the nucleus (Figure 10). At the posterior margin of the apical cap, three to five concentric lamellae extend outward at 90 to the microgamete plasmalemma (Figures 10 and 11). Electron-dense granules are present in the cytoplasm of the mid region of the microgamete.

Macrogamonts are unique from all other stages by having a relatively large, centrally-located nucleus with a prominent nucleolus, lipid bodies, amylopectin granules, and wall-forming bodies (Figure 12).

Little is known about fertilization of macrogamonts by microgametes. Microgametes attach to the plasmalemma of host cells harboring macrogamonts and schizonts. Portions of microgametes have been observed within macrogamonts (Figure 12), but the actual penetration of a microgamete into a macrogamont has not been observed. Microgametes attach via their apical cap to the plasmalemma of host cells harboring macrogamonts (Figure 12). Evidently, only the nucleus and closely associated microtubules of the microgamete enter the macrogamont during fertilization (Figure 12). Penetration of the microgamete into the macrogamont and the fusion of the gamete nuclei apparently occur rapidly.

After fertilization, the zygote develops into an oocyst with either a thin or thick wall (Figure 13). Oocysts with thin walls evidently lack the wall-forming bodies characteristic of oocysts with thick walls. Those that form thick walls have types 1 and 2 wall-forming bodies similar to other coccidia (i.e., *Eimeria* spp.). The first step in oocyst wall formation is separation of the sporont from the feeder or attachment organelle (Figure 8) with the simultaneous appearance of two unit membranes external to the zygote plasmalemma (Figure 12).

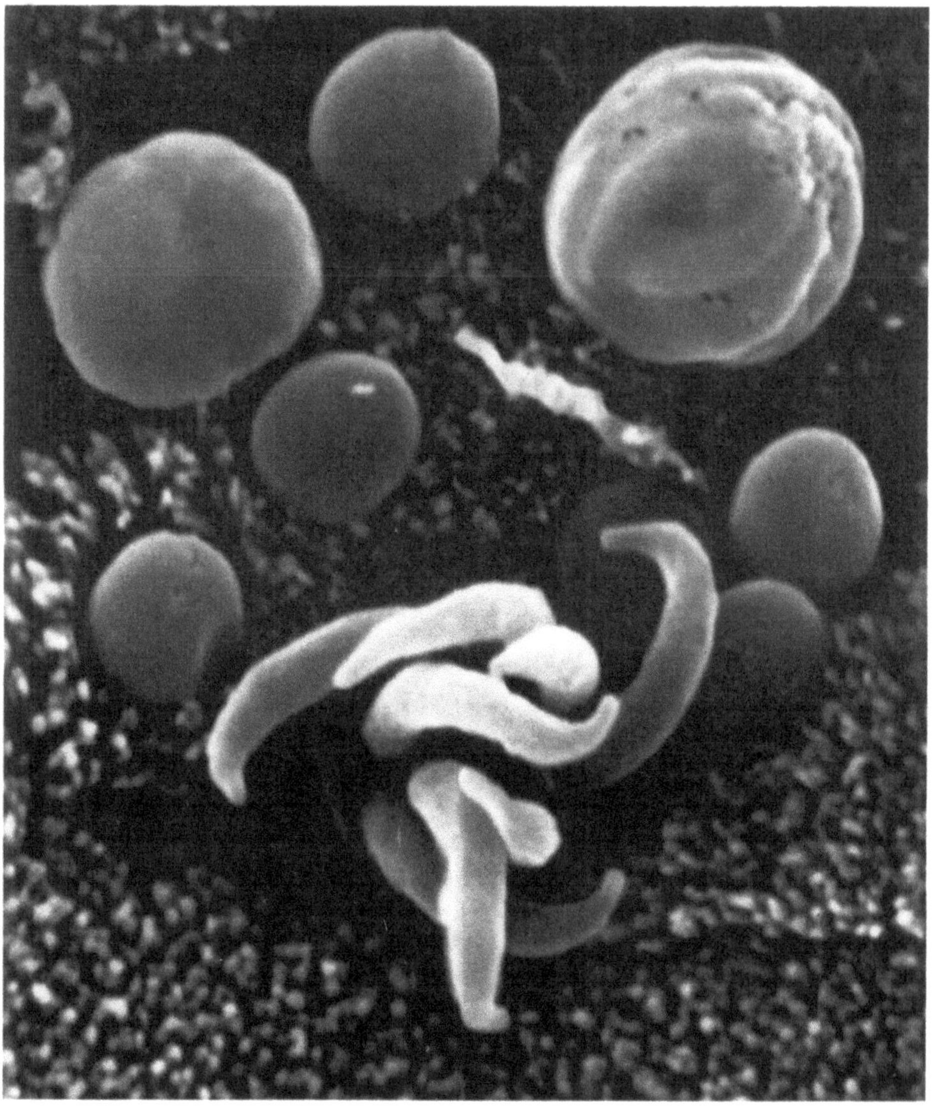

FIGURE 9. SEM of several cryptosporidia on sheep intestinal epithelial cells. Eight merozoites are escaping from one host cell, and the impressions of merozoites are visible in host cell at upper right. (Magnification × 15,750.)

Wall-forming body material is transported or exocytosed across the oocyst pellicle (i.e., plasmalemma and inner membrane) where it forms a thin, moderately coarse outer layer and a finely granular inner layer (Figure 14). The two layers are separated by an electron-lucent zone that consists of the two oocyst membranes sandwiched between the outer and inner layers of the oocyst wall (Figure 14). The outer layer of the oocyst wall is continuous and of uniform thickness, whereas the inner layer contains a suture at one pole which spans 1/3 to 1/2 the circumference of the oocyst (Figures 5, 13, and 14).

VI. CLINICAL SIGNS, PATHOGENESIS, AND LESIONS

It is not known how cryptosporidia cause disease. However, there is an apparent association between the presence of organisms at a specific site in the body and clinical features of disease.

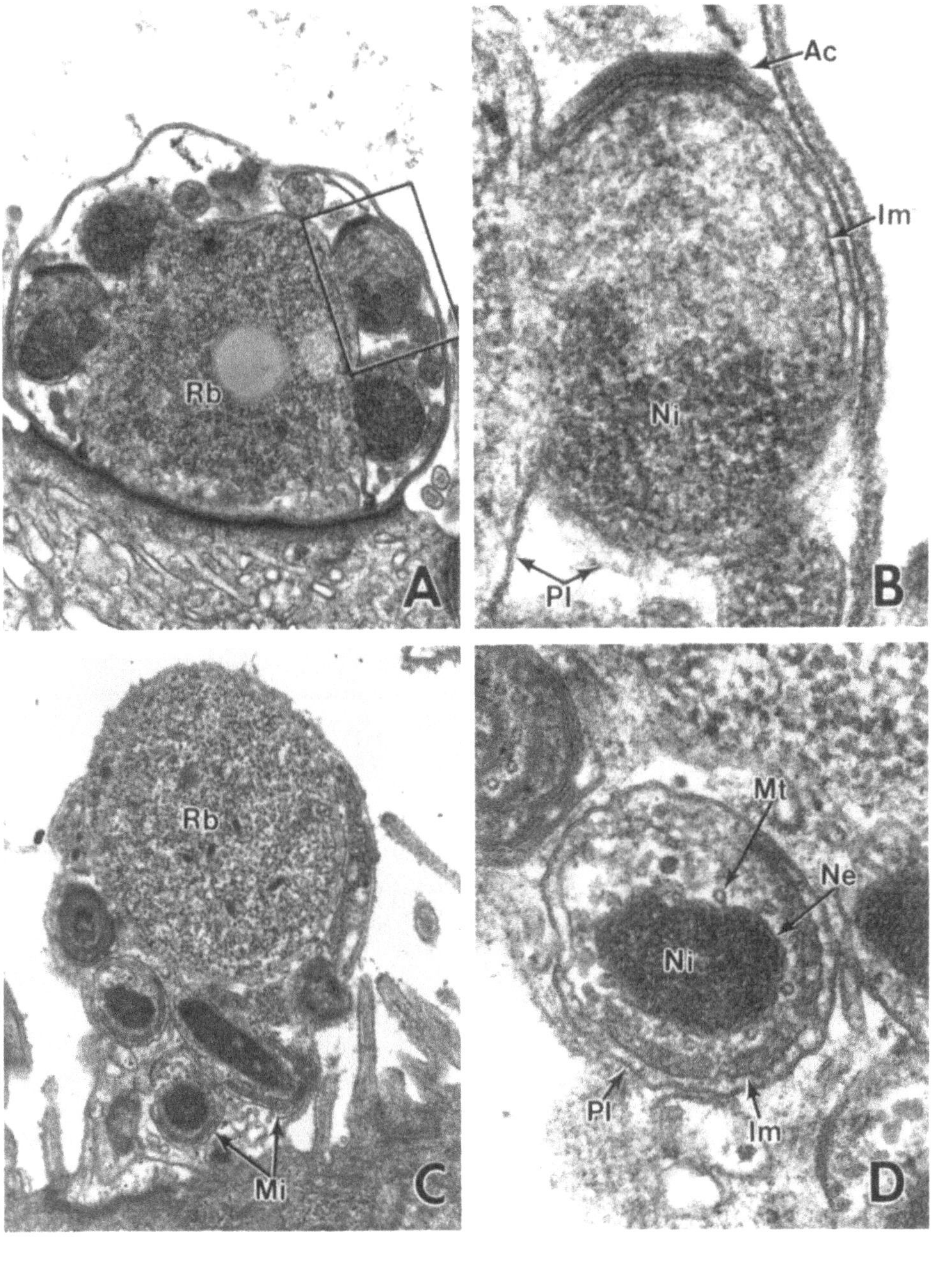

FIGURE 10. TEMs of microgamonts and microgametes. (A) Microgamont with microgametes budding from a residual body (Rb). (Magnification × 16,000.) (B) High magnification of area in rectangle in A showing budding microgamete; note the apical cap (Ac) of the microgamete and that the plasmalemma (Pl) of the microgamont is continuous with that of the microgamete; Im, inner membrane of microgamete; Ni, nucleus of microgamete. (Magnification × 80,000.) (C) Mature microgamont showing portions of six microgametes that have separated from the residual body (Rb). (Magnification × 19,000.) (D) Cross-section of microgamete showing microtubules in close proximity to the nuclear envelope (Ne), microgamete nucleus (Ni), plasmalemma (Pl), and inner membrane of microgamete. (Magnification × 66,000.)

For the most part, sites of infection include the respiratory tree, the gastrointestinal tract, and associated organs. Specificity, which might be a reflection of the species of *Cryptosporidium,* varies from one group of animals to another. For example, in fish and reptiles, the stomach and possibly the intestine are most often infected. In birds, most infections are respiratory, with many

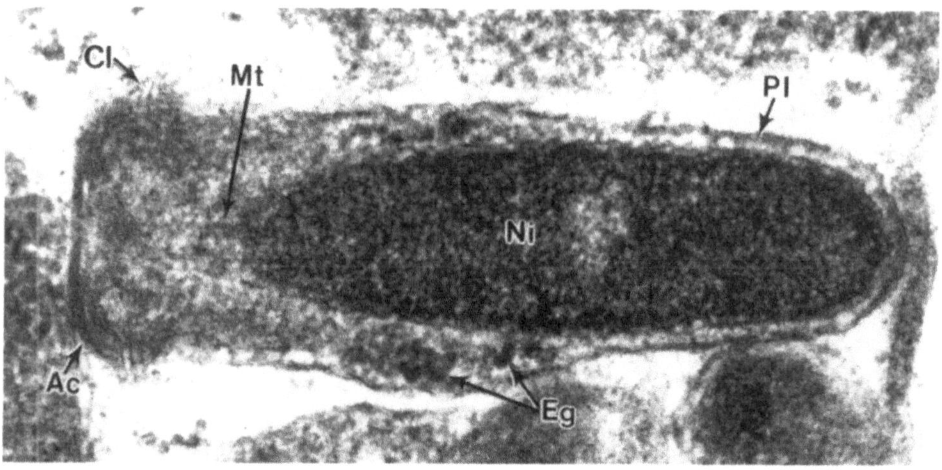

FIGURE 11. TEM of longitudinal section of microgamete of *C. parvum*. Note concentric lamellae (Cl) just posterior to the apical cap (Ac). Eg, electron-dense granules; Mt, microtubule; Ni, nucleus of microgamete; Pl, plasmalemma of microgamete. (Magnification ×75,600.)

fewer infections reported for the intestine and associated organs such as the bursa or cloaca. In immunocompetent mammals, most infections are intestinal, whereas immunocompromised mammals might also have infections of virtually the entire length of the gastrointestinal tract from esophagus to rectum including the pancreas, liver, and gallbladder, as well as the respiratory tree.

Depending on the species and isolate of the parasite and the age and immunologic status of the host, the severity of infection can range from subclinical to severe. Generally, young animals are more susceptible to infection and have the most severe clinical signs. Most reports of infections for domesticated animals, companion animals, and poultry are for individuals less than 1 month of age. Most reports of infections for immunologically healthy humans are children less than 2 years of age. However, of all the animals known to acquire infection with *Cryptosporidium,* the only immunologically normal adults that are routinely susceptible to infection are humans, pigs, and guinea pigs. Immunocompetent animals and humans recover spontaneously from clinical signs lasting a few days to a few weeks, whereas immunodeficient individuals can become chronically and terminally ill.

The most prominent signs of infection in reptiles are a chronically swollen stomach accompanied by anorexia and regurgitation (see Chapter 10). Chickens most often exhibit acute signs which include watery eyes, sneezing, coughing, and respiratory distress. Other birds usually have diarrhea (see Chapter 9).

In various mammals including humans the most prominent sign is watery, cholera-like diarrhea associated with anorexia, weight loss, dehydration, and abdominal discomfort (see Chapter 4). The severity of illness can vary greatly among individuals. In addition to the aforementioned clinical effects of cryptosporidiosis, the location of the organisms at the cell surface, concurrent loss of microvilli, and decreased levels of microvillar disaccharidases might interfere with absorption of nutrients, leading to malabsorption and malnutrition (Figures 15 to 17).[342] Based on study of pulmonary hydrogen gas levels in calves experimentally infected with 1×10^8 oocysts of *Cryptosporidium* spp., it was suggested that the accompanying diarrhea was most likely due to accumulation of nonabsorbed hypertonic nutrients within the lumen of the large intestine.[356b]

Abdominal X-ray has revealed large amounts of gas in the colon and fluid in the large and small bowel.[591] In one case, toxic dilation of the colon to a transverse diameter of 8 cm was

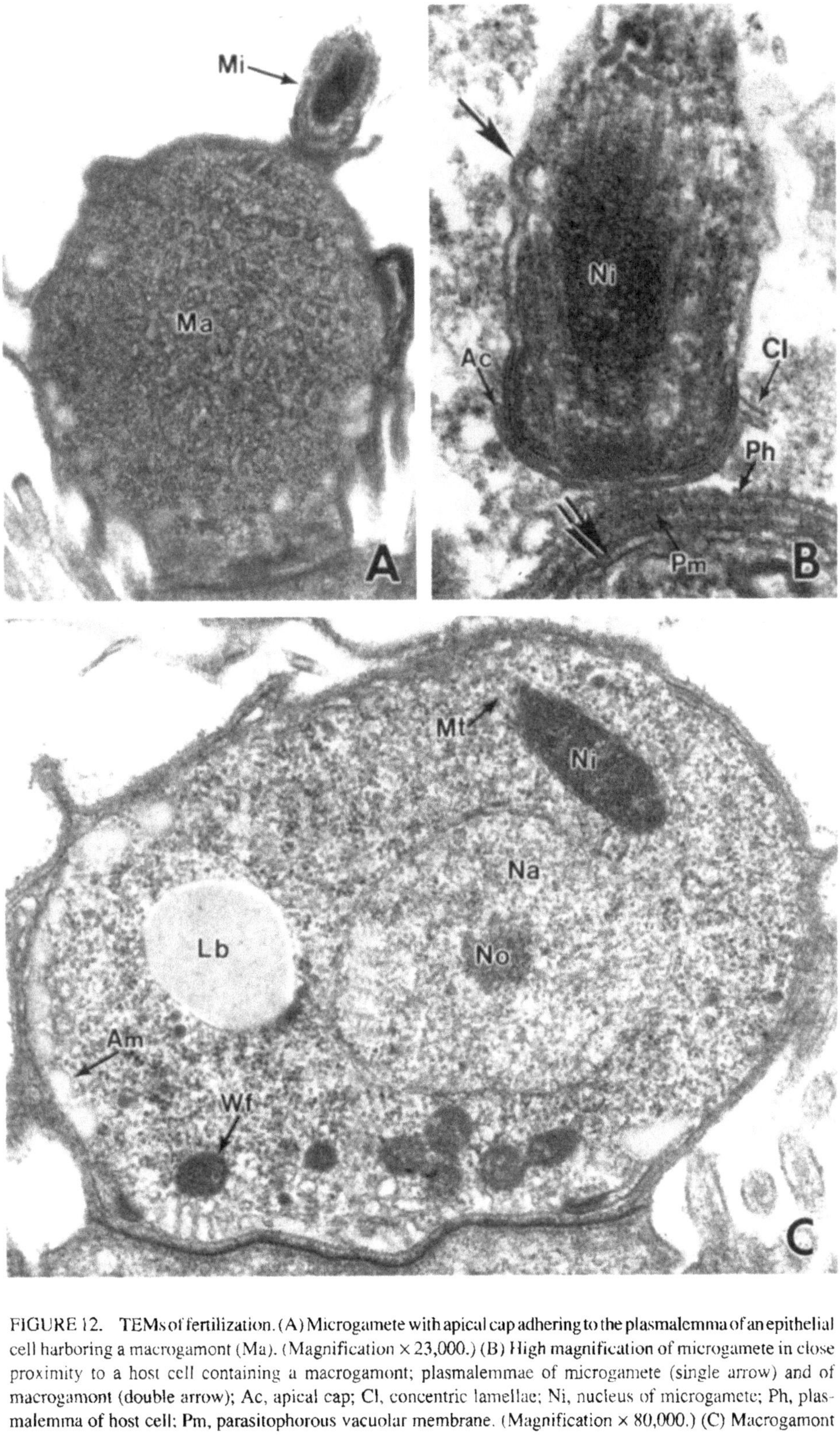

FIGURE 12. TEMs of fertilization. (A) Microgamete with apical cap adhering to the plasmalemma of an epithelial cell harboring a macrogamont (Ma). (Magnification × 23,000.) (B) High magnification of microgamete in close proximity to a host cell containing a macrogamont; plasmalemmae of microgamete (single arrow) and of macrogamont (double arrow); Ac, apical cap; Cl, concentric lamellae; Ni, nucleus of microgamete; Ph, plasmalemma of host cell; Pm, parasitophorous vacuolar membrane. (Magnification × 80,000.) (C) Macrogamont containing nucleus (Ni) and associated microtubules (Mt) of microgamete; Am, amylopectin; Lb, lipid body; Na, nucleus of macrogamont; No, nucleolus of macrogamont; Wf, wall forming body. (Magnification × 33,600.)

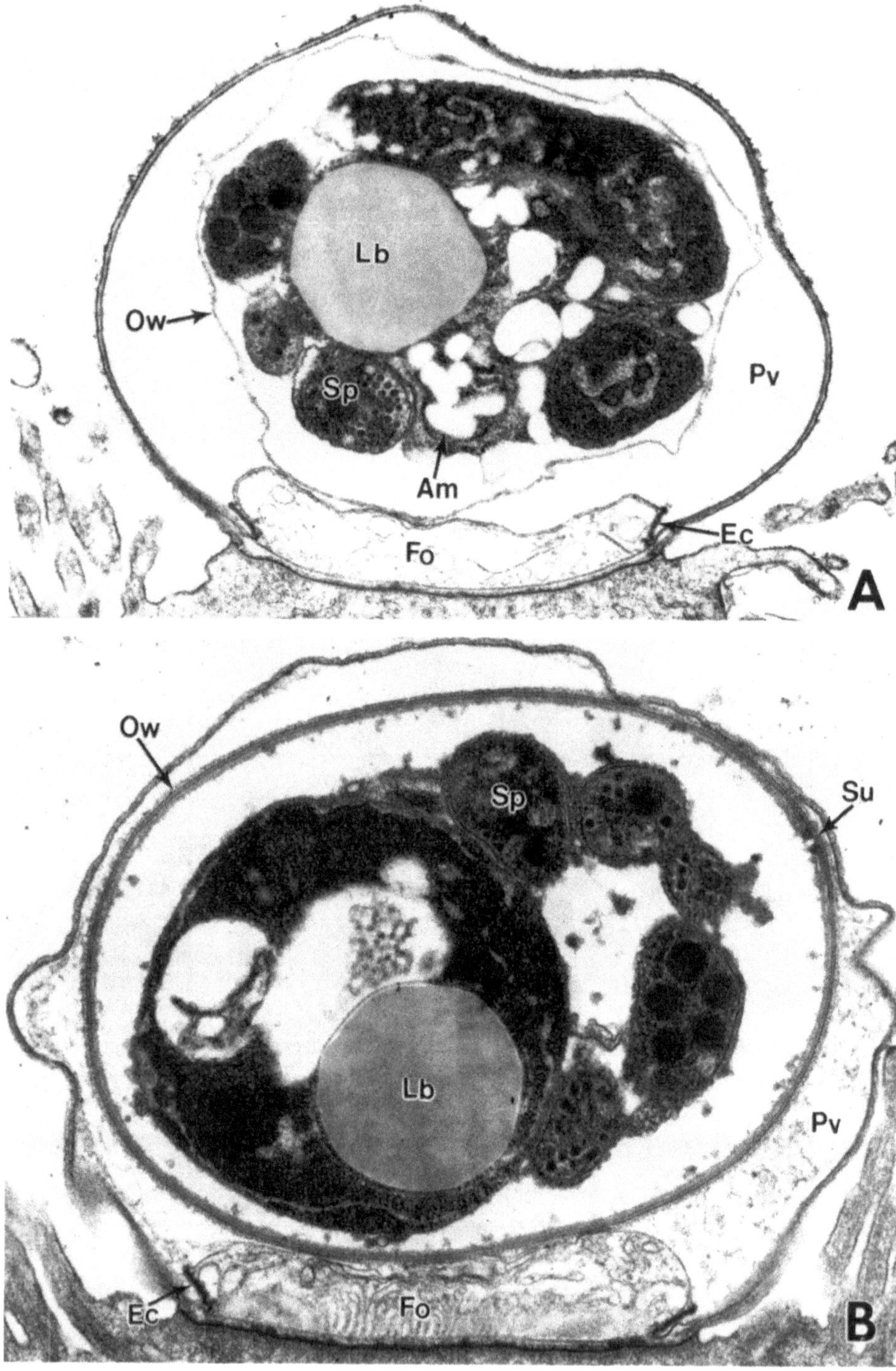

FIGURE 13. TEMs of *C. parvum* oocysts. (A) Thin-walled oocyst; Am, amylopectin; Fo, feeding organelle; Lb, lipid body; Ow, thin oocyst wall; Pv, parasitophorous vacuole; Sp, sporozoite. (Magnification ×20,000.) (B) Thick-walled oocyst with suture (Su); Lb, lipid body; Ow, thick oocyst wall; Pv, parasitophorous vacuole; Sp, sporozoite. (Magnification ×23,240.)

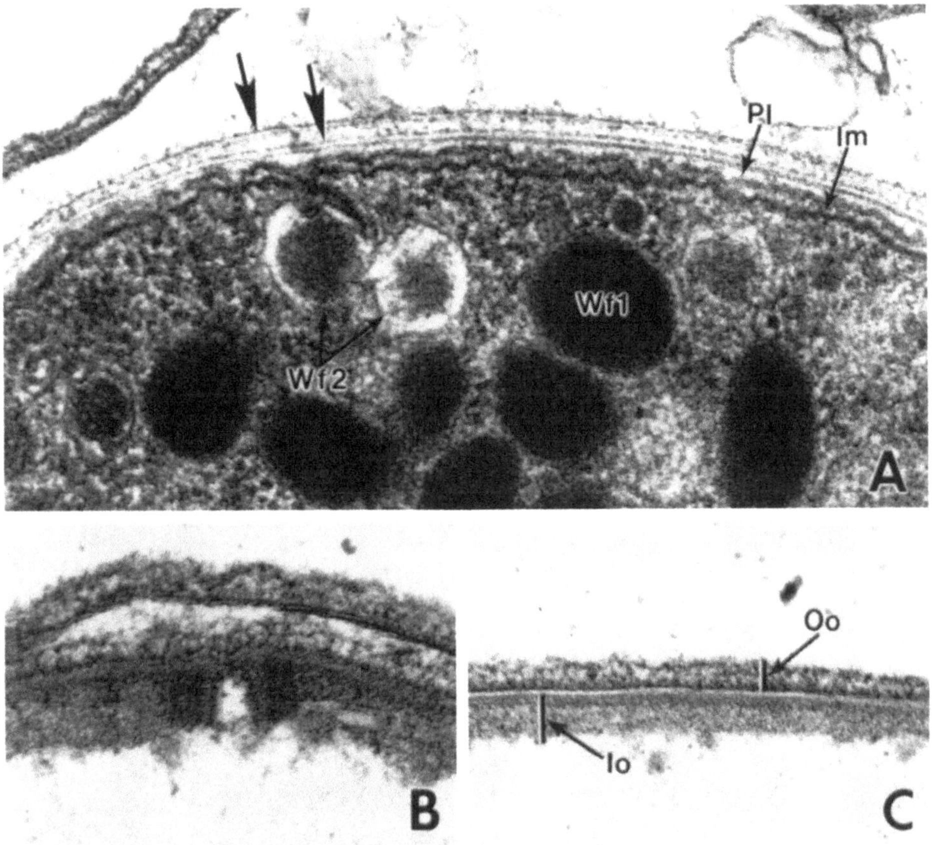

FIGURE 14. High magnification TEMs. (A) Margin of zygote showing two membranes (arrows) external to the parasite plasmalemma (Pl) and two types of wall forming bodies (Wf1 and Wf2); Im, inner membrane of zygote. (Magnification × 80,000.) (B) Portion of Figure 11B showing suture in oocyst wall. (Magnification × 118,800.) (C) Cross-section of completely formed oocyst wall showing inner (Io) and outer (Oo) layers. (Magnification × 118,800.)

reported.[172] Barium enema has shown fine spiculated ulcerations producing a markedly abnormal mucosal pattern, edema of the descending and sigmoid colon and rectum, consistent with acute colitis.[591] Proctosigmoidoscopy has revealed edematous mucosa and yellow plaques thought to be fibrinous exudate.[591]

Although lesions can be present throughout the gastrointestinal tract, they are usually most severe in the lower jejunum and ileum.[540] Mesenteric lymph nodes are sometimes swollen. Intestines appear unremarkable or hyperemic. Microscopic lesions follow a similar pattern among almost all host species infected. Cryptosporidia are found in the epithelium, usually without obvious damage to host cells except that cells may be low columnar, cuboidal, or squamous, and microvilli may be absent at the site of parasite attachment.[540] At the electron microscopic level, the cytoplasm of infected cells sometimes contains mitochondria that appear swollen and vacuolated, and smaller smooth membraned vacuoles containing clear fluid may be scattered throughout the cytoplasm around the cell nucleus.[98] Usually, mild to severe villous atrophy is apparent at the light microscope level. Villi appear blunt, shorter, and wider than normal and are sometimes fused with other villi.[769,871] Some fused villi have desmosomal junctions between enterocytes of adjacent villi, visible with electron microscopy. Inflammatory changes are marked by cellular infiltration, usually plasma cells and neutrophils, but also macrophages and lymphocytes, into the subepithelial lamina propria.[540,769] Peyer's patches

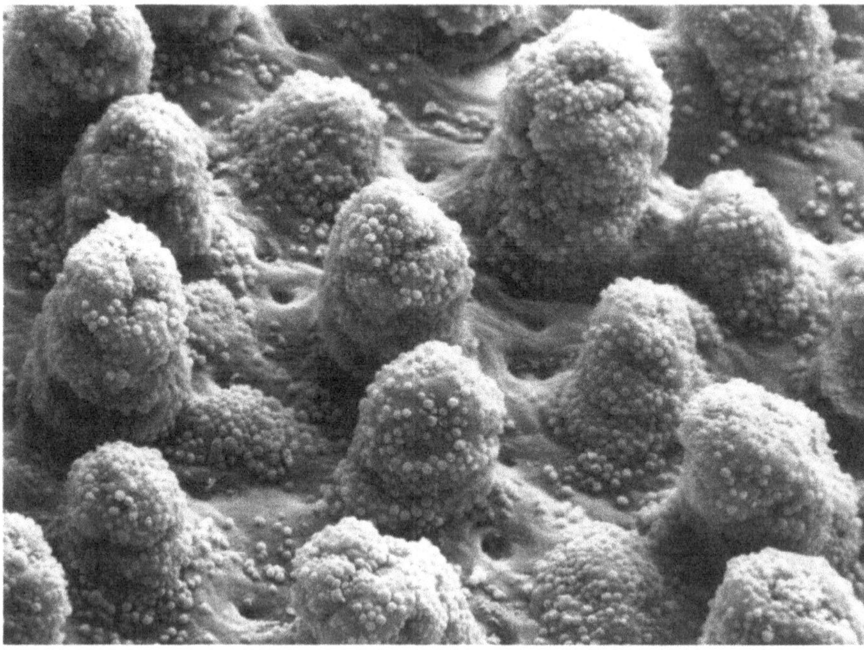

FIGURE 15. *C. parvum* covering the villar and intervillar surface of small intestine of experimentally infected mice. (Magnification × 225.) (From Vítovec, J. and Koudela, B., *J. Vet. Med.*, 35, 515, 1988. With permission.)

appear reactive.[250] Crypts are enlarged, hyperplastic, and may contain dead cells or pink-staining gelatinous material.[636]

In the respiratory tree there may be excess mucus, swollen mucosa, and epithelium that is deciliated, hyperplastic, or hypertrophied. Inflammatory cells may infiltrate into the subepithelial lamina propria.

Hemolytic uremic syndrome, a rare disease in infants, was presented as hemolytic anemia with massive hemolysis, thrombopenia, and renal failure in children with cryptosporidiosis.[95,433] The trigger mechanism for this syndrome is unknown, but infective agents such as viruses have previously been associated with it. The relationship, if any, to cryptosporidiosis is not clear.

VII. IMMUNITY AND IMMUNOPROPHYLAXIS

Age-related susceptibility and resistance to cryptosporidiosis has been reported. Much of the present state of knowledge comes from the observation of clinical cases rather than direct experimentation. In general, young, immunologically immature mammals have a greater prevalence of infection and experience more severe illness than adults. In fact, most adult mammals appear asymptomatic and parasite free. In human populations, although the age range for infection is from 3 d to 95 years, most infections are reported for children under 2 years of age (see Chapter 4). Where sanitation and nutrition are poor, there is an especially high prevalence of cryptosporidiosis in children. In populations where good sanitation and nutrition normally exist, there are reports of outbreaks of cryptosporidiosis in children in special settings such as day care centers where children become infected first and then adults associated with the children (see Chapter 4). Populations of adults and children become infected after drinking contaminated water (see Chapter 3). These episodic infections indicate that susceptibility might

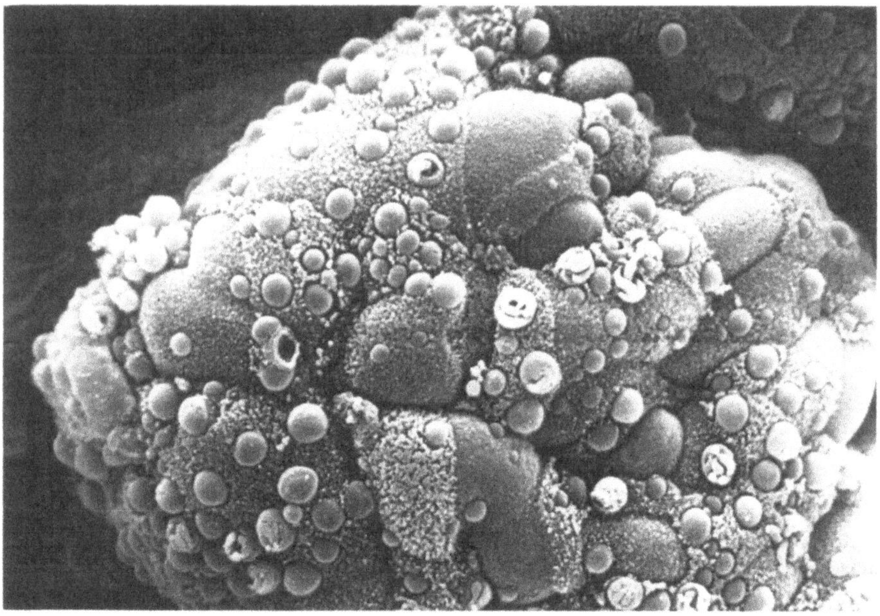

FIGURE 16. Erosion of villi in mice experimentally infected with *C. parvum*. (Magnification × 925.) (From Vítovec, J. and Koudela, B., *J. Vet. Med.*, 35, 515, 1988. With permission.)

be influenced more by previous exposure than by age. Most infected bovines are neonates, although the role of exposure resulting in immunity has not been ruled out (see Chapter 5). Virtually all laboratory-raised strains of suckling rodents develop intestinal cryptosporidiosis of short duration after oral inoculation with *C. parvum* oocysts, whereas previously unexposed mice older than 3 weeks of age are usually not susceptible to infection (see Chapter 8). Even administration of immunosuppressant drugs does not appreciably change the resistance in adults.[734] Exposure of neonatal nude mice with greatly reduced T cell function results in persistent cryptosporidiosis.[328] Still, suckling nude mice develop heavier infections than adult nude mice after oral infection with oocysts.[328] Although the adults of some wild mice might be susceptible to infection,[418] this has not been tested under controlled conditions. Unlike mice, adult rats immunosuppressed with hydrocortisone, cyclophosphamide, or dexamethasone become susceptible to intestinal cryptosporidiosis after oral inoculation with oocysts of *C. parvum*,[123,678] indicating that acquired immunity might play a role in resistance to infection in rats. Malnutrition, through suppression of immunity, might affect the susceptibility to infection or the severity of the illness.[180]

Distinct differences are seen in the clinical course of cryptosporidiosis in immunocompetent vs. immunodeficient humans. In the former group, clinical signs are of shorter duration than in the latter group, in which infection can be life threatening. This latter group has been represented in the literature by persons receiving immunosuppressive drugs, AIDS patients, and a person with hypogammaglobulinemia, but with normal T cell function.[187] A humoral response to infection has been demonstrated by a wide range of mammals[236] including immunologically healthy and immunodeficient persons.[838]

Specific bovine IgA, IgG, and IgM antibodies to *Cryptosporidium* sporozoites and oocysts have been identified by enzyme-linked immunosorbant assay (ELISA).[239,838,889] Specific anti-*Cryptosporidium* human IgA, IgG, IgE, and IgM also have been identified in sera by ELISA.[148,838,840] Despite the demonstration that sporozoite infectivity can be neutralized by exposure to immune serum,[689,690] it is apparent that a measurable humoral response does not necessarily provide immunological protection since some persons with HIV infection produce

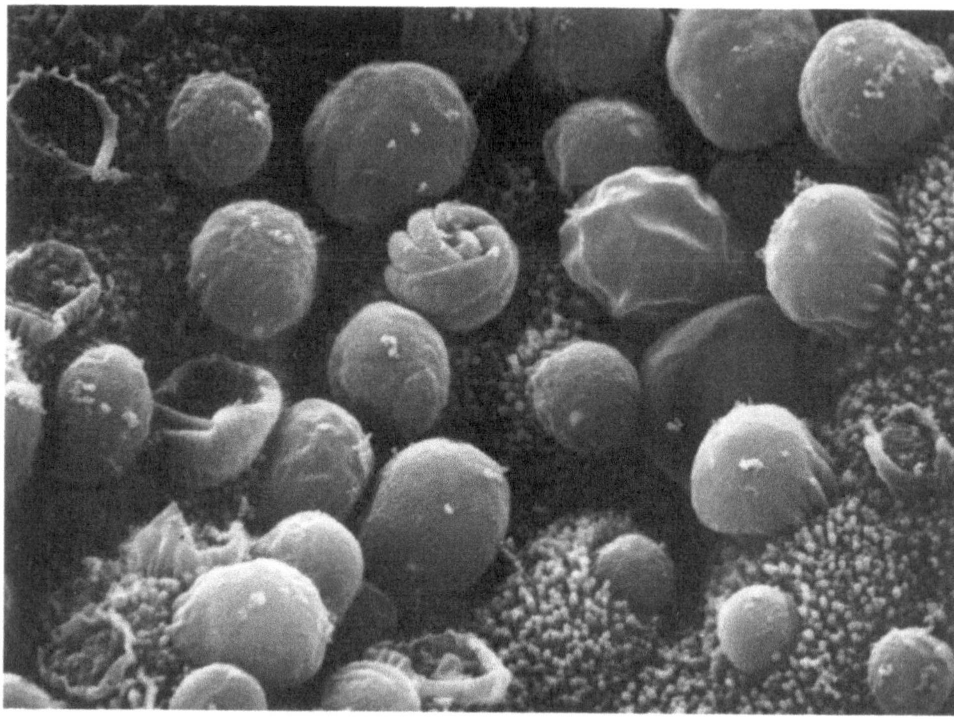

FIGURE 17. Cloaca of chicken experimentally infected with *C. baileyi*. Some villi are fused near the surface of the parasites. The membrane covering one of the schizonts (S) has been removed exposing the merozoites. Note the craters left after the parasites have ruptured and left the attachment site. (Courtesy of W.L. Current and S.L. White, Lilly Research Laboratories, Greenfield, IN.)

IgM and all produce IgG during the course of chronic cryptosporidiosis. It has become apparent from studies of other coccidia that immunity to coccidiosis is T cell dependent.[700] Not surprisingly, humans as well as mice with specific T cell dysfunctions are unable to develop an effective immunity against cryptosporidiosis. A protective immune response might require a T cell-dependent induction that results in specific humoral and cellular mechanisms.[199]

The nature or identity of any antigens that stimulate a protective response is not known. Many antigens have been recognized by hyperimmune rabbit serum, monoclonal antibodies, and sera from humans, calves, and horses.[444,490,492,538,690] Monoclonal antibodies recognize the same epitope on oocyst proteins from different species of *Cryptosporidium*. A monoclonal antibody against *C. parvum* reacts with *C. muris* and *C. meleagridis*, but not *C. baileyi* (Meridian Diagnostics, Cincinnati, OH). *Cryptosporidium muris* and *C. baileyi* share four proteins of different molecular weight, *C. parvum* and *C. baileyi* share one, and *C. parvum* and *C. muris* share one. Of over 75 electrophoretically differentiated antigens from oocysts and sporozoites that react with hyperimmune rabbit serum, less than 20 also react with immune human sera.[199] One antigen, a 23-kDa protein from the oocyst stage, reacts with sera from 93% of persons with cryptosporidiosis. Studies with ConA and lectins to probe electrophoretic profiles transferred to nitrocellulose indicate that 14 to 19 antigen bands are glycosylated.[798] Other studies have shown that *C. parvum* antigens recognized by mice after intestinal exposure were carbohydrates alone or associated with lipids or proteins.[490]

Attempts to provide protection or therapy through passive immunity have yielded mixed results. In an initial uncontrolled study, bovine transfer factor prepared from lymph node lymphocytes of calves immunized with *Cryptosporidium* provided a spectrum of response for five of eight AIDS patients with cryptosporidiosis.[488] In a controlled study, six of seven patients

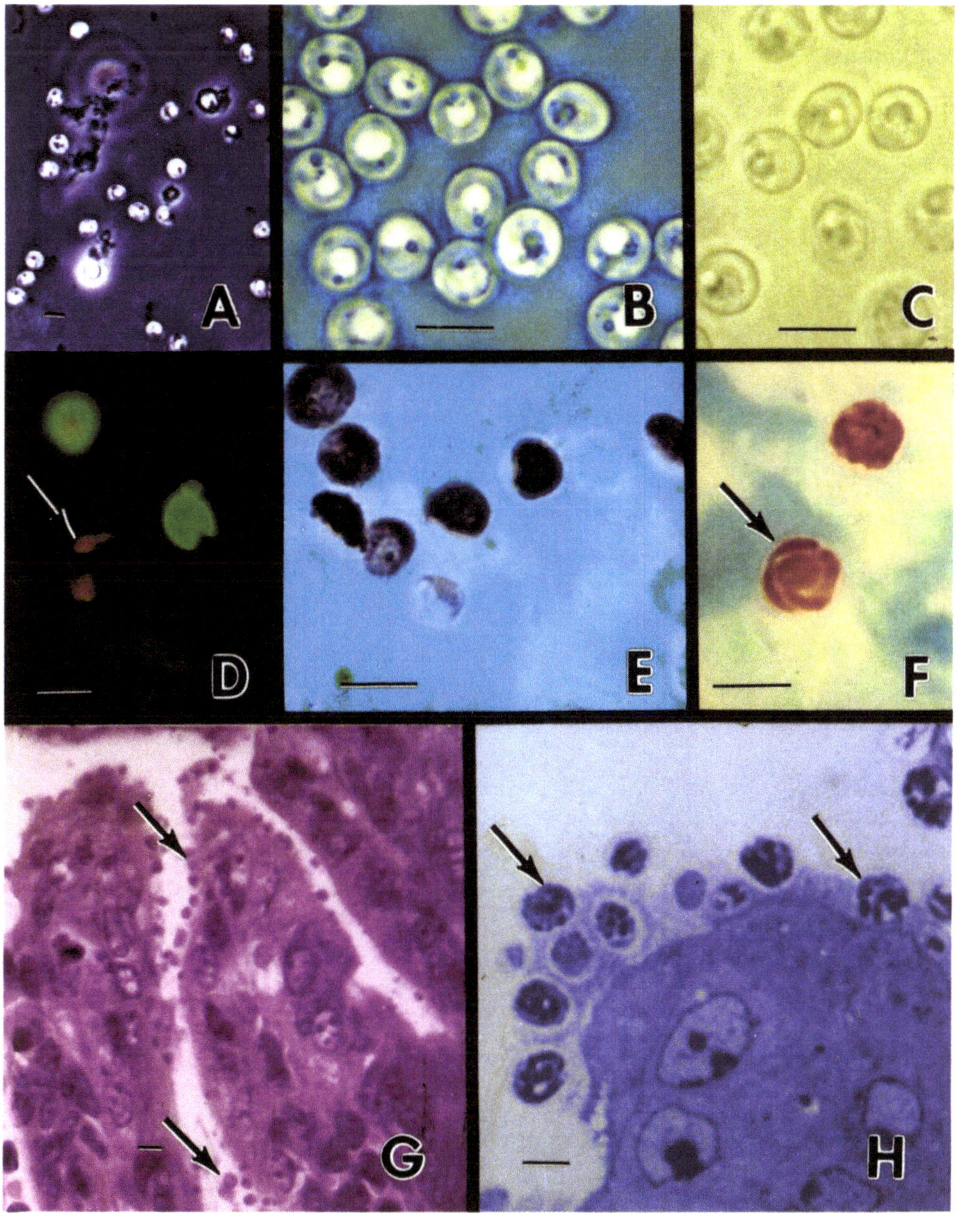

PLATE 1. (A to C) Oocysts of *C. parvum* in a Sheather's sugar flotations of stool specimens. (A and B) Oocysts viewed with phase-contrast microscopy are bright and refractive and contain one to four dark granules. (A viewed with 40× objective, B viewed with 100× oil immersion objective.) (C) Oocysts viewed with bright-field microscopy are translucent. (D) Oocysts of *C. parvum* stained with fluorescein by indirect immunofluorescent antibody procedure. The suture of one fluorescing oocyst has opened, allowing escape of the sporozoites (arrow). (E) Oocysts of *C. parvum* in an acid-fast stained smear of a fecal specimen. Oocysts stain red; fecal debris and yeast cells stain green or blue, depending on counterstain. (F) Note four sporozoites (arrow) and a prominent residual body in an acid fast stained oocyst. (G) Histologic section of human small intestine. Note the numerous parasites (arrows) within the brush border of the enterocytes. (H. & E. stain.) (H) A thin plastic section stained with toluidine blue showing numerous developmental stages of *C. parvum* (arrows) in the microvillous region of ileal enterocytes of an experimentally infected suckling mouse. Bar = 5 µm. ([A to D and H] Courtesy of W. C. Current, Lilly Research Laboratories, Indianapolis; [F] Courtesy of B. C. Anderson, University of Idaho, Caldwell.)

given immune transfer factor responded with decreased frequency of bowel movements and weight gain; five patients had eradication of oocysts from stools.[535a] However, the same transfer factor did not provide prophylaxis for any of nine calves challenged with oocysts of *Cryptosporidium*.[237] Epidemiologic studies on diarrhea and *Cryptosporidium* in Costa Rica, Ecuador, Guatemala, Haiti, and Liberia indicated that breast-fed children rarely or never had cryptosporidiosis.[236] These studies did not provide the type of data to enable investigators to determine whether antibody or other biologically active factors in milk or whether reduced exposure to contaminated food, water, or utensils were responsible for the lower prevalence of cryptosporidiosis in the breast-fed children. However, the concept that passive lacteal immunity might be beneficial was investigated in studies utilizing colostrum. In a mouse model, dams that recovered from enteric cryptosporidiosis and were immunologically boosted by oocyst challenge were unable to passively protect suckling pups from experimental cryptosporidiosis.[561] Clinical cure of cryptosporidiosis was reported for three immunodeficient persons that received hyperimmune bovine colostrum.[830,833] In contrast, no cure was obtained for three other immunodeficient persons that received nonhyperimmune bovine colostrum.[720] In a series of studies, hyperimmune bovine colostrum was found to provide partial prophylaxis for calves against oral challenge with oocysts,[239] neutralization of sporozoites and oocysts inoculated into mice,[238] and therapeutic treatment of an AIDS patient.[842] Although the concentrations of anti-*Cryptosporidium parvum* IgA, IgG, and IgM in the colostrum were determined to be high,[239] the factors actually responsible for this passive lacteal protection in calves, mice, and the humans had not been determined.

VIII. TREATMENT

A total of 94 therapeutic and preventive modalities have been tested for efficacy against cryptosporidiosis in humans or animals (Table 4). Supportive care with oral or intravenous hydration, possibly with parenteral nutrition, is a therapeutic intervention that offers a clear benefit for most immunocompetent patients.[236] For these individuals the disease is self-limiting, and hydration alone or no treatment may suffice. For immunocompromised individuals the disease may be fulminating and life threatening unless any exogenous cause of immunosuppression can be removed. Few chemotherapeutics have shown any efficacy. Anecdotal success has been reported for diloxanide furoate,[42] furazolidone,[42] quinine plus clindamycin,[44] amprolium,[852] and interleukin-2.[405] Inconclusive results have been obtained with spiramycin, a macrolide antibiotic similar to erythromycin and clindamycin. An evaluation of 37 patients with cryptosporidial diarrhea who were treated with spiramycin led investigators to conclude that 28 had a favorable response in the reduction of the number of daily bowel movements to less than 50% of baseline and 12 stopped shedding oocysts.[567] Unfortunately, details regarding other concurrent medications and specific clinical data were lacking. Of ten immunodeficient patients with cryptosporidiosis, five had complete resolution of diarrhea and four had symptomatic improvement after treatment for 1 week; one patient required 30 d of treatment.[656] One AIDS patient with cryptosporidiosis involving the pancreas, biliary tree, bowel, and respiratory tract resolved his diarrhea, but continued to shed oocysts after spiramycin treatment for 4 weeks.[299] Five AIDS patients with cryptosporidiosis treated with spiramycin had significant reductions in stool volumes, but side effects limited the length of treatment; after treatment stopped, cryptosporidia were still found in stools.[173] In a study of cryptosporidial diarrhea of infants, 21 of 39 patients received spiramycin and 18 received a placebo with no difference in outcome between the two groups.[894] In a report of the Centers for Disease Control, 7 of 14 patients showed no improvement.[44] These studies fail to clearly establish spiramycin efficacy for treatment of cryptosporidiosis. Erythromycin has had less testing than spiramycin. Of ten AIDS patients with cryptosporidial diarrhea, six showed a significant decrease in stool volumes in response to erythromycin therapy, but treatment was discontinued due to adverse side effects.[173] Some

TABLE 4
List of Therapeutic and Preventive Modalities Used for *Cryptosporidium* Infections

Alborixin[b]
Amikacin[a]
Amphotericin B[a]
Ampicillin[a,b]
Amprolium[a,b]
Arpinocid[b]
Bismuth salicylate[a]
Bleomycin[b]
Bovine colostrum[a,b]
Bovine transfer factor[a,b]
Carbenicillin[a]
Cefamandole[a]
Chloramphenicol[b]
Chloroqine[a]
Cholestyramine[a]
Cimetadine[a]
Clindamycin[a]
Clonidine[a]
Clopidol[b]
Cloxacillin[a]
Colistin[a,b]
Cotrimoxazole[a]
Cyclosporin A[b]
Decoquinate[b]
Diclazuril[a]
Difluoromethyl-ornithine[a,b]
Di-iodohydroxy-equinoline[a,b]
Diloxanide furoate[a]
Dimetridazole[b]
Dinitolmide[b]
Diphenoxylate HCl[a]
Doxycycline[a]
Emtryl[b]
Enterolyte N[b]
Erythromycin[a]
Ethopabate[b]
Framycetin[a]
Furaltadone[b]
Furazolidone[a,b]
Gamma globulin[a]
Gentamicina,[b]
Gluten-free diet[a]
Halofuginone[b]
Indomethacin[a]
Interleukin-2[a]
Iodoquinol[a]
Ipronidazole[b]
Ivermectin[b]
Kaolin-pectin[a,b]
Ketoconazole[a]
Lasalocid[b]
Levamisol[a]
Lincomycin[b]
Loperamide[a]
Maduramycin[b]
Mebendazole[a]
Mepracrine[a]
Methylbenzoquate[b]
Metronidazole[a,b]
Monensin[b]
Morphine[a]
Naproxyn[a]
Neomycin[b]
Nicarbazin[b]
Nitrofurantoin[a]
Nystatin[a,b]
Octreotide[a]
Oxytetracycline-metronidazole[b]
Paramomycin[a]
Penicillin[a]
Pentamidine[a]
Phenamidine[b]
Piperazine[a]
Polymixin-Furazolidone[b]
Primaquine[a]
Pyrimethamine[a]
Quinacrine[a,b]
Quinine[a]
Quinine-clindamycin[a]
Robenidine[b]
Salinomycin[a,b]
Septrin[a]
Somatostatin[a]
Spectinomycin[b]
Spiramycin[a]
Streptomycin[b]
Sulfonamides[a,b]
Tetracycline[a]
Thiabendazole[a]
Trimethoprim[a,b]
Trimethoprim-sulfamethoxazole[a,b]
Trinamide[b]
Tincture of opium[a]
Vancomycin[a]
Zidovudine[a]
Zoaquin[b]

[a] Humans.
[b] Other mammals.

Modified from Reference 236.

TABLE 5
Groups at Risk of Acquiring Cryptosporidiosis

Occupational groupings	
	Veterinary workers
	Farmers
	Animal handlers and caretakers
	Medical/health care workers
	Day care center employees
Social groupings in contact with infected persons/places	
	Family
	Sexual partner
	Nosocomial setting
	Day care center attendees
	Pet owner
Travelers to selected areas	
Immunodeficient and immunosuppressed persons	
	Congenital deficiency
	Acquired deficiency
	Chemically induced suppression
	Malnourished

Adapted from Reference 180.

antidiarrheal drugs have provided symptomatic improvement. Somatostatin analogue SMS-201-995 was administered subcutaneously to an AIDS patient who had cryptosporidial diarrhea for 10 months which was refractory to conventional therapy.[175] Although the cryptosporidiosis remained unchanged, the number of bowel movements and stool volume dramatically decreased with therapy, which was still being administered when the manuscript was submitted 8 months after it was initiated. Diphenoxylate treatment was effective until adverse side effects necessitated discontinuation.[173] In all seven patients given long-acting morphine sulphate, stool volumes were reduced more than 50% and stools became formed in six patients.[173] Treatment of three AIDS patients having cryptosporidial diarrhea by zidovudine (AZT) resolved both the diarrhea and oocyst shedding.[173] Testing of the relatively new anticoccidial drug diclazuril is now underway in AIDS patients with cryptosporidiosis.[756a]

Most drug testing with animal models has been conducted using prophylactic regimens. Of 15 anticoccidials tested in suckling mice, none prevented infection even at high doses.[35] However, amprolium, arprinocid, dinitolmide, salinomycin, and sulfaquinoxaline reduced the number of oocysts excreted compared with untreated controls. The most efficacious of these drugs, arprinocid, reduced but did not completely inhibit development of *C. parvum* in neonatal hamsters,[410] was ineffective in mice when given therapeutically,[820] and failed to control cryptosporidiosis in lambs.[35] Lasalocid, an ionophore polyether antibiotic, was found effective for prophylaxis against experimental *C. parvum* in calves, but was thought to be toxic at the levels administered.[557] However, when administered at nearly twice that level it effectively interrupted the life cycle of the parasite in 12 experimentally infected calves and in 320 naturally infected calves without toxic effects.[281] Initial screening studies utilizing a suckling mouse model indicate cyclosporin A,[867] alborixin, and maduramycin to be efficacious in suppressing oocyst development when administered prophylactically.[105b] Thorough evaluations of arprinocid, lasalocid, cyclosporin A, alborixin, and maduramycin should be undertaken.

IX. CONTROL OR PREVENTION

The key to control and prevention is elimination or reduction of the exogenous oocyst stage from the environment or avoidance of contact with known sources. Certain groups of people are

TABLE 6
Disinfectants Tested for Killing of *Cryptosporidium* Oocysts

Agent	Concentration
Ammonia	1%, 5%
Benzylkonium chloride	5%
Campden solution	0.5%
Chloramine B	3%
Cresylic acid	3%
Ethanol	90%
FAM 30	1%
Formaldehyde	1%, 5%
Formol saline	10%
Glutaraldehyde	2%
Hydrogen peroxide	10 vol
Hypochlorite	1%
Hypochlorite pH 7.6	10%
Iodophore	4%
Isopropanol	90%
Lysol	4%
Paracetic acid	0.2%
Phenol	1%
Potassium permanganate	1%
Propanol	90%
Providone iodine	10%
Commercial products	
Buraton	—
Dettol	5%
Dikonit	3%
Exspor	—
Formula H	—
Hycolin	—
Joctonal A	3%
Lastanox Q	0.2%
Mycolastanox	0.2%
Oo-cide	5%
Presept	—
Sporocidin	—
Tegodor	—
Vircon	1%

Compiled from References 32, 75a, 133, and 621.

at greater risk of exposure than others (Table 5). Under favorable conditions of moisture and moderate temperature, oocysts can remain viable and infectious for a relatively long time. Oocysts for research purposes are usually stored in aqueous suspensions at approximately 4 C. Although oocysts begin to lose infectivity as early as 2 months,[556] some have been found viable at 6 to 9 months[824] and infectious for cultured cells at 12 months.[189] The best storage solution seems to be 2.5% aqueous potassium dichromate.[189,673] Because chromium is a heavy metal and requires special handling to prevent environmental contamination, another storage medium should be sought.

Temperature extremes affect oocyst viability. Freezing, even in the presence of cryoprotectants, kills oocysts. Infectivity decreased after warming to 45 C for 5 to 20 min,[21] and viability was lost after 65 C for 30 min.[824]

Of the various chemical agents assayed for disinfection by *in vitro* excystation technique or mouse bioassay, few have been found to be effective within a relatively short time of exposure (Table 6). The four most effective agents for killing oocysts are hydrogen peroxide, a chlorine dioxide-based cold sterilant (Exspor), a two phase product producing ammonia (Oo-cide),[109] and ozone.[278]

BIBLIOGRAPHY

Information pertinent to the subject matter of this chapter may be found in References 7, 11, 12, 16, 17, 21, 32, 33, 35, 36, 37, 42, 44, 46, 61, 70, 95, 98, 102, 105b, 109, 119, 123, 148, 161, 172, 173, 175, 180, 186, 187, 189, 190, 194, 195, 197, 199, 200, 217, 234 to 238, 239a, 250, 278, 279, 281, 286, 299, 314, 319, 322, 325, 326, 328, 342, 349, 355, 356, 370, 375, 389, 390, 403, 405, 410, 418, 419, 434, 444, 457, 458, 468, 473, 474, 490, 492, 512, 530, 539, 540, 556 to 558, 561, 591, 597, 602, 607, 621, 636, 644, 648, 656, 664, 669, 673, 674, 679, 689, 690, 700, 704, 720, 734, 740, 743, 756a, 769, 774, 782, 789, 795, 798, 806 to 809, 811, 813, 814 to 816, 820, 821, 824, 825, 830, 833, 838, 840, 842, 852, 854, 867, 871, 882, and 894

Chapter 2

TECHNIQUES AND LABORATORY MAINTENANCE OF *CRYPTOSPORIDIUM*

W. L. Current

TABLE OF CONTENTS

I. EXPERIMENTAL INFECTIONS

A. CALVES, GOATS, LAMBS

The three best laboratory hosts for propagating large numbers of *C. parvum* oocysts are suckling calves, goats, and lambs that are less than 3 d of age at the time of inoculation.[27,40,336] Obtaining and maintaining calves, goats, and lambs completely free of contaminating isolates of *C. parvum* is sometimes difficult and requires special isolation facilities. All three of these hosts are highly susceptible to *C. parvum* infections, and inoculation with most isolates of the parasite results in the production of diarrheic feces containing many oocysts.[12,13,18,110] It is easier to purify oocysts from diarrheic feces than from formed stools. The following procedures should provide the most consistent results for producing large numbers of oocysts that can be isolated and purified from feces.

1. Place 1-d-old animals in isolation in pens (such as a metabolic crate) equipped with a pan to collect feces.
2. Mix 10^6 to 10^7 oocysts with a small volume of fluid (warm water, colostrum, milk, or milk replacer) and feed by bottle. Prior to inoculation oocysts can be purified by the Sheather's sugar flotation procedure described below and then counted with a hemacytometer.[110] Counting oocysts is much easier if one uses phase-contrast or Nomarski interference contrast microscopy and a thin hemacytometer (S/P phase counting chamber, Scientific Products) designed for platelet count by phase-contrast microscopy.
3. Most isolates of the parasite will produce diarrhea beginning 3 to 4 d after inoculation. At the onset of diarrhea, the animals should be fed a commercial oral electrolyte solution instead of milk or milk replacer. After oocyst collection is complete, the animals can again be placed on milk or milk replacer. The use of oral electrolyte solutions reduces the amount of fat that is often in high concentration in diarrheic feces of infected animals fed milk. A high concentration of fat interferes with oocyst recovery and purification.
4. Feces of infected animals should be collected in pans containing 2 to 3% (w/v) potassium dichromate ($K_2Cr_2O_7$); it is best to store each daily collection separately and refrigerated. The largest numbers of oocysts are usually found on days 5 to 8 after inoculation. Daily collections should be examined microscopically, and those with the highest numbers of oocysts should be used first for isolation and purification of the parasite. Collections containing fewer oocysts are suitable for reinfecting animals.

B. SUCKLING RODENTS

Suckling mice or rats are suitable hosts for maintaining pure isolates of *C. parvum* in the laboratory. Most rodents are supplied as *Cryptosporidium*-free animals and can be maintained easily in strict isolation. However, the number of oocysts that can be recovered from these suckling rodents is extremely small compared to the number that can be recovered from a calf.[110] The following procedures may be useful for maintaining pure isolates in suckling mice or rats. Additional aspects of using suckling rodents as models to study *Cryptosporidium* infections are discussed in Chapter 8.

1. Separate *C. parvum* oocysts from fecal debris by the Sheather's sugar flotation method described below. If this procedure does not result in a sufficiently clean suspension of oocysts (10^7/ml), additional purification can be achieved by the Percoll® isopycnic or the cesium chloride density gradient centrifugation techniques (below). Count with a hemacytometer and resuspend in PBS or water at 10^7 oocysts per milliliter.
2. Orally inoculate 10^5 oocysts of *C. parvum* (10 μl of the purified 10^7 oocyst per milliliter suspension) into 3- to 5-d-old mice or rats by placing a disposable tip fitted to a micropipet into the back of the animal's throat and slowly delivering the oocyst suspension.

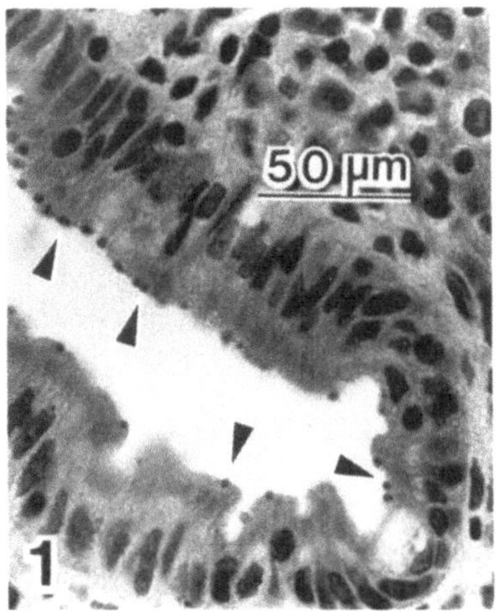

FIGURE 1. Photomicrograph of a histologic section of a small bowel biopsy obtained from a patient with congenital hypogammaglobulinemia and persistent cryptosporidiosis.[186] Arrowheads point to four of the many developmental stages of the *C. parvum* in the microvillous border of the enterocytes. (H. & E. stain.)

Swallowing reflexes can be stimulated by gently rubbing the back of the suckling rodent's tongue. Since the ID_{50} (dose at which half of the inoculated animals become infected) is between 100 and 500 oocysts,[233] it is not necessary for each animal to swallow all 10 μl of the oocyst suspension.

3. Maintain the suckling animals as litters with their dams until oocysts are harvested 6 to 8 d postinoculation (DPI) at which time the animals should be euthanized by CO_2.
4. The largest numbers of oocysts are usually present in the gut 6 to 8 DPI. It is impractical to collect feces from suckling mice and rats because fecal output is minimal and the dam usually eats the feces. The lower half of the small intestine, the cecum, and the entire colon of each animal in the litter should be pooled, placed in approximately 10 ml of deionized water containing 0.02% Tween®-20 (dH_2O+T), and homogenized to free oocysts present in the intestinal contents and enterocytes. Homogenization is best achieved by the use of a high-speed tissue homogenizer (e.g., Brinkmann Polytron).
5. Gut homogenates containing oocysts should be washed three times by centrifugation in dH_2O at $1000 \times g$ for 10 min. The pellet containing oocysts should then be resuspended in 2.5% $K_2Cr_2O_7$ or balanced salt solution containing antibiotics (see below) and stored at 4°C. Most oocysts stored this way will remain viable for 3 months, and some may remain viable for up to 12 months. Since oocysts begin to loose viability after 3 months of storage, it is advisable to generate fresh inoculum every 3 to 4 months. Oocysts harvested from suckling mice or rats can be purified using the same techniques to purify oocysts from the feces of suckling calves or goats (see below). Oocysts recovered from suckling mice or rats can also be used to inoculate suckling calves or goats for the production of large numbers of parasites.

II. IDENTIFICATION OF OOCYSTS AND ENDOGENOUS STAGES

Before 1980, human cryptosporidiosis was diagnosed by histologic examination of biopsy of the intestinal mucosa.[192] Routine hematoxylin and eosin can be used to visualize the developmental stages of the parasite as dark, basophylic, spherical bodies 2 to 5 μm (depending on stage in life cycle) in the brush border of mucosal epithelium (Figure 1). Transmission

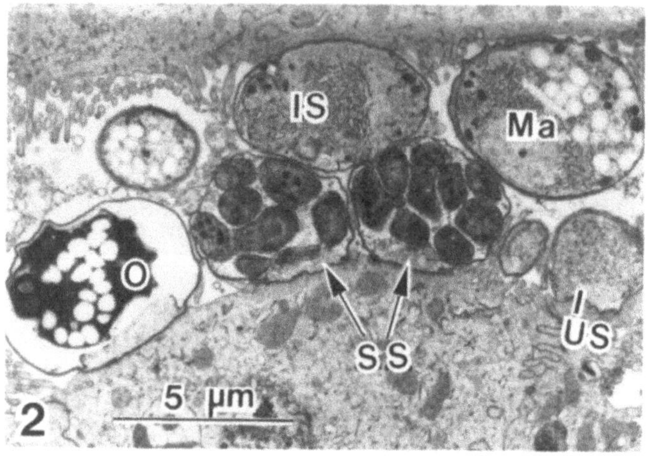

FIGURE 2. Transmission electron micrograph of ileal enterocytes of a suckling mouse experimentally infected with *C. parvum*. Developmental stages of the parasite are within parasitophorous vacuoles confined to the microvillous border of the host cells. Developmental stages shown include mature schizonts (S) containing merozoites, an immature schizont (IS) in the process of merogony, a macrogamont (Ma), a sporulated oocyst (O), and several uninucleate schizonts (US).

electron microscopy[195] can be used to confirm the diagnosis (Figure 2). Such invasive, expensive, and time-consuming procedures are no longer required for the diagnosis of cryptosporidiosis since a variety of techniques have been developed to identify *Cryptosporidium* spp. oocysts in feces and other body fluid specimens.[261] Oocysts, usually released in large numbers from the infected epithelium, can be identified in stool specimens representing a sampling of the entire intestinal tract or in sputum or respiratory aspirates representing a sampling of the entire respiratory tract. Biopsy specimens may be of some value in determining the site of infection and lesions associated with the parasite.

A. PRESERVATION AND STORAGE OF SPECIMENS CONTAINING OOCYSTS

1. Preservation of Specimens

Most diagnostic specimens examined for oocysts will be stool samples; however, other fluids such as sputum should be handled the same way. For the diagnosis of cryptosporidiosis, stool specimens should be submitted as fresh material or in 10% formalin or sodium acetate-acetic acid-formalin (SAF) preservatives. If parasites are not going to be used for subsequent experimental purposes, fixed specimens are recommended because of biohazard considerations. Fresh or preserved stool specimens can be examined as wet mounts or they can be concentrated or stained as outlined below to aid in the visualization of *Cryptosporidium* spp. oocysts.

2. Storage of Specimens

Potassium dichromate solution ($K_2Cr_2O_7$, 2 to 2.5% w/v in water) is used routinely as a storage medium to preserve oocyst viability; it is not a fixative. When stored at 4°C in $K_2Cr_2O_7$, *Cryptosporidium* spp. oocysts remain viable for at least 3 months, and some may retain infectivity for up to 12 months.[192] Since the percentage of viable oocysts begins to decrease after 3 months of cold storage, it is advisable to generate fresh oocysts every 3 to 4 months. Some researchers prefer to store oocysts of *Cryptosporidium* spp. in balanced salt solutions supplemented with antibiotics because chromium is an environmental contaminant requiring special disposal and handling procedures. Hank's balanced salt solution (HBSS) containing penicillin, 10,000 U; streptomycin,10 mg; fungizone, 0.05 mg; and mycostatin, 500 U/ml is recommended for the storage of *Cryptosporidium* spp. oocysts and also for the storage of sporocysts of *Sarcocystis* spp.[225]

B. IDENTIFICATION OF OOCYSTS

Two general approaches have been widely used to identify *Cryptosporidium* spp. oocysts in stools or body fluid specimens. Concentration techniques are used routinely in the research laboratory, whereas staining techniques, before or after concentration of oocysts, are used more often in the clinical microbiology laboratory.[259,261] Because *Cryptosporidium* spp. oocysts are similar in size and shape to some yeast cells, considerable experience is required to obtain accurate results with concentration and/or staining techniques. Useful stool concentration techniques include flotation of oocysts in Sheather's sugar solution, in zinc sulfate solution (1.18 or 1.2 sp gr), or in saturated sodium chloride solution (1.27 sp gr). Stool concentration techniques using sedimentation include formalin-ether and formalin-ehtyl acetate. Some studies have found no differences among these methods, whereas others have found formalin-ether sedimentation and sodium chloride flotation to be the most sensitive procedures.[146,259] Some researchers feel that the Sheather's sugar flotation gives results equal to or better than those obtained with formalin-ether or formalin-ethyl acetate sedimentation.[498,536,913] Sheather's sugar flotation, perhaps the most widely used concentration technique, is outlined below.[12,186]

1. Sheather's Sugar Flotation

1. If the stool specimen is not fluid, it may be necessary to dilute it with water before straining it through cheesecloth to remove large particulate material. Mix thoroughly approximately 1 ml of fecal (sputum) suspension with 10 ml of Sheather's sugar solution (sugar, 500 g; water, 320 ml; phenol, 6.5 g) in a 15-ml centrifuge tube. A cap should be placed on the tube to prevent aerosol formation.
2. Centrifuge at 500 × *g* for 10 min, obtain a sample from the surface of the flotation medium (this can be done easily by using a wire loop or Pasteur pipet), place it on a microscope slide, add a coverslip, and examine microscopically (see 3).
3. Sheather's sugar flotation combined with phase-contrast microscopy is an excellent procedure to identify oocysts and to distinguish oocysts from contaminating yeast cells.[186,192,194] *C. parvum* oocysts appear as bright, birefringent, 4.5 to 5.5 μm spherical bodies containing one to four dark granules (Figure 3). With good-quality, phase-contrast, oil immersion objectives the four sporozoites surrounding the centrally located oocyst residuum can often be seen. Oocysts of other species of *Cryptosporidium,* e.g., *C. muris* or *C. baileyi,* will be of a different size and shape, but will otherwise appear similar to *C. parvum.*[194,844] Contaminating yeast cells do not appear as bright, birefringent bodies, do not contain dark granules or sporozoites, and sometimes have small budding yeasts on their margin. When viewed with bright-field microscopy, oocysts are often difficult to distinguish from yeast cells.[12,259,498] When viewed with bright-field microscopy, *C. parvum* oocysts appear as 4.5- to 5.5-μm translucent, spherical bodies containing one to four dark granules (Figure 3). With some microscopes, these oocysts often appear light pink. Yeast cells do not have the characteristic granules and are not pink in color. After approximately 15 min in the hyperosmotic Sheather's solution, some oocysts begin to collapse, and the percentage of collapsed oocysts increases with time. Collapse of the oocysts usually involves a folding in of the oocyst wall along the oocyst wall suture.

2. Acid-Fast Staining

Most recommended stains for *Cryptosporidium* spp. oocysts cannot be performed on stools

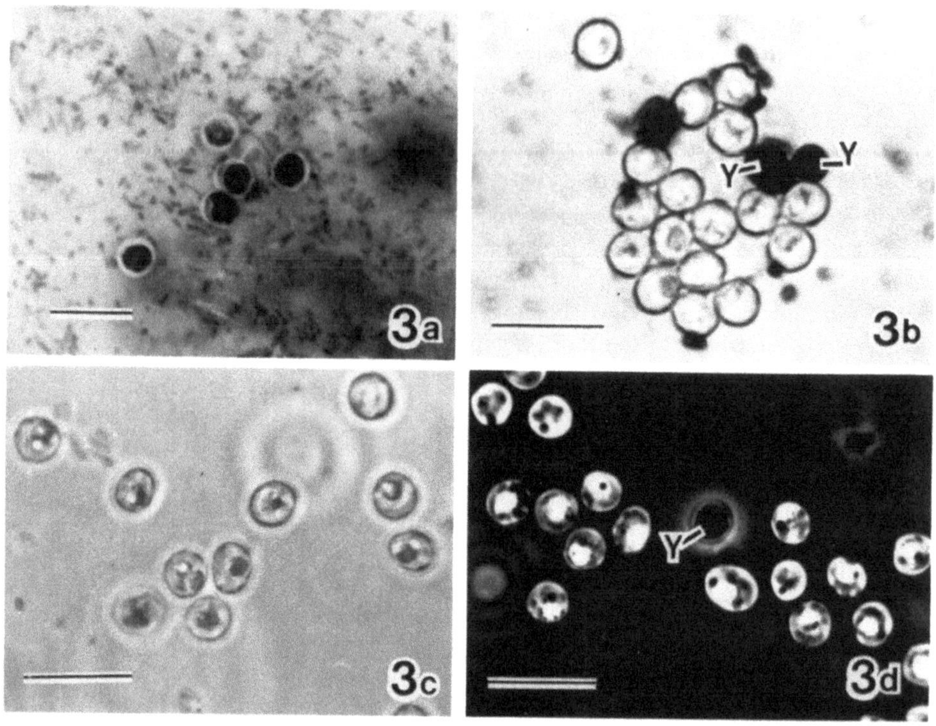

FIGURE 3. Photomicrographs of *C. parvum* oocysts as they appear using several staining and concentration procedures developed to demonstrate *Cryptosporidium* spp. oocysts in stool specimens. (a) Acid-fast stain of a fecal smear viewed with bright-field microscopy. Oocysts stain red against a blue or green (depending on counterstain) background. Fecal debris, bacteria, and yeast cells do not stain red, they take up the counterstain. (b) Negative stain of fecal smear viewed with bright-field microscopy. Oocysts appear bright and birefringent, whereas fecal debris and yeast cells (Y) stain darkly. (c) Oocysts concentrated by Sheather's sugar flotation and viewed with bright-field microscopy. Oocysts appear as translucent, spherical bodies containing one to four dark granules. (d) Oocysts concentrated by Sheather's sugar flotation and viewed with phase-contrast microscopy. Oocysts appear bright and birefringent, whereas yeast cells and most fecal debris are dark. One to four dark granules can be seen within the oocysts.

preserved in polyvinyl alcohol (PVA) fixative. The routine stains (trichrome, iron hematoxylin) used for stool diagnosis of other parasites are not helpful for the identification of *Cryptosporidium* spp. oocysts. The most widely used staining techniques for demonstrating *Cryptosporidium* spp. oocysts in fecal specimens are based on the acid-fast staining properties of the oocyst wall. Several acid-fast stains[261,498] have been used in hot or cold procedures[259,495] and modified by incorporation of dimethyl sulfoxide.[127,643] It appears that reported differences in acid-fast stains and methods represent personal preference rather than marked differences in sensitivity and specificity. The most commonly used acid-fast procedure is the modified Kinyoun cold technique[495] outlined below.

1. With a wooden applicator stick or a disposable culture loop, spread the specimen on a microscope slide and allow to air dry.
2. Fix in absolute methanol for 5 to 10 min and air dry.
3. Cover the smear with modified Kinyouin acid-fast stain (fuchsin-carbol, Harleco, Gibbstown, NJ) for approximately 2 min.
4. Rinse briefly in tap water (some prefer 50% ethanol), cover the smear with decolorizing solution (10% H_2SO_4), and incubate for approximately 1 min and then rinse briefly in tap water.

5. Cover the smear with counterstain (light green SF yellowish stain, Harleco; or methylene blue, 0.3 g/100 ml deionized water), incubate for approximately 1 min, rinse in tap water, air dry, and observe microscopically.
6. For optimal staining with this procedure, staining, decolorizing, and counterstaining times are determined empirically. Also, it is important that fecal smears are not too thick to ensure adequate penetration of the reagents.
7. When stained properly, oocysts are 4.5- to 5.5-μm red, spherical bodies against a green or blue (depending on counterstain) background. Oocysts stain red with varying degrees of intensity, whereas most fecal debris and yeast cells take up the color of the counterstain. Acid-fast stained oocysts in a fecal smear are shown in Figure 3.

3. Negative Staining

The negative staining procedure originally described by Heine[327] is also useful for detecting oocysts in fecal specimens. This procedure has the advantage of requiring less preparation time and fewer procedures than the Kinyouin technique, but does not result in the production of a permanent slide. The procedure works best on fresh fecal specimens, but it will give satisfactory results with specimens fixed in 10% formalin. The author does not know if specimens preserved in other fixatives lend themselves to negative staining. A modification[187] of the original negative-staining procedure[327] is outlined below.

1. With a wooden applicator stick, mix thoroughly on a microscope slide an equal volume (5 to 10 μl) of feces and modified Kinyoun's carbol fuchsin stain. (The stain is prepared by dissolving 4.0 g of basic fuchsin in 100 ml deionized water plus 20 ml 95% ethanol. After the stain is dissolved, slowly add 8 ml of liquified phenol.)
2. As soon as the smear has air dried, add a drop of immersion oil directly to the smear. Place a coverslip on top of the immersion oil, and observe with a bright-field microscope.
3. The characteristic bright, birefringent appearance of *Cryptosopridium* spp. oocysts against a dark-staining background is due to an oocyst wall that is not stained during this procedure and that is impervious, preventing entry of the stain and loss of water from the interior of the oocyst. Fecal debris, yeast, and bacteria all stain darkly. Often the stain collects around the outside of the oocyst as the smear dries, resulting in dark rings around bright oocysts (Figure 3). After approximately 15 min in immersion oil, the oocysts begin to collapse.

4. Immunofluorescent Antibody Stains

Most humans and animals with cryptosporidiosis pass enough oocysts in their stools so that most of the concentration and/or staining techniques described in the literature (several are outlined above) are adequate for detection and diagnosis. However, more sensitive techniques are sometimes needed to detect oocysts in specimens that contain few parasites and large amounts of debris. These specimens include fecal samples from asymptomatic carriers or filtrates from surface or drinking water samples. Both polyclonal and monoclonal antibodies specific to *C. parvum* can be used in direct or indirect immunofluorescent assays to detect oocysts in these specimens.[260,502,602] An immunofluorescent diagnostic kit is available commercially (Merridian Laboratories, Cincinnati, OH) which utilizes a monoclonal antibody specific to the oocyst wall of *C. parvum*.[260] Polyclonal antibodies can also be used as described below.

Preparation of polyclonal antisera —

1. For immunizing rabbits, suspend oocysts (10^8/ml) in 2 m*M* TRIS buffer (pH 8.0) containing 10 m*M* phenylmethyl-sulfonylfluoride (PMSF) and disrupt by three freeze-thaw cycles, involving incubation in liquid nitrogen for 10 min followed by incubation at

37°C until the suspension is thawed. Disruption of oocysts and sporozoites can be monitored microscopically.

2. Inoculate rabbits every 4 weeks with approximately 10^7 disrupted oocysts. Choices concerning the use of adjuvants and route of administration are numerous and a matter of preference. Intravenous administration of disrupted oocysts without adjuvant and subcutaneous boosting (two to three times) with the antigen mixed in Freund's incomplete adjuvant has been used in our laboratory and results in sera with high titers of antibody specific to *Cryptosporidium*.[196]

Indirect immunofluorescent antibody procedure using polyclonal antibodies —

1. With a wooden applicator stick or a disposable culture loop, spread the specimen on a microscope slide and allow to air dry. To save reagents, a 12-well multispot microscope slide (Shandon, Inc., Pittsburgh, PA, Catalog No. 99910090) can be used. Approximately 4 µl of specimen can be placed in each well, air dried, and processed as described below.
2. Fix the smear in cold 95% ethanol for 5 min and then incubate in phosphate buffered saline (PBS, pH 7.2) at room temperature for 2 min.
3. Cover the smear (4 µl if 12-well slides are used) with rabbit antiserum (diluted to appropriate concentration in PBS) and incubate for 30 min at 37°C. The rabbit antiserum can be diluted 1:100 to 1:1000, depending on titer.
4. Wash 5 times by incubating in PBS for 3 min.

4a. If background fluorescence is a problem, one can use an appropriate blocking agent at this point to prevent nonspecific binding of the FITC-conjugated antibody added in step 5. Appropriate blocking agents include 1% bovine serum albumin or 1% gelatin, both in PBS.

5. Cover the smear (4 µl if 12-well slides are used) with goat anti-rabbit IgG conjugated to FITC (1:100 to 1:1000 dilution) and incubate for 30 min at 37°C. Addition of 0.01% (w/v) Evans blue stain to the diluted FITC-conjugated antibody helps distinguish oocysts from fecal debris and yeast cells. Oocysts will fluoresce bright green against a dark background, and contaminating debris and yeast cells will appear red as a result of the Evans blue counterstain.

C. DETECTION AND IDENTIFICATION OF ENDOGENOUS STAGES

Routine histologic methods can be used to detect endogenous stages of *Cryptosporidium* spp. in specimens of mucosal epithelium, but identification of different life cycle stages can be difficult (Figure 1). Transmission electron microscopy facilitates identification of developmental stages of *Cryptosporidium* spp. that reside within parasitophorous vacuoles confined to the microvillous region of epithelial cells (Figure 2).[192,195] To identify stages of *Cryptosporidium* spp. in fresh mucosal scrapings,[195] cultured cells,[189] or the chorioallantoic membrane of chicken embryos,[188] a light microscope equipped with high-quality phase-contrast or Nomarski interference contrast (NIC) optics is best (Figure 4). The author prefers NIC optics for this purpose. If NIC or phase-contrast optics are not available, one can use bright-field microscopy to identify some of the developmental stages of *Cryptosporidium* spp. in stained mucosal smears.[441]

D. SERODIAGNOSIS

The use of serodiagnostic techniques for monitoring exposure of humans and animals to *Cryptosporidium* sp. has thus far been limited to a few laboratories. Indirect immunofluorescent antibody procedures have been used to detect *Cryptosporidium*-specific antibodies in the sera of humans,[134,148,200,420] in sera obtained from chickens,[196] and in sera obtained from eight other species of animals.[819] Specific anti-*Cryptosporidium* IgG and/or IgM can also be detected by an

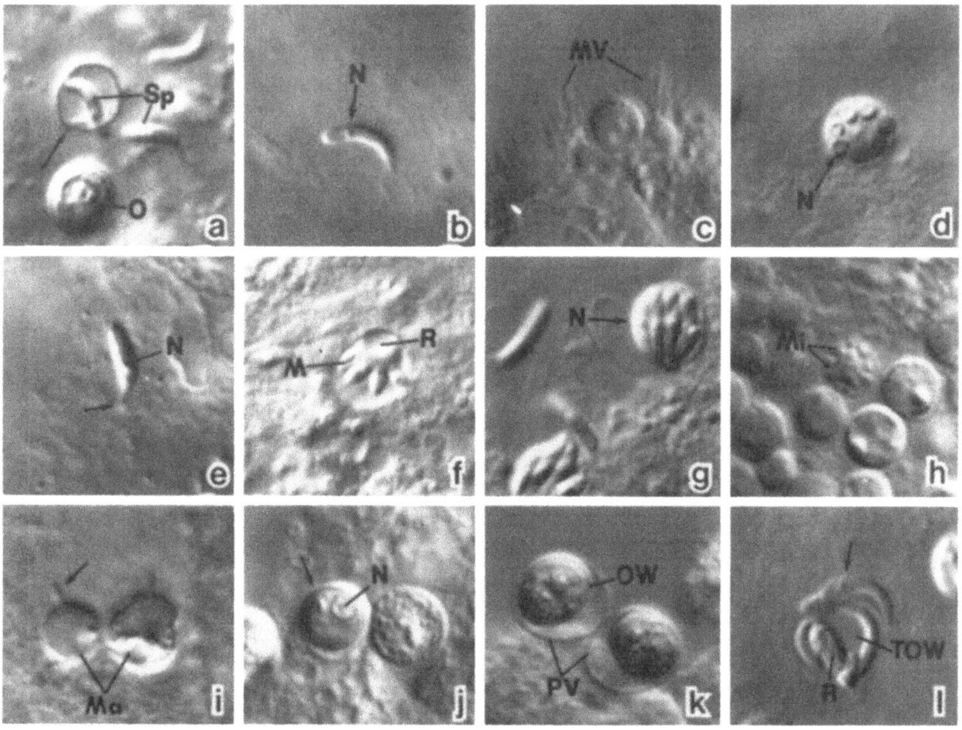

FIGURE 4. Nomarski interference-contrast photomicrograph of developmental stages of *C. parvum* in a mucosal scraping obtained from the ileum of experimentally infected suckling mice. (a) Sporozoites (Sp) free and excysting from the opening (arrow) in an oocyst, and an intact oocyst (O). (b) Free sporozoite showing the posterior location of the nucleus (N). (c) Uninucleate schizont (meront) surrounded by hypertrophied microvilli (MV). (d) Immature type I schizont (meront) with peripherally located nuclei (N), six of which are in focus. (e) Merozoite attached to host cell (arrow points to attachment site). Note the centrally located nucleus (N). (f) Immature type II schizont (meront) with four merozoites (M) budding from the residuum (R). (g) Mature type II schizont (meront) showing the four merozoites arranged like the segments of an orange. Nuclei (N) of all four merozoites are aligned in the center of the schizont. (h) Microgamont with microgametes (Mi) budding from the surface of the residuum. (i) Two macrogamonts (Ma) each with an attached microgamont (arrow points to one that is in focus). (j) Macrogamete containing an eccentric nucleus (N) and what appears to be a microgamete within the cytoplasm (arrow). (k) Two sporulated oocysts with thick walls (OW), both within parasitophorous vacuoles (PV). Most of the thick-walled oocysts are passed unaltered in the feces. (l) An autoinfective, thin-walled oocyst that has ruptured under coverslip pressure releasing the four sporozoites from the thin oocyst wall or membrane (TOW). Note the granular oocyst residuum (R) and the posterior location of the sporozoite nuclei (N). (Adapted from Current, W. L. and Reese, N. C., *J. Protozool.*, 33, 289, 1986.)

enzyme-linked immunosorbent assay (ELISA).[196,838,839] Additional evaluations are needed to confirm the utility of these serologic procedures for diagnosing and monitoring infections, for determining the prevalence of prior exposure in selected study populations, and to determine if there is any correlation between the presence of *Cryptosporidium*-specific serum antibodies and resistance to reinfection.

1. Indirect Immunofluorescent Antibody Procedure[134]

The same basic indirect fluorescent antibody (IFA) procedure has been developed using histologic sections of infected mouse intestine[134,819] or purified oocysts.[148,200] Infected mouse intestine can be used to detect antobodies to all life cycle stages of the parasite, whereas intact oocysts are used for detection of antibodies to the outer oocyst wall. The procedure using intact oocysts is outlined below.

1. Purify *Cryptosporidium* sp. oocysts as described below and suspend approximately 10^7 in deionized water. Add 4 µl of the oocyst suspension to each 6-mm well of a 12-well multispot microscope slide (Shandon, Inc., Pittsburgh, PA, Catalog No. 99910090) and air dry. These air-dried antigen preparations remain useful for IFA procedures for up to 12 months if the slides are stored frozen (–70°C) in a desicator.

1a. If multispot microscope slides are not available, 6-mm diameter wells can be made on a microscope slide with a wax pencil.

2. Fix oocysts in cold 95% ethanol for 5 min and then incubate in PBS (pH 7.2) at room temperature for 2 min.

3. Blot dry the Teflon® coating around each well, add 4 µl of the appropriate concentration of test serum (diluted in PBS) to each well, and incubate for 30 min at 37°C. An appropriate initial dilution of test serum to initiate a titration series is 1:10.

4. Wash five times by incubating in PBS for 3 min.

4a. If background fluorescence is a problem, one can use an appropriate blocking agent at this point to prevent nonspecific binding of the FITC-conjugated antibody added in step 5. Appropriate blocking agents include 1% bovine serum albumin or 1% gelatin, both in PBS.

5. Blot dry the Teflon® coating as in step 3, add 4 µl of appropriately diluted (in PBS) FITC-conjugated antibody that will recognize the host species and class of antibody of interest and incubate for 30 min at 37°C. Addition of 0.01% (w/v) Evans blue counterstain to the diluted FITC-conjugated antibody results in contaminating debris and yeast cells fluorescing red. Oocysts fluoresce bright green against a dark background (Figure 4g).

2. ELISA Procedure[196,839]

1. The antigen for coating 96-well plates is prepared from highly purified *Cryptosporidium* sp. oocysts obtained by the procedures outlined below. Oocysts (10^8/ml) should be suspended in 2 m*M* TRIS buffer (pH 8.0) containing10 m*M* phenylmethyl-sulfonylfluoride and subjected to three freeze-thaw cycles, involving incubation in liquid nitrogen for 10 min followed by incubation at 37°C until the suspension is thawed. Add sodium dodecyl sulfate to the ruptured oocyst-sporozoite suspension (1% w/v final concentration), centrifuge at 10,000 × *g* for 20 min at 4°C, and recover the supernatant. The stock antigen is prepared by diluting the recovered supernatent 1:10 in coating buffer (0.1 *M* carbonate buffer in deionized water, pH 9.6). Store the stock antigen as 1-ml frozen (–70°C) aliquots.
2. Coat 96-well assay plates overnight by adding (100 µl per well) antigen diluted appropriately (1:500 should be appropriate) in coating buffer.
3. The following day, remove the antigen solution and coat the wells with 1% gelatin (or other appropriate blocking agent) for 1 h at 37°C.
4. Rinse the wells (150 µl per well) three times with PBS containing 0.05% Tween® 20 (PBS-T).
5. Add 100 µl of test serum diluted appropriately in PBS-T containing 0.5% bovine serum albumin (PBS-T-BSA) and incubate for 1 h at 37°C. Rinse three times as in step 4.
6. Add 100 µl of appropriately diluted (in PBS-T-BSA) antibody conjugated to peroxidase that will recognize the host species and class of antibody of interest and incubate for 1 h at 37°C. Rinse three times as in step 4.
7. Add 100 µl of the appropriate substrate (ABTS, Kirkgard Perry Laboratories, Gaithersburg, MD), incubate for 30 to 60 min, and read absorbance on an ELISA reader with a 405-nm absorbance filter.

III. OOCYST PURIFICATION

Several techniques are used to isolate and purify oocysts from feces of infected hosts. With all of these techniques it is important to start with fecal specimens that have large numbers of oocysts. Diarrheic feces from heavily infected calves is the best source of oocysts to be purified (procedures for infecting calves and collecting and storing feces containing oocysts are presented above). However, efficiency of oocyst recovery varies from one specimen to another. The Sheather's sugar flotation procedure can serve as a stand-alone purification technique or it can be combined with isopycnic Percoll® centrifugation[56] or with cesium chloride step gradient centrifugation[409] if further purification of oocysts is required. Prior to initiation of the Sheather's sugar flotation procedure it is also important to minimize the amount of fecal debris in the specimen, and in some cases it may be necessary to remove some of the fecal fat. The following procedures have proven useful for isolation and purification of *Cryptosporidium* spp. oocysts from feces.

A. INITIAL REMOVAL OF FECAL DEBRIS

1. Mix the fecal specimen with an equal volume of 2.5% $K_2Cr_2O_7$ or antibiotic solution.
2. Strain the mixture through a series of wire-screened sieves, usually 100, 200, and then 300 or 400 mesh, rinsing the retained debris several times with water to flush through any trapped oocysts.
3. Allow the screened fecal suspension to stand in a tall glass cylinder for 1 h to sediment the heavier fecal debris. Remove and save the upper layer of fluid which contains most of the oocysts and continue with the Sheather's sugar flotation procedure described below.

B. REMOVAL OF EXCESS LIPID OR FECAL FAT

If the specimen contains excess lipid (fecal fat), treatment with chloroform may be helpful. Suggestions are also presented above (Section I.A) that will minimize the amount of lipid in feces from infected calves.

1. Vigorously mix equal volumes of the fecal specimen and chloroform and then allow two distinct layers to form. Ethyl ether can be substituted for chloroform.
2. Oocysts will remain in the aqueous phase and the lipid material will be partitioned into the upper chloroform layer.
3. Oocysts should be washed by centrifugation ($1000 \times g$ for 15 min) in water to ensure the residual chloroform is removed and then store at 4°C in 2.5% $K_2Cr_2O_7$ or antibiotic solution (described above).

C. SHEATHER'S SUGAR FLOTATION

After the large fecal debris is removed and the specimen does not contain an excessive amount of fat, a modification of the Sheather's sugar flotation procedure used to detect oocysts[186] can be used alone or in combination with the Percoll® isopycnic centrifugation or cesium chloride step gradient centrifugation techniques described below.

1. Adjust Sheather's solution (sugar, 500 g; water, 320 ml; phenol, 6.5 g) to a specific gravity of 1.2 (1.2 g/ml).
2. Add Tween®-20 (0.2% w/v) to the fecal suspension containing oocysts and mix thoroughly.
3. Mix thoroughly in a 50 ml centrifuge tube 30 ml of the adjusted Sheather's sugar solution (sp gr 1.2) with 5 ml of fecal suspension containing 0.2% Tween®-20.

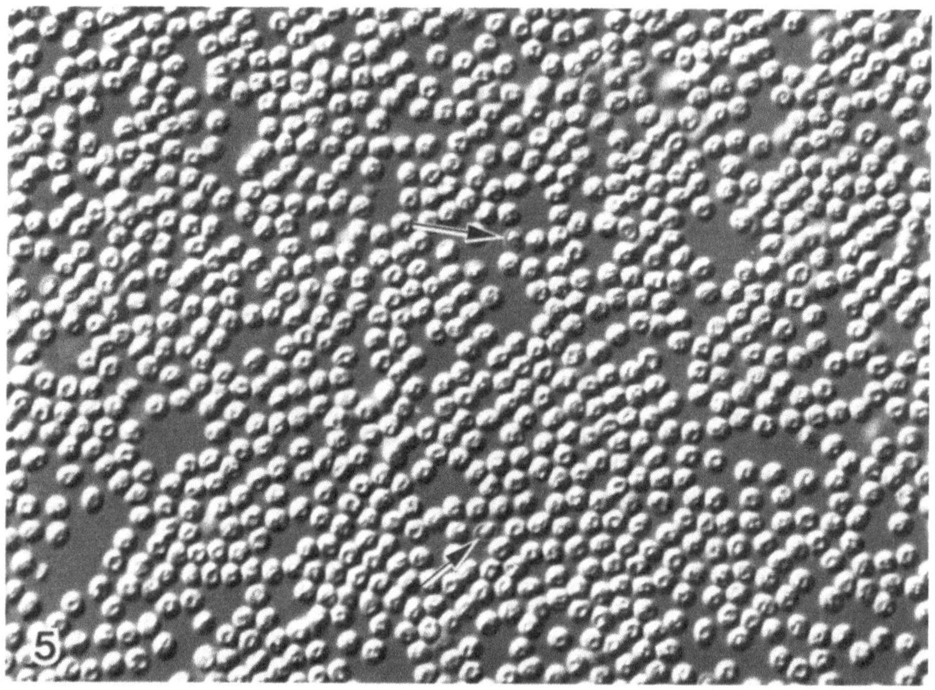

FIGURE 5. Nomarski interference-contrast photomicrograph of oocysts of *C. parvum* (approximately 5 µm in diameter) purified from calf feces by Sheather's sugar flotation. Note the absence of fecal debris. Two of the oocysts have released their sporozoites and are seen as oocyst walls surrounding the oocyst residuum (arrows).

4. Centrifuge at 1000 × *g* for 10 min and then layer carefully 5 ml of water containing 0.2% Tween®-20 on top of the flotation medium.
5. Stir the top layer of water with the tip of a pipet. By watching carefully one can see a wisp of oocysts being washed off of the surface of the flotation medium into the uppermost water layer.
6. Aspirate carefully the water layer. The 5 ml of water containing oocysts from each harvest can be pooled into the same tube(s). The number of oocysts recovered each time can be monitored microscopically.
7. Thoroughly remix the remaining flotation medium and repeat steps 4 through 6 two to four times.
8. Add an equal volume of water containing 0.2% Tween®-20 to the tube(s) containing the harvested oocysts and centrifuge at 1000 × *g* for 15 min. The pellet often has two layers, the whitish oocysts on the bottom and a darker upper layer containing fine fecal debris and bacteria. The upper layer can often be removed by careful aspiration and discarded with the supernatant. Recentrifugation and removal of the top layer of the pellet can result in suspensions containing highly purified oocysts (Figure 5).
9. Recovered oocysts should be stored in 2.5% (w/v) $K_2Cr_2O_7$ or in antibiotic solution (described above) at 4°C.

D. ISOPYCNIC PERCOLL GRADIENT CENTRIFUGATION

If the Sheather's sugar flotation procedure outlined above or the Sheather's sugar step gradient procedure described elsewhere[56] does not result in an oocyst suspension of high enough purity, oocysts can be purified further by isopycnic Percoll® gradient centrifugation.[56] Details of this secondary purification procedure are outlined below.

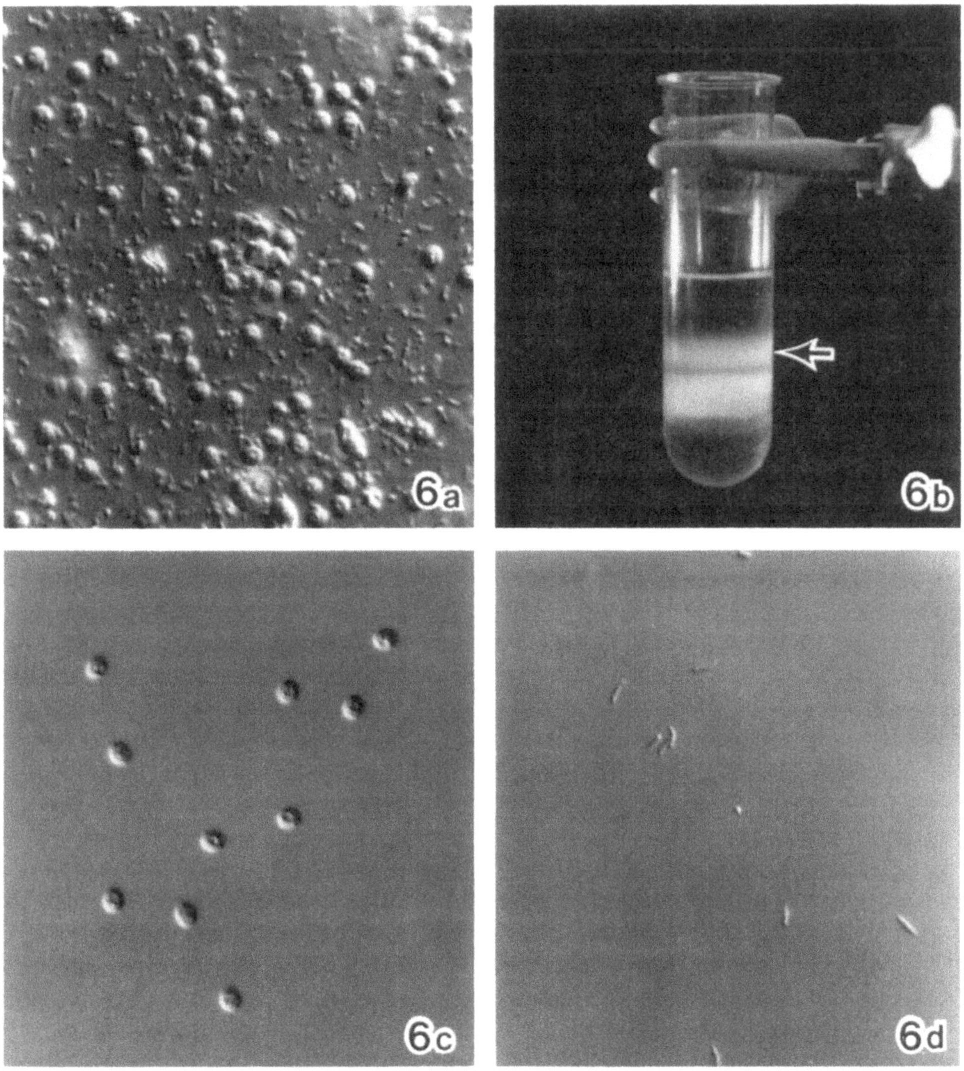

FIGURE 6. Nomarski interference-contrast photomicrographs (a, c, and d) and a photograph (b) demonstrating different steps in the purification of *C. parvum* oocysts by the cesium chloride step gradient centrifugation and anion exchange procedures (see Section III). (a) Oocysts (approximately 5 μm in diameter) concentrated from feces by Sheather's sugar flotation. (b) Following centrifugation, most oocysts can be recovered from the uppermost band (arrow). Fecal debris remains in the bands below the one containing most of the oocysts. (c) Oocysts recovered from the band denoted by an arrow in b. (d) Sporozoites recovered after excystation and purification by the anion exchange procedure described in Section IV. B. (Courtesy of M. Tilley and S. J. Upton, Division of Biology, Kansas State University, Manhattan, KS).

1. Prepare a Percoll® (Pharmacia, Piscataway, NJ) solution of 1.09 g/ml (320 mOsm, pH 7.4). The solution is prepared by mixing thoroughly nine parts Percoll®, one part 10x Alsever's solution, and nine parts 1x Alsever's solution. The Percoll® solution should be prepared fresh each time and used immediately. Hank's balanced salt solution (10x and 1x, pH 7.2) can be substituted for Alsever's solution.
2. Place 9 ml of the Percoll® solution in a 15-ml, high-speed centrifuge tube and then layer on the top 1 ml of an oocyst suspension (2 to 4×10^8/ml in Alsever's) obtained from the Sheather's sugar flotation procedure.

3. Centrifuge at 22,000 × *g* for 30 min at room temperature.
4. Oocysts form a dense band approximately one third the distance from the bottom, below the upper layer of bacteria.
5 Remove the band of oocysts and wash three times by centrifugation (1000 × *g* for 10 min) in PBS (pH 7.2) or water.
6. If the recovered oocysts are not going to be used immediately, they should be stored in 2.5% $K_2Cr_2O_7$ or antibiotic solution (described above) at 4°C.
7. If the recovered oocysts are not of sufficient purity, the entire procedure can be repeated using oocysts recovered from the first attempt.

E. CESIUM CHLORIDE STEP GRADIENT CENTRIFUGATION

The cesium chloride step gradient procedure, derived from the one described by Kilani and Sekla,[409] is outlined below and certain aspects are shown in Figure 6. Some feel that this procedure is superior to the Percoll® isopycnic centrifugation method because it results in more highly purified oocysts that are easily separated from the purification medium.

1. Prepare a stock solution of CsCl with a specific gravity of 1.8 (1.8 g/ml deionized water). If a refractometer is used, first dissolve approximately 10.0 g CsCl in 5 ml deionized water. Check the density with a refractometer and add water (in most cases) or CsCl to obtain a proper density of 1.8 g/ml (the refraction index at 25°C should read about 1.409 for 1.805 g/ml).
2. Prepare a stock TRIS buffer solution of 50 mM TRIS and 10 mM EDTA in deionized water (6.055 g/l TRIS; 2.923 g/l EDTA; pH 7.2).
3. From the stock solutions of CsCl and TRIS buffer prepare three solutions as follows:
   Solution 1 (1.4 g/ml) — mix 1 part stock CsCl solution with 1 part stock TRIS buffer.
   Solution 2 (1.1 g/ml) — mix 1 part stock CsCl solution with 7 parts stock TRIS buffer.
   Solution 3 (1.05 g/ml) — mix 1 part stock CsCl solution with 15 parts Stock TRIS buffer.
4. Into a 15-ml high-speed centrifuge tube carefully layer 3 ml of each solution; bottom 3 ml, 1.4 g/ml; middle 3 ml, 1.1 g/ml; top 3 ml 1.05 g/ml.
5. Layer on the top of the three-step gradient 1 ml of oocysts in the stock TRIS buffer (approximately 10^8/ml) obtained by the Sheather's flotation procedure. These oocysts should have been washed twice by centrifugation (1500 × *g*) in deionized water and once in the stock TRIS buffer.
6. Centrifuge at 16,000 × *g* for 60 min at 4°C.
7. Three bands form within the step gradient as well as a pellet at the bottom. The top band, formed at the interface between solution 2 and solution 3 contains the purified oocysts. The other two bands contain varying amounts of debris and a few oocysts. If the oocysts are not of sufficient purity, the purification process (steps 4 through 7) can be repeated using oocysts recovered from the first attempt.
8. The recovered oocysts should then be dialyzed overnight against 10 m*M* TRIS/1.0 m*M* EDTA buffer to remove the CsCl. The dialysis buffer is prepared by mixing one part stock TRIS buffer with nine parts deionized water.
9. Oocysts can be used immediately or stored in 2.5% $K_2Cr_2O_7$ or antibiotic solution (described above) at 4°C.

IV. EXCYSTATION AND SPOROZOITE PURIFICATION

A. *IN VITRO* EXCYSTATION

One of the key events in the life cycle of any coccidian parasite is excystation, the release of infective sporozoites from environmentally resistant oocysts (Figure 4a and b). Excystation usually occurs in the lumen of the intestinal tract following ingestion of oocysts by a suitable host. In most coccidian parasites studied to date, excystation can also be triggered *in vitro* by exposing oocysts to conditions that simulate the gastrointestinal environment of the host; i.e., reducing conditions, CO_2, host body temperature, pancreatic enzymes, and bile salts. Studies of *C. parvum* [109,192,673,900] and of *C. baileyi* [782] demonstrate that *in vitro* excystation is readily achieved in the laboratory. One procedure which results in a high, consistent percentage of excystation is outlined below.

1. Oocysts previously purified by one or more of the methods described above should be washed free of $K_2Cr_2O_7$, suspended in 3 ml of PBS (pH 7.2) in a 15-ml centrifuge tube, and placed in an ice bath. Hank's balanced salt solution can be substituted for PBS in this or any subsequent step.

1a. Bleach treatment can increase the percentage of oocysts that excyst *in vitro* and is necessary if sterile sporozoite preparations are needed. If sterile sporozoites are not required, steps 2 to 4 may be skipped.

2. Add an equal volume of cold, 40% commercial laundry bleach (full strength bleach is 5.25% sodium hypochlorite) to the oocyst suspension (final suspension is 1.05% sodium hypochlorite) and incubate on ice for 10 min. It may be necessary to mix or stir the suspension periodically.
3. At the end of 10 min, add approximately 6 ml of cold PBS to the suspension and centrifuge at $1500 \times g$ for 15 min.
4. Resuspend the pellet of oocysts in cold PBS and recentrifuge. This washing procedure should be repeated 4 times or until all of the bleach is removed.
5. The percentage of sporozoites excysting can be increased by preincubating oocysts (bleach-treated or nontreated) in PBS for 1 h at 37°C. This step is optional.
6. Add an equal volume of prewarmed 2x excystation fluid, consisting of 1.5% (w/v in PBS) sodium taurocholate with or without 0.5% (w/v in PBS) trypsin, and incubate for 30 to 45 min at 37°C (on a shaker).
7. The excystation process can be monitored microscopically. Depending on the age of the oocysts and the excystation solution, one should expect to get 60 to 90% excystation. Sporozoites should be washed free of excystation fluid and resuspended in HBSS.
8. An alternative excystation procedure used by Blewett[109] also results in a high percentage of excystation. Oocysts are incubated for 1 h at 37°C in HBSS (adjusted to pH 4 to 5) containing 1% (w/v) trypsin. The solution is then neutralized using HBSS containing 2.2% (w/v) sodium bicarbonate. Following neutralization to pH 7.2, HBSS containing 1% (w/v) sodium deoxycholate (SDC) is added to make a final concentration of 0.1% SDC. After 30 min at 37°C, excystation should be complete.

B. PURIFICATION OF SPOROZOITES

Several techniques are available for the purification of sporozoites, and the one of choice may depend on the volume of sporozoites required, the amount of time and effort one is willing to spend, and how easily the particular procedure is adapted into a laboratory. Three such procedures are outlined below. The first is a simple, one-step filtration that results in suspensions of sporozoites that are usually free from oocysts and oocyst walls. However, the percentage of excysted sporozoites recovered by this filtration procedure may not be as high as that for the other two procedures which involve Percoll® isopycnic centrifugation or the use of an anion exchange resin.

1. One-Step Filtration

1. Excyst sporozoites *in vitro* as described above. When excystation is complete (after 30 min to 1 h) place the warm excystation fluid containing free sporozoites, oocyst walls, and

intact oocysts into a syringe and gently force it through a polycarbonate membrane with a pore size of 2 µm (Nucleopore Corp., Pleasanton, CA). Then, force warm HBSS (twice the volume of the excystation fluid used) through the membrane.

2. Sporozoites which pass through the membrane can then be washed free of excystation fluid by centrifugation at 1000 × *g* for 10 min in HBSS.
3. Because sporozoites are thin (1 to 2 µm in diameter) and motile, many will pass through the 2 µm pores of the membrane; the 4.5- to 5.5-µm oocysts and oocyst walls are retained on the membrane. The percentage of sporozoites that pass through the membrane depends on the amount of debris (oocysts and oocyst walls included) that plug the pores of the membrane; thus, better yields are usually obtained when the number of oocysts (excysted and intact) forced onto the filter is below $10^7/cm^2$ of membrane surface.

2. Isopycnic Percoll Centrifugation[56]

1. Following excystation, wash the free sporozoites, oocysts, and oocyst walls in 1x Alsever's solution (HBSS, pH 7.2 can be substituted). Resuspend the pellet in 1x Alsever's solution (up to the equivalent of 5×10^8 oocysts per milliliter) and carefully layer 1 ml on top of 9 ml of the Percoll® solution described in Section III.D.
2. Centrifuge as described in Section III.D. Oocyst walls band at the top of the gradient, and the sporozoites form a band approximately one third from the bottom. Sporozoites recovered from the gradient may be contaminated with 2 to 4% intact oocysts; most intact oocysts sediment below the band of sporozoites.

3. Anion Exchange

The procedure described below is modified from the one described by Riggs and Perryman.[689]

1. Add 5 g of Watman® DE-52 preswollen matrix (Chemtrix, Hillsboro, OR, Catalog Number 4057-050) to 50 ml of column buffer. Column buffer is prepared by adding the following to 1 l of deionized water: Na_2HPO_4 (anhydrous), 10.78 g; $NaH_2PO_4.H_2O$, 0.62 g; NaCl, 3.4 g; glucose, 10.0 g. Adjust to pH 8.0. Ionic strength should be 0.290.
2. Wash the matrix in column buffer three times by stirring gently, letting it settle to the bottom, and decanting the supernatant. Do not use a stir bar because it can crush the matrix into fine particles. Store overnight in 50 ml of column buffer at 4°C.
3. Treat purified oocysts with cold 10% commercial bleach (full strength bleach is 5.25% sodium hypochlorite) for 5 min, wash three times with deionized water by centrifugation 1000 × *g*, 20 min) followed by two times in PBS, and then excyst sporozoites as described above in Section IV.
4. After excystation, wash sporozoites, oocyst walls, and oocysts in column buffer two times by centrifugation. Check the matrix buffer slurry and, if necessary, readjust to pH 8.0.
5. Add 1 ml of matrix to 10 ml of column buffer containing sporozoites, oocyst walls, and oocysts recovered from the excystation of approximately 10^8 oocysts. Swirl gently for 1 to 2 min and then add to a column prepared from a 3-ml syringe barrel. Columns are prepared by placing flexible tubing equipped with clamps on the end of a 3-ml syringe barrel containing a plug of nylon wool and 1 ml of the DE-52 matrix.
6. Elute with 30 to 50 ml of column buffer (flow rate 2 to 3 ml/min), collect sporozoites by centrifugation into a pellet, and resuspend sporozoites to desired concentration in HBSS or PBS. Sporozoites recovered from the column should be almost free of contaminating oocysts; less than 1% contamination should be expected (Figure 6).

V. *IN VITRO* CULTIVATION

A. CULTIVATION IN CHICKEN EMBRYOS

Both *C. parvum*[188] and *C. baileyi*[477] can complete their entire life cycle, from sporozoite to infective oocyst, in chicken embryos. The parasite isolated from chickens, *C. baileyi,* grows readily in endoderm cells of the chorioallantoic membrane (CAM) of chicken embryos from virtually any flock. Large numbers of *C. baileyi* oocysts can be harvested from the chorioallantoic fluid, making it possible to pass the parasite from embryo to embryo without the need of supplemental oocysts harvested from adult chickens. *C. baileyi* has been transferred from embryo to embryo every week for more than 3 years without any noticeable changes in its infectivity for embryos nor its infectivity for or pathogenicity to chickens.[196]

In contrast to *C. baileyi,* the mammalian parasite (*C. parvum*) will grow only in embryos obtained from certain flocks. For example, all embryos obtained from a flock of White Leghorn chickens maintained by the University of Gerogia supported prolific growth of *C. parvum,* whereas only one of every five to ten of the embryos from White Leghorn chickens maintained by Auburn University supported limited growth of the parasite.[196] The reason(s) for this marked difference in the suitability of embryos as hosts for *C. parvum* is (are) not understood. Therefore, if one is going to grow *C. parvum in ovo,* a reliable supply of suitable chicken embryos must be available. At the time of this writing, trial and error is the only way to determine if the embryos from a certain flock are suitable for the culture of *C. parvum.* Also, a reliable supply of oocysts must be available. Unlike *C. baileyi,* oocysts of *C. parvum* are often not released from endoderm cells of the CAM into the allantoic fluid and are therefore difficult to recover in large numbers. If sufficient numbers of *C. parvum* oocysts are recovered, embryo to embryo passage is possible. Below is a description of the procedures for cultivation of *C. baileyi* or *C. parvum* in chicken embryos.[188,477]

1. Candle 10-d-old embryos and mark an inoculation site approximately 0.5 cm below the air sac, being sure to avoid blood vessels in the CAM. Sterilize the inoculation site with disinfectant (betadine solution) and then punch a hole in the egg shell with a 23-gage needle.
2. To prevent bacterial contamination, one may wish to inject antibiotics into the allantoic fluid prior to introducing the parasite. Using a 1-ml syringe with a 26-gage needle, inject 0.1 ml containing 1000 U of penicillin and 1.0 mg of streptomycin. Other suitable antibiotics can be substituted.
3. Obtain purified oocysts from feces by using the procedures outlined earlier in this chapter. Purified oocysts can also be obtained from infected chicken embryos for embryo to embryo transfer by following steps 6 through 8 below.
4. Excyst sporozoites by the procedures outlined earlier in this chapter, wash by centrifugation in HBSS ($1000 \times g$ for 10 min), count with a hemacytometer, and resuspend in HBSS at 10^7/ml. Using a 1-ml syringe with a 26-gauge needle, inoculate 0.1 ml of the sporozoite suspension into each embryo.
5. Seal the inoculation site with melted dental wax and hold the embryos in an egg incubator for 7 d.
6. Using scissors, cut away the egg shell covering the air sac at the top of the embryo and moisten the surface of the exposed CAM with sterile HBSS so that the blood vessels in the CAM are readily visualized when the embryo is exposed to a bright light (candler). Using a sterile needle, puncture the CAM (avoid large blood vessels) in an area near the shell, insert a sterile Pasteur pipet into the hole and then aspirate the allantoic fluid. Approximately 10 to 15 ml of fluid can be obtained from each embryo.
7. Allantoic fluid should be pooled into centrifuge tubes and centrifuged at $1000 \times g$ for 15

min. Discard the supernatant and resuspend the pellet containing oocysts in deionized water plus 0.02% Tween®-20 (H_2O-T). Wash two times in H_2O-T by centrifugation and store the oocysts in H_2O-T overnight in the refrigerator (4°C). The next day, oocysts can be used or they can be placed in 2.5% $K_2Cr_2O_7$ and stored at 4°C.

8. Purification of oocysts harvested from chicken embryos is accomplished by incubating the oocysts in HBSS with cold 20% commercial bleach for 10 min and then washing with HBSS four times by centrifugation. The pellets that are formed may have two layers, a darker upper layer with some cellular debris from the CAM, and a bottom white layer of oocysts. Removal of the top layer by aspiration results in highly purified oocysts (Figure 5). These purified oocysts are then ready for excystation.
9. Developmental stages of the parasite can be monitored and photographed at various times after inoculation of sporozoites by examining the endoderm surface of the CAM. Observing fresh CAM preparations with Nomarski interference-contrast (NIC) microscopy allows one to identify different developmental stages. The appearance of different developmental stages of *C. parvum* viewed with NIC microscopy is shown in Figure 4.

B. DEVELOPMENT IN CULTURED CELLS

Complete development, from sporozoite to infective oocyst, in cell culture has been reported for human and calf isolates of *C. parvum*.[189] From these studies it is apparent that the parasite can complete its life cycle in a variety of cultured cells; however, the extent of multiplication and the number of oocysts produced is much lower than in experimentally infected animals or chicken embryos. The developmental stages of *C. parvum* observed in human fetal lung (HFL) cells during the first 3 d after introduction of sporozoites was similar to that observed in mice during the first 3 d after oral inoculation of oocysts. However, on days 4 to 8 after inoculation, the number of parasites in HFL cells decreased, whereas the number in ileal enterocytes of infected mice increased.[189] This difference was attributed to the presence of thin-walled oocysts in mice[195] and to the absence of these autoinfective forms in cultured cells.[189] If this proposal is correct, then the key to obtaining large numbers of oocysts and other developmental stages may require manipulation of the infected cell cultures so that the parasite will produce this autoinfective stage. At the time of this writing, this author was not aware of anyone who had been able to achieve parasite development in cultured cells comparable to that in the gut of the mouse or calf. In fact, most researchers cannot recover enough oocysts to pass the parasite from one culture to another. However, there is one report[206a] that CACO2 cells, a human colon carcinoma cell line that expresses some characteristics of enterocytes in culture, supports endogenous development of *C. parvum*. Enough oocysts were obtained from the culture fluid of the infected CACO2 cells to initiate infections in another cell culture. Attempts to grow other species of *Cryptosporidium* in cultured cells has been more disappointing. Several attempts to grow *C. baileyi* in cultured cells have been unsuccessful,[477] and at least one attempt in our laboratory to grow *C. muris* (bovine isolate) has also failed. The procedures, outlined below, for monitoring the development of *C. parvum* in cultured cells are similar to those used for growing the parasite in chicken embryos.

1. Select an appropriate cell line (or cell for a primary culture) and grow to approximately 75% confluency on 13-mm round coverslips in the bottom of 24-well plates.
2. Use the procedures outlined above to obtain purified oocysts and excysted sporozoites. Tissue cultures can be inoculated with a mixture of free sporozoites, oocyst walls, and intact oocysts or with purified sporozoites obtained by one of the procedures described above. By removing oocyst walls and intact oocysts from the inoculum, it is easier to monitor different developmental stages of the parasite that occur in the cell layer.
3. Resuspend the excysted sporozoites in tissue culture medium supplemented with 5 to 10%

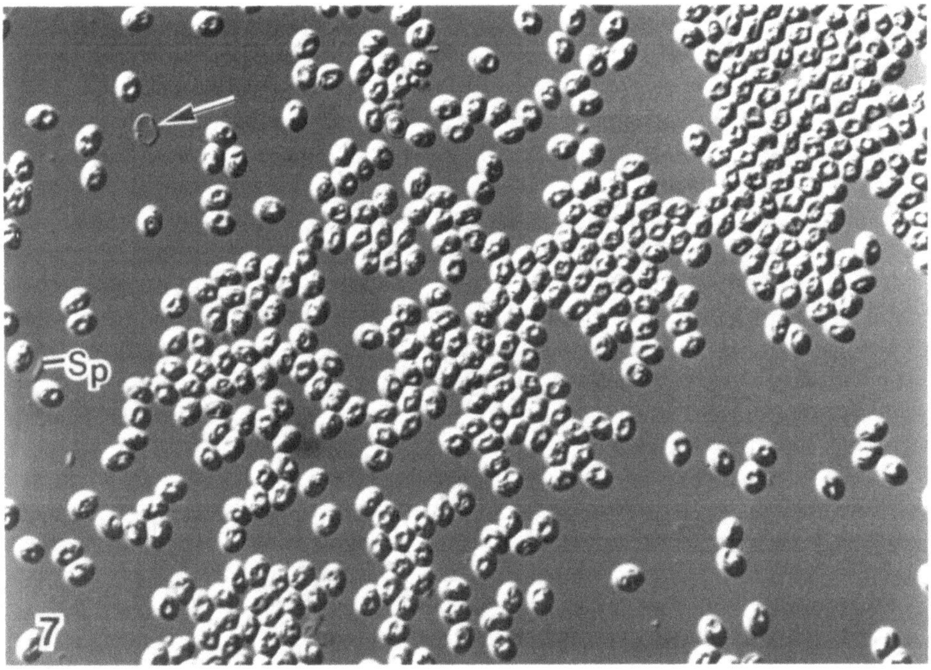

FIGURE 7. Nomarski interference-contrast photomicrograph of oocysts of *C. baileyi* (approximately 6 × 4.5 μm) purified from the allantoic fluid of experimentally infected chicken embryos. Note the absence of contaminating debris. One empty oocyst (arrow) and a sporozoite (Sp) are seen.

fetal bovine serum, 2 m*M* *l*-glutamine, 50 mg/ml dihydorstreptomycin, and 50 U/ml penicillin G. Count the sporozoites with a hemacytometer, inoculate 2 to 5 × $10^4/cm^2$ of cell monolayer, and incubate at 37°C in 5% CO_2, 95% air.

4. After 4 to 12 h, the monolayer should be washed gently by aspiration with a pipet and the culture medium should be changed. This procedure removes most of the sporozoites that are free in the medium or on the surface of the host cells. Culture medium should be changed every 2 to 4 d or when needed. Once the cell layer becomes confluent, it may be necessary to reduce the fetal bovine serum to 1 to 2% to prevent overgrowth and cell sloughing.
5. Parasite growth and development can be monitored and photographed at various times after sporozoite inoculation by examining double-coverslip preparations with a compound microscope equipped with Nomarski interference contrast or with phase-contrast objectives. Place the 13-mm round cover slip on a slide with the cell layer facing up, add a drop of culture medium, and then place a 22-mm coverslip on top. Appearance of different developmental stages of *C. parvum* viewed with NIC microscopy is shown in Figure 4.

BIBLIOGRAPHY

Information pertinent to the subject matter of this chapter may be found in References 12, 13, 18, 27, 40, 56, 109, 110, 127, 134, 146, 148, 186 to 189, 192, 194 to 196, 200, 225, 233, 259 to 261, 327, 336, 409, 420, 441, 477, 495, 498, 502, 536, 602, 643, 673, 689, 782, 819, 838, 839, 844, 900, and 913.

Chapter 3

WATERBORNE CRYPTOSPORIDIOSIS

Charles R. Sterling

TABLE OF CONTENTS

I. BACKGROUND

Waterborne disease outbreaks due to parasites and other etiologic agents have increased in the U.S. for the past 20 years despite regulations and measures taken to ensure safe drinking and recreational water for all. More waterborne disease outbreaks were reported during 1971 to 1985 than during any 15-year period since 1920. During this interval, 502 outbreaks involving 111,228 cases of illness were reported to the Centers for Disease Control (CDC) and the U.S. Environmental Protection Agency (USEPA).[178] Of these, 92 (18.3%) involving 24,365 individuals were directly attributable to the protozoan parasite, *Giardia lamblia*, making it the predominant cause of identifiable waterborne disease. For all the outbreaks during the 1971 to 1985 period, the responsible agent was identified only 50% of the time.

Cryptosporidium recently joined the list of diseases transmitted via water. A disease outbreak in San Antonio, TX, in 1984 was linked to sewage leakage into potable well water.[200] *Cryptosporidium* in surface drinking waters in Sheffield, England, was postulated to cause cases of cryptosporidiosis reported in 1986.[704] In New Mexico, an outbreak of cryptosporitiosis in day care centers during the summer of 1986 was statistically linked to consumption of surface waters.[258] In January 1987, Carrollton, GA, experienced a waterborne outbreak of cryptosporidiosis involving an estimated 13,000 individuals in a county of 64,900 residents in western Georgia.[322] This unconfirmed number of cases represents the single largest waterborne disease outbreak in the U.S. attributable to an identified agent. Smaller suspected waterborne outbreaks of cryptosporidiosis have subsequently been reported to Ayrshire, Scotland, and Oxfordshire-Swindon, England.[47,743]

The increase in reported waterborne disease outbreaks due to *Cryptosporidium* and other infectious agents is of great concern to the water industry. In the U.S., 56 million individuals daily drink water receiving disinfection only.[270] Such waters are acknowledged to be the leading cause of waterborne disease outbreaks. To address this concern, the USEPA proposed the Surface Water Treatment Rule in 1987, requiring communities to filter and disinfect all surface water.[843a] Criteria are also specified for operating under a variance, i.e., not being required to utilize filtration technology if the absence of regulated agents in the source water can be verified. The intent is to have systems meet design and operating criteria which would effect a 99.9% removal and/or inactivation of *Giardia* cysts and 99.99% removal/inactivation of enteric viruses. Treatment techniques are specified because the USEPA believes it is neither economically nor technologically feasible to measure certain contaminant levels in drinking water. *Giardia* is included as a regulated contaminant for which the USEPA believes that analytical methods, validation, and monitoring are beyond the capabilities of all but a few highly specialized laboratories. *Cryptosporidium* is under consideration for inclusion on that list.

In this chapter outbreaks attributable to *Cryptosporidium*, methods for recovering and identifying the parasite, distribution in the environment, methods potentially useful in removing or inactivating the parasite, and potential for waterborne transmission are reviewed.

II. WATERBORNE DISEASE OUTBREAKS ATTRIBUTABLE TO *CRYPTOSPORIDIUM*

During the summer of 1984, Braun Station, a suburb of San Antonio, TX, experienced two outbreaks of gastroenteritis caused by contamination of the community's well water supply.[200] Limited serologic data implicated Norwalk virus as the pathogen of the first outbreak, whereas presence of oocysts in the stools of 47 of 79 residents established *Cryptosporidium* as the pathogen of the second. In both cases, the incidence of reported gastrointestinal illness was approximately 12 times greater than in surrounding communities utilizing a different water supply. *Cryptosporidium* oocysts were not recovered from the source water, although untreated water samples contained fecal coliforms. Subsequent water testing using dyes showed that the likely source of the contamination was the community's sewage system. Water from the well serving the residence of Braun Station was chlorinated but unfiltered.

In Sheffield, England, a peak in cryptosporidiosis was recorded during the summer of 1986 among individuals whose only epidemiological source link was drinking water supplied from a common reservoir.[704] In this instance, oocysts were identified from the second source water, cattle grazing on adjacent agricultural land, and from the intestinal contents of brown trout within the reservoir using modified Ziehl-Neelsen and monoclonal immunofluorescent staining techniques. It was suggested the reservoir may have been contaminated by surface runoff following heavy rain.

Consumption of untreated surface water appeared to be a predominant risk factor associated with cryptosporidial illness among 78 laboratory-confirmed cases during the summer of 1986 in New Mexico.[258] In this case control study, 5 of 24 case patients who had cryptosporidiosis drank untreated surface water, while none of 46 age, sex, and neighborhood matched controls had consumed untreated surface waters. Matched set analysis also identified attendance at day care centers as a risk factor associated with acquisition of cryptosporidial illness in this study. Surface waters sampled from New Mexico and Arizona, where the case-patients might have been exposed, tested negative for *Cryptosporidium* oocysts. Testing, however, was conducted approximately 2 months after the reported illnesses.

In January and February 1987, an estimated 13,000 cases of gastroenteritis occurred among the residents of Carroll Country, GA.[322] *Cryptosporidium* oocysts were identified in 39% of patients tested during the outbreak. A randomized telephone survey indicated an attack rate of 54% within the city of Carrollton and 40% overall for the county. The only significant risk factor associated with illness was exposure to the public water supply which was filtered and chlorinated. Records kept during the outbreak indicated that the water treatment facility serving Carrollton was operating within established EPA guidelines. Samples tested negative for coliform bacteria, were in the range of 0.07 to 0.18 nephelometric turbidity units (NTU), and had chlorine residuals as high as 1.5 mg/l at the treatment facility. Despite these better than normal operating values, an advisory to boil water was issued on January 30 and kept in effect until March 2. Low numbers of oocysts were ultimately identified from water samples collected at the water treatment plant and in dead-end mains of the system in late January and early February. Oocysts were also identified in streams draining into the main river serving the treatment plant, but not in the river itself. A small number of dairy cattle grazing on land adjacent to the watershed also tested positive for *Cryptosporidium*. Investigations into the treatment plant's operation revealed that filters were frequently restarted without undergoing backflushing. Turbidity measurements of water from these filters, taken within three hours of restarting, showed higher than normal values (up to 3.2 NTU). It was postulated that the restarting of a higher than usual number of such filters during this period, without appropriate backflushing, may have discharged a bolus of infectious oocysts which had been trapped on the filters into the drinking water supply. In addition, mechanical agitators used in flocculation had been removed in anticipation of the arrival of new ones and the efficiency of filtration had been impaired by the equipment used to control water flow through the filters.

Two additional *Cryptosporidium*-related waterborne outbreaks have been reported in Ayrshire, Scotland, and Oxfordshire-Swindon, England, in 1988 and 1989.[47,548a,743] Both were remarkably like the Sheffield incident. The latter is particularly noteworthy because of the boil water advisory issued to 600,000 people of the Oxford-Swindon area. Once again, *Cryptosporidium* oocysts were identified first from affected people and secondly from the water supply.

The increase in reported waterborne disease outbreaks due to *Cryptosporidium*, as exemplified by the cases just discussed, is no doubt related to improvements in providing positive identification of the responsible agent. The foregoing also demonstrates the difficulty in assessing risk factors which might have contributed to the outbreak. In each instance, the organism was first identified from infected individuals before being isolated from the suspected water source. Even then, proof for the involvement of *Cryptosporidium* as the responsible agent in the majority of these outbreaks could be made only through epidemiological "guilt by association". Acknowledging the difficulties attendant with assigning risk factors which might contribute to waterborne disease outbreaks, it is perhaps pertinent to assess the methods which are useful for recovering and identifying *Cryptosporidium* from water sources used for drinking purposes. In doing so, it may provide an alternative approach to assessing the true potential for *Cryptosporidium* transmission via the waterborne route.

III. RECOVERY AND DETECTION METHODOLOGIES

The development of methods for detecting parasites in water has largely paralleled the rise in interest in *Giardia* as a waterborne disease entity. The first method developed for detecting *Giardia* was the "EPA method".[378] The method involves filtration of large volumes of water through a spun polypropylene cartridge filter having a standard nominal porosity, elution and concentration of the sample from the filter, and microscopic examination of the concentrated sediment for detection of cysts. With an efficiency of organism recovery by this method reported to average 6.3%, the possibility of conducting viability studies on the few organisms likely to be recovered from environmental samples is precluded. Recent improvements upon the EPA method, involving more efficient concentration and detection methodologies, however, have resulted in a dramatic increase in organismal recovery and identification.

A. SAMPLE PROCUREMENT

Recovery of the few organisms likely to be present in an environmental sample requires an adequate sample size. This is most easily accomplished by running water through a filter cartridge housing unit containing a 25-cm polypropylene filter with 1-μm nominal porosity and an on-line flow meter. In field situations, collection can be facilitated by a portable gasoline-powered water pump. The drawback of this collection method is the quality of the water being sampled. Extremely turbid water can reduce the sample volume to 37.8 l (10 gal) before the filter clogs. In most instances, 378 to 3785 l can be filtered. When extensive field sampling is conducted on the same day, it is advisable to flush the pump, filter housing unit, and flow meter at each new site with 189 to 378 l of water to reduce the possibility of contamination carry-over from a previous collection site. Filters recovered from each sampling site are placed on ice for transport back to the laboratory and processing. When in extremely remote sites, it is best to fix the filters with 10% neutral buffered formalin.

B. SAMPLE PROCESSING

Elution and concentration procedures are utilized to recover as much of the sample trapped on the filter as possible. Initially, filters were cut apart, repeatedly washed, and all collected washings were centrifuged to concentrate the debris eluted from the filters.[378,379] More recently this procedure has become more efficient by improvements to both the filter elution and specimen concentration procedures. Elution has improved by backflushing the filter with a nitrogen-driven pressure vessel, cutting apart the wound filter fibers, and mechanically agitating the cut fibers in the backflush eluent.[570] All eluted material is then concentrated by centrifugation. This procedure has been modified by filtering 20-l water samples sequentially through 293-mm to 5- and 1-μm polycarbonate filters.[602] The final filter is inverted in a pan containing 200 ml of distilled water and vibrated to elute retained material. Eluted material is concentrated by centrifugation. Two drawbacks remain to this modification: the need to transport 20 l of water back to the laboratory and the inability to sample relatively turbid waters.

Samples can be concentrated further by several other gradient centrifugation procedures. One procedure involves fixation of the eluted material in 10% formalin; 1% sodium dodecyl sulfate, 0.1% Tween® 80 (SDS/Tween®), and a few drops of Antifoam A are added to break up of large aggregates and prevent sample foaming during vortexing. Small sample portions are then layered onto sucrose gadients (1.23 g/ml) and centrifuged at 1000 × *g*.[502,696] Other useful gradients include potassium citrate (1.195 g/ml) and Percoll®-sucrose (1.09 g/ml) for sample separation.[448,602] The entire supernatant is collected, washed with phosphate buffered saline (PBS), and pelleted by centrifugation. The pellet can be stored in 10% formalin or directly evaluated for the presence of oocysts.

C. DETECTION OF *CRYPTOSPORIDIUM* OOCYSTS

Initially, the few parasites in environmental water samples were identified by conventional staining techniques and microscopy which was very labor intensive.[378,379] Identification required highly trained technicians who could differentiate parasites from environmental debris. The development of polyclonal antisera and monoclonal antibodies (MAbs) specific for the cyst and oocyst stages of *Giardia* and *Cryptosporidium*, respectively, has led to detection of these parasites in both clinical and environmental samples by immunofluorescent microscopy.[260,688,719,743a,759,773,775,776]

Most laboratories interested in quantitative counts of parasites recovered from environmental samples filter a portion of the final concentrated sample through a syringe or manifold-mounted filter of cellulose nitrate or acetate and then apply appropriate MAb with subsequent incubations and washes in the same unit.[448,502,602,696,699,699a,719,775a] Filters can then be counterstained, cleared, and mounted on slides for microscopic examination and quantitation of oocysts.

D. COMMENTS ON SAMPLE RECOVERY AND DETECTION METHODOLOGIES

While a consensus seems to have been reached on how to best obtain a water sample which may contain parasites, there appears to be divergent opinion on how to best concentrate and subsequently detect organisms. The scheme outlined above is extremely labor intensive and costly, and there is an acknowledged wide range of recovery efficiencies attributable to the quality of water being tested. Only a handful of laboratories are set up to conduct water monitoring for parasites as outlined above. A single sample can take 8 h or more to process. Where strict quantitation is desired, it is mandatory to run parallel seeded samples to determine efficiency of recovery so one can correct the final count on the unknown sample.

Simplification of methodology without sacrificing proficiency has met with a certain degree of success for *Cryptosporidium*. First, a slide smear was made from each sample recovered from the sucrose gradient and examined for oocysts by immunoflourescence.[775a] With this modification, oocysts were detected in every instance where they were evaluated quantitatively. Additional time-saving measures included performing slide-smear quantitative evaluations on the initial concentrated eluent from the backflushed filter and on material subsequently concentrated by ethyl acetate using a commercially available parasite-fecal concentration kit.[774] Based on the results of these modifications, no single slide-smear examination technique appeared preferable, although experiments were not performed to determine how many slides would have to be examined to find a sample positive.

Another factor complicating accurate identification of *Cryptosporidium* infections for humans is the specificity of the reagents used to identify the parasite. A monoclonal antibody, thought initially to react only with oocysts from mammals, recently has been shown to react with *Cryptosporidium* from turkeys and quail, albeit with a different fluorescent staining pattern.[743b] An inexperienced technician might not be able to discern between the two. In addition, prevalence of the avian species in comparison to the mammalian species in any water source is not known. Nor is the infectivity potential for recovered oocysts to humans known. Based on the attendant difficulties in defining infection risk associated with identifying the parasite, as well as the highly variable and frequently poor parasite recoveries from these samples, it does not appear practical to quantitatively assess oocyst numbers encountered in water samples. No wonder the USEPA has taken the stance it has in promulgating treatment regulations.

IV. DISTRIBUTION OF *CRYPTOSPORIDIUM* IN WATER

Despite sampling and diagnostic difficulties, information regarding the prevalence of *Cryptosporidium* in a variety of waters is beginning to emerge. In the first study, *Cryptosporidium* oocysts were identified in concentrations of 5 and 17/l in secondarily treated sewage.[570] Subsequently, high concentrations of oocysts were recovered and identified in raw (5180/l) and treated (1063/l) sewage effluents.[502] Oocysts were recovered in much lower numbers (10/l) in sewage effluents discharged from facilities employing sand filtration in the final treatment procedure. In this same study, oocysts were encountered in highly variable numbers (0.8 to 5800/l) in surface waters likely to be contaminated by some type of sewage or fecal discharge. In yet another study, conducted on water from 11 rivers in Washington and California, all contained oocysts at levels between 2 and 112/l.[602]

Of 101 surface water sites in Arizona, 24 (23.8%) were positive for *Cryptosporidium* oocysts in numbers ranging from 0.005 to 252.7/l.[774,774b,775a] Sites with high oocyst counts included streams adjacent to agricultural runoff areas, streams and lakes of high recreational use, and rivers flowing intermittently from Mexico into Arizona. Of 107 sites throughout the western U.S., 77% were positive for *Cryptosporidium* oocysts in numbers ranging from 4.1 to 1732/l for treated sewage and from 0.04 to 18/l for stream samples.[699] Other data suggest that *Cryptosporidium* is widely distributed in western Pennsylvania.[448]

Of ten filtered drinking water samples tested for the presence of *Cryptosporidium* oocysts in the western U.S., two were positive with a geometric mean of 0.001 oocysts/l.[699] One came from a treatment plant having problems with the coagulation and disinfection system, and the other from a plant providing filtration but not coagulation. Two of four plants using chlorination as the only treatment tested positive for oocysts, but in low numbers (0.006/l).

In many third world countries, drinking water usually receives no treatment. In Peru, *Cryptosporidium* oocysts have been found in high numbers from drinking water cisterns serving individual residences (13.2/l).[774a] They have been identified in low numbers in water delivery trucks and in high numbers in outflow from oxidation ponds (198.2/l) which serve as a water source for people too poor to purchase water from the delivery trucks. Because water for personal consumption in many of these communities often sits in containers or cisterns for a week or more, additional contamination to the water could come from animals or from humans dipping for water. That such contamination occurs is supported by the finding of high fecal coliform counts in these residential waters. The presence of fecal coliforms in such water, however, may not always be an indicator of parasite presence.

It is frequently difficult to assess the relevance of reported oocyst numbers. Recovery efficiencies were not reported in every instance where oocyst counts were given. That would indicate an actual underreporting of the actual number present. The oocyst numbers reported are very likely to vary as do the conditions of the water sampled and of the techniques used for their recovery and identification.

V. WATER TREATMENT — REMOVAL AND INACTIVATION

The primary role of water treatment is the prevention of infectious disease transmission. Despite existing regulations and treatment techniques designed to eliminate this possibility, waterborne disease outbreaks continue to occur. Only a limited number of outbreaks have been directly attributed to *Cryptosporidium*, but one has to wonder what percentage of undefined outbreaks of the past might have been caused by this parasite. The mainstays of the most effective type of water treatment to reduce the number of microorganisms passing into finished drinking water include some type of water pretreatment to effect coagulation and sedimentation of suspended particles followed by filtration and disinfection.[482] Limited data on the zeta potential (a measure of surface charge in relation to pH conditions) of the *Cryptosporidium* oocyst indicates that it should respond to coagulation in a fashion similar to *Giardia* and, therefore, should be removed by filtration in proportion to other particles of similar size (4 to 6 μm).[483]

A. FILTRATION

Data for removal efficiencies of *Cryptosporidium* in potable water by filtration are now being collected. Preliminary data, based on sampling of source, waste, filtered, and backflush water, however, have provided a measure of information about the efficiency of filtration as a treatment process.

In studies of wastewater treatment, oocyst counts (based on averages) were reduced by 74 and 79% in plants relying on secondary treatment without filtration and by a further 87 and 99% in plants employing sand filtration.[502,696] The apparent overall efficiency of those plants having sand filtration, based on the average number of oocysts present in raw sewage from the studies, was 96.6 and 99.8%, respectively. These figures must be viewed cautiously, however, because they are short term, at a single plant, and reflect only averages of a random survey.

Similar information has emerged from drinking water treatment facilities. Examination of backflush water from rapid sand filters of drinking treatment facilities has shown oocyst to be present in great numbers (2906 to 24,306/l), indicating that the filters were effectively removing oocysts. 696,697 A comparison with numbers present in the source water compared to the finished water indicated a 91% removal efficiency. In another study involving a slow sand filtration plant, *Cryptosporidium* oocysts were detected in the source and finished water, but were not quantified.[697] The experience at Carrollton, GA, where the water treatment facility has rapid sand filtration and was meeting federal EPA guidelines, demonstrates how important it is for a facility to continually monitor operational efficiency. At present we really know little about *Cryptosporidium* removal efficiency as it relates to the operational efficiencies of different types of filters operating under different conditions of water pretreatment and with variable types of source waters.

B. DISINFECTION

Early clues to resistance to disinfection came from studies in which *Cryptosporidium* oocysts were exposed to various disinfectants commonly used in hospitals and laboratories.[32,133] Only 10% formaldehyde, 5 to 10% ammonia, and 70% to full strength commercial bleach were reported to destroy oocyst infectivity (see Chapter 1, Table 6). Exposure to temperatures below –20°C and above 65°C for 30 min will also destroy infectivity.[21,824] *Cryptosporidium* oocysts are routinely stored in a 2.5% solution of potassium dichromate at 4°C, where they retain their infectivity for 2 to 6 months or longer.[189] In two studies, oocysts were exposed to the action of chlorine as sodium hypochlorite.[133,673] Viability of oocysts as measured by either excystation or *in vivo* infectivity was reportedly not affected by exposure of oocysts to sodium hypochlorite concentrations of from 1.05 to 3% or by exposure times of from 10 s to 18 h at temperatures of either 4 or 37°C. It is difficult to evaluate such data in relation to water disinfection because the chlorine bleach was utilized in the first instance to enhance excystation and in the second was mixed with whole intestinal homogenates in which case disinfection demand due to organic materials present was high.

Studies employing disinfectants commonly used to treat water have so far been limited to oocysts exposed in a demand-free situation. Under these circumstances, oocyst viability as measured by either excystation or infectivity is abolished following exposure to 80 ppm chlorine, 25°C, and pH 7 for 2 h.[422b] These data give a C.t′ (concentration × time) value of 9600 for *Cryptosporidium* compared to a C.t′ of <15 for *Giardia* exposed to chlorine at 25°C and pH 7.[382] Exposure to chlorine dioxide at 0.43 ppm significantly reduced but did not eliminate

TABLE 1
Resistance of Different Microorganisms to Chlorine and Ozone for 99% Inactivation

	Chlorine				Ozone			
Organism	Temp. (°C)	pH	C.mg/l[a]	C.t ′ value[b]	Temp.	pH	C.mg/l[a] (°C)	C .t′ value[b]
Escherichia coli	5	6.5	0.02—0.10	0.02[124]	1	7.2	0.065	0.022[400]
	23	7.0	0.1	0.014[687]				
Poliovirus	20	6.0	0.3—0.7	0.5—0.7[730]	5	7.2	0.15	0.22[701]
					20	7.2	0.15	0.075[701]
Giardia muris	5	7.0	23.8—78.5	449—1012[446]	5	7.0	0.15—0.7	1.94[884]
	25	7.0	2.8—7.1	25.5—44.8[446]	25	7.0	0.03—0.15	0.27[884]
G. lamblia	5	7.0	2.0—8.0	56—152[382]	5	7.0	0.11—0.048	0.53[885]
	25	7.0	1.5	<15[382]	25	7.0	0.03—0.15	0.17[885]
Naegleria gruberi	25	7.0	0.64—3.42	12.1[703]	5	7.0	0.55—2.0	4.23[884]
					25	7.0	0.3—1.2	1.29[884]
Cryptosporidium parvum[c]	25	7.0	80	9600[422b]	25	7.0	1.0	<10.0[422b]
					N.A.	N.A.	1.11—2.25	6.66—20[634]

[a] Concentration in milligrams per liter.
[b] Concentration of disinfectant (milligrams per liter) times contact time (minutes).
[c] Inactivation measured at the 99.9% confidence level.

infectivity within 15 and 30 min of exposure.[634] Ozone has also been tested for its ability to abolish oocyst infectivity. Ozone is increasingly becoming popular as a method of water treatment not only because it is more efficient in terms of killing microorganisms, but also because it increases coagulation during sedimentation and can be used to eliminate bad taste in water.[278] Treatment of demand-free water containing 10^4 oocysts per milliliter with 1.11 ppm ozone for 6 min totally eliminated infectivity.[634] In another study, the same effect was achieved by treating oocysts at a concentration of 10^5/ml with 1 ppm ozone for 10 min.[422b] A comparison of the resistance of various microorganisms to chlorine and ozone is shown in Table 1. Under normal operating conditions, water utilities using ozone or chlorine usually maintain a residual activity of 0.4 and 1.0 ppm, respectively.[634,793] In the case of ozone, this activity cannot effectively be maintained since ozone is extremely unstable.[278] These preliminary studies show that *Cryptosporidium* is more likely to survive disinfection practices, such as those used by the water industry, than other microorganisms.

VI. THE REAL POTENTIAL FOR WATERBORNE TRANSMISSION

The real potential for waterborne transmission depends on many factors including determining the infectious dose of *Cryptosporidium* for humans, strain variability that may influence virulence, individual susceptibility to infection, conditions under which *Cryptosporidium* can survive in the environment, and which treatment processes most effectively remove *Cryptosporidium* from drinking water. Water sampling has demonstrated that *Cryptosporidium* is ubiquitous in the environment and always likely to be present as a waterborne pathogen. Its presence is linked to prevailing sanitary conditions and water treatment technology, hence the higher incidence of disease in third world countries.

BIBLIOGRAPHY

Information pertinent to the subject matter of this chapter may be found in References 21, 32, 47, 124, 133, 178, 189, 200, 258, 260, 270, 278, 322, 378, 379, 382, 400, 422b, 446, 448, 482, 483, 502, 548a, 570, 602, 634, 673, 687, 688, 696, 697, 699, 699a, 701, 703, 704, 719, 730, 743 to 743b, , 759, 773 to 776, 793, 824, 843a, 884, and 885.

Chapter 4

CRYPTOSPORIDIOSIS IN HUMANS (*HOMO SAPIENS*)

Beth L. P. Ungar

TABLE OF CONTENTS

I. HISTORIC PERSPECTIVE

The first case of *Cryptosporidium* infection in a human was described in 1976 in a previously healthy 3-year-old child from a farm community in rural Tennessee.[591] The child was vomiting and had severe watery diarrhea, malaise, and a dry cough, but recovered within 2 weeks; diagnosis of cryptosporidiosis was made by electron microscopic examination of rectal biopsy tissue. Within 2 months, the second human case was reported, this time in a 39-year-old college administrator who lived on a cattle farm, had an ileostomy for ulcerative colitis, and was treated with maintenance cyclophosphamide and prednisolone for bullous pemphigoid.[540] He had severe watery diarrhea accompanied by crampy abdominal pain which stopped 2 weeks after discontinuation of cyclophosphamide, but not prednisolone; diagnosis was again made by electron microscopic examination of biopsy tissue, this time from proximal jejunum and distal ileum.[540]

Over the next 6 years, until the acquired immunodeficiency syndrome (AIDS) epidemic gave human cryptosporidiosis a new importance, additional sporadic case reports appeared in the world literature. These linked the severity of illness associated with *Cryptosporidium* infection in humans to varying levels of immune function. Among the cases of immunologically healthy individuals are described a 25-year-old man with abdominal cramping and diarrhea lasting approximately 11 d and beginning 3 weeks after starting a survey in calves for *Cryptosporidium*;[674] a 25-year-old veterinary student with fever, nausea, diarrhea, abdominal pain, chills and sweating that began five days after caring for infected calves;[17] a 27-year-old man who developed headache, lethargy, weakness, abdominal cramping, and watery diarrhea approximately 1 week after his 8-year-old daughter and a number of her classmates experienced similar symptoms (of note, this is the first case in which cryptosporidiosis was diagnosed by identification of Giemsa-stained oocysts in a fecal specimen);[812] a 24-year-old woman with a 5-d history of fever, nausea, abdominal pain, and watery diarrhea who underwent hemicolectomy for presumed (not confirmed) Crohn's disease;[250] and a 31-year-old major league baseball player who also worked as a horse trainer with a 1-month history of anorexia, nausea and diarrhea.[64]

Among the cases of immunologically compromised individuals are reported a 9-year-old child with congenital hypogammaglobulinemia and diarrhea for 3 years which continued even after concurrent giardiasis was successfully treated;[440] a child with congenital hypogammaglobulinemia, recurrent cholera-like diarrhea, vomiting, and abdominal pain from 6 to 12 years of age;[41,741] a 58-year-old male renal transplant recipient with IgA deficiency on long-term prednisone and azathioprine with a 7-week history of diarrhea;[871] a 35-year-old immunodeficient homosexual man with intractable diarrhea who died after 2 months of hospitalization;[911] a 48-year-old homosexual man with a 9-month history of diarrhea and a 19 kg weight loss;[870] a 52-year-old woman with multiple immunologic abnormalities, disseminated toxoplasmosis, and a 22-month history of severe diarrhea and abdominal pain.[769] All cases in immunlogically healthy individuals recovered within approximately a month, while all cases with immune dysfunction in which outcome was reported died regardless of treatment, unless the immunosuppression was corrected.

In 1982, the first outbreak of cryptosporidiosis in immunologically healthy persons and the first statistics from the Centers for Disease Control (CDC) on cryptosporidiosis and AIDS were published. The latter reported that 14 of 21 patients with AIDS and *Cryptosporidium*-associated diarrhea had died foreshadowing a large-scale problem in human immunodeficiency virus (HIV)-infected patients.[42] The former gave details of *Cryptosporidium* infection in 12 immunocompetent persons, all of whom had direct contact with feces from calves with cryptosporidiosis and 9 of whom experienced up to 10 d of abdominal cramps and diarrhea.[41a,186] This recognition of *Cryptosporidium* as a new and potentially dangerous human enteric pathogen stimulated increased activity in the medical and scientific community, particularly in refining diagnostic techniques, defining the epidemiology, and searching for a cure for cryptosporidiosis.

Within the next few years, because of the difficulty in diagnosing *Cryptosporidium* infection except by intestinal biopsy, more than a dozen stool concentration and staining techniques were developed allowing direct visualization of *Cryptosporidium* oocysts in stool specimens[19,30,57,59,67,71,72,74,79,106,144,146,213,259,260,262,269,334,361,395,495,498,535,536,545,642,643,646,705,773,776,913] (see Chapter 2). By 1986, the CDC reported that 697 (3.6%) of the first 19,187 AIDS patients were infected with *Cryptosporidium* with a case fatality rate of 61% ($p < 0.01$ compared to AIDS patients without cryptosporidiosis);[587] on a worldwide basis, *Cryptosporidium* infection had been described in more than 20 countries on 6 continents.[236]

Now *Cryptosporidium* has been generally accepted as an agent of diarrhea throughout the world. Certain groups are considered to be high risk such as immunocompromised (particularly HIV-infected persons), household or family members and sexual partners of infected patients, health care workers and their patients, children and caretakers in day-care centers, travelers, and animal handlers (in farm, veterinary, laboratory, and household domestic settings). Waterborne outbreaks have been well-documented. In immunologically deficient patients and perhaps ultimately in others, *Cryptosporidium* is also starting to be recognized as a probable extra-intestinal pathogen. This chapter will describe cryptosporidiosis in humans, specifically clinical manifestations, immunologic response to infection, putative patterns of transmission, and global epidemiology. For details on pathogenesis, pathologic features, diagnosis, and treatment, see Chapters 1 and 2.

II. CLINICAL PRESENTATION

Our understanding of the clinical manifestations of cryptosporidiosis comes from three sources: individual case reports, descriptions of outbreaks, and geographically based epidemiologic surveys. Based on these, the demographic picture, incubation period, variety of clinical symptoms, and duration of oocyst shedding are described.

A. DEMOGRAPHIC INFORMATION

The age range reported for *Cryptosporidium* infection is from 3 d in an infant vaginally delivered to a mother with cryptosporidiosis[116] to 95 years of age.[359] Epidemiologic surveys, however, and reports of numerous outbreaks in child care facilities (see section on transmission) suggest that young children are most susceptible to symptomatic infection, likely reflecting immunologic immaturity, some degree of postinfection protective immunity in later life, and the increased fecal-oral transmission of pathogens which occurs in the toddler age group (lack of toilet training, frequent hand-to-mouth behavior, and child-to-child contact).

In geographically diverse surveys for fecal *Cryptosporidium* oocysts in both adults and children or in children only, significantly higher prevalences were found in the majority of studies in young children, usually less than 2 years of age[2,76,114,138,140,177,181,184,264,317,354,364,369,501,505a507,512,526,583,681,682,693,728,729,732,742,797,823,831,869,873,875,876,886.] In a single serologic survey performed in a community near Lima, Peru, where housing and sanitation are very poor, seropositivity for specific anti-*Cryptosporidium* immunoglobulin G (IgG) was 69% at birth (umbilical cord blood), declined to 33% during the first 2 years of life, and peaked at 73% between 2 and 5 years of age;[840] detectable anti-*Cryptosporidium* IgG represents infection sometime previously in life, again targetting the toddler age as a likely time for infection.

Exclusively breast-fed children have been reported to have fewer infections: cryptosporidiosis was not found or found significantly less often in breast-fed children in some studies.[184,524,525,609,869] Others, however, have found a similar prevalence of infection in bottle- and breast-fed children.[138,139,177,183,526] Bottle feeding may be the actual risk: 28% of bottle-fed children less than 18 months of age, compared to 9.1% of children never bottle fed, were infected with *Cryptosporidium* in a study in Liberia;[354] in India, all infants (five) excreting *Cryptosporidium* were bottle fed.[507]

TABLE 1
Clinical Features of Human Cryptosporidiosis Compiled from Case Reports[a]

Clinical feature	Number of infected persons		
	HIV-infected patients[b]	Other immunodeficient patients	Immunologically healthy patients
Diarrhea			
Diarrhea	70	33	73
Abdominal pain	36	11	55
Anorexia/nausea/vomiting	36	12	29
Fever	39	10	22
Malaise/fatigue/weakness	21	7	24
Respiratory problems	20	11	3
Duration of diarrhea (days)			
<4	0	1	11
4—10	0	0	17
11—20	1[c]	1	29
21—30	0	0	10
>30	56	24	4
Unknown	12	6	2
Outcome			
Recovery	1	14	81
Persistent infection	6	2	0
Death	42	10	0
Unknown	29	8	1

[a] These 194 patients were described in the following references: (1) HIV-infected patients (29, 85, 92, 96, 111, 119, 120, 159, 163, 164, 166, 172, 176, 186, 252, 268, 272, 285, 296, 299, 300, 316, 348, 356, 391, 396, 401, 402, 419, 450, 497, 514, 516, 527, 549, 595, 636 to 638, 721, 751, 761, 895, 911, and 912); (2) other immunodeficient patients (165, 168, 186, 209, 271, 314, 357, 404, 408, 422, 440, 464, 509, 527, 537, 540, 546, 505, 600, 660, 741, 769, 777, 870, 871, 872, and 874); (3) immunologically healthy patients (17, 64, 65, 70, 95, 101, 102, 126, 168, 186, 201, 224, 230, 240, 316, 320, 321, 355, 385, 420, 435, 499, 527, 532, 591, 644, 665, 685, 733, 751, 752, 772, 812, 851, 856a, and 866). Cases reported in the context of therapeutic drug trials are not included.

[b] These patients were seropositive for HIV or presumed infected based on clinical symptoms.

[c] This patient died from other causes early in his *Cryptosporidium* infection.

Like children, the elderly may also be at increased risk for *Cryptosporidium* infection, perhaps related to normal immunosenescence.[65] Males and females appear to be equally susceptible to infection.[236]

B. INCUBATION PERIOD

Cryptosporidium infects humans after oral ingestion of oocysts. Potential sources of organisms are other infected humans, animals, or the environment, particularly contaminated water. The best estimates of the prepatent or incubation period necessary before symptoms develop are from instances where initial exposure can be most precisely timed. For example, four Finnish veterinary students were all exposed to the same infected calves on the same day and all developed symptoms 6 or 7 d later, although the duration and severity of infection varied.[644] In other veterinary or laboratory animal exposures, time lapse from presumed oocyst ingestion to symptoms ranges from 5 to 28 d, most often 7 d.[17,41a,102,186,355,462,674] In circumstances where the moment of exposure might be less clear, such as during travel or contact with an infected patient, the average incubation period is also 7 to 10 d.[70,168,388,390,420,499,685,772]

C. CLINICAL FEATURES

Cryptosporidium infection was first recognized in humans because of its association with severe diarrhea (Table 1); more recently other clinical profiles including asymptomatic and extra-intestinal manifestations have received attention. Because the diagnostic proficiency varies with experience and frequency of cryptosporidiosis in a given population, and because oocysts are more easily identified in persons with diarrhea than those without, less apparent clinical presentations may be underrecognized.

Diarrhea is the most noteworthy symptom. Characteristically, it is voluminous and watery, often called cholera-like. As many as 71 stools and 12 to 17 l/d have been reported.[29,42,638] However, less fulminant diarrhea also occurs, even in HIV-infected persons.[721] Mucous may be associated with diarrhea, but blood or leukocytes are rarely reported. As much as a 25-kg weight loss has been noted.[285,751] Crampy abdominal pain, sometimes quite severe, may occur. Less frequent symptoms accompanying diarrhea include low grade fever (<39°C) perhaps actually caused by other concomitant infections, particularly in immunologically deficient hosts; generalized malaise, weakness, or fatigue; and loss of appetite, nausea, and vomiting (Table 1). Toxic megacolon has been associated with cryptosporidiosis in one HIV-infected patient with severe diarrhea.[172] On biopsy or autopsy of immunocompromised patients with diarrhea, organisms have been located throughout the gastrointestinal tract from the esophagus and stomach to rectum and even the appendix;[85,111,268,299,300,301,402,440,498,540,591,639,769,870,871] *Cryptosporidium* has been found in the appendix of an immunologically healthy child as well,[665] but is normally limited to the small and large intestine in these patients.

The clinical course and severity of diarrhea-associated illness varies considerably from individual to individual, which may relate to host factors or to intensity of exposure. Symptoms may wax and wane or be relentlessly persistent in any patient, although the latter is more common in immunologically impaired patients; in one group of 50 immunologically healthy patients, the range of illness was 2 to 26 d (mean duration 12 d).[390] Whether or not clinical symptoms correlate with number of oocysts shed is uncertain.

Ultimate outcome of illness and duration of diarrhea typically depend on the immune status of the host, and any immunocompromised patient must be considered at risk (Table 1). Treatment, particularly with spiramycin, somatostatin, interleukin-2, azidothymidine, or bovine colostrum hyperimmune to *Cryptosporidium* may modify symptoms in HIV-infected or other immunocompromised patients and prolong life.[44,153,173,175,399,405,488,567,656,708,709,830,833,842,852,881,894]

In HIV-infected persons (either presumed, based on clinical picture in some of the earlier cases, or serologically diagnosed), cryptosporidiosis lasting more than 30 d is ususally followed by death, although spontaneous clinical recovery has been reported.[92] In patients with other immune deficiencies, the length of illness generally varies with the ability to reverse the immunosuppression. Included here are individuals on exogenous immunosuppressive agents, especially patients with cancer;[116,168,271,408,464,509,527,537,540,546,600,660,777,863,871] hypo- or a-gammaglobulinemic patients;[41,54,165,186,208,209,422,440,505,741,863,871] malnourished persons, particularly children;[114,139,184,318,404,501,524,525,681,710,728,729,742,790,796,874,875,894] and patients with intercurrent viral infections such as measles or perhaps chickenpox.[214,314,369,681,766,790] Pregnancy may also predispose to *Cryptosporidium* infection.[201,435,637]

Immunosuppressive therapy may activate *Cryptosporidium* infection: an asymptomatic *Cryptosporidium* infection occurred in a hospitalized patient who had received prednisone and azathioprine only while in the hospital for treatment of skin lesions and in whom prior fecal examination for *Cryptosporidium* during hospitalization had been negative; the authors postulated reactivation of a latent infection.[357] One child with leukemia on immunosuppressive therapy had two episodes of cryptosporidiosis and diarrhea associated with active infection in a pet cat; both episodes resolved when immunosuppressive therapy was discontinued.[464] An adult on chemotherapy for lymphoma and another leukemic child, however, had disappearance of both symptoms and fecal oocysts, although anticancer chemotherapy was not stopped.[537,600] One adult leukemic patient with disseminated *Candida* infection in whom chemotherapy was continued, at autopsy had *Cryptosporidium* not only in the intestinal mucosa and respiratory parenchyma, but also in the lumen of submucosal colonic vessels, suggesting the possibility of hematogenous dissemination in this setting.[271]

Malnutrition may contribute to increased length of diarrhea, hospitalization, and even to fatality associated with cryptosporidiosis,[113,114,404,501,710,742,790,894] although some authors feel nutritional status is not important.[139,609,729] Experience in Haiti suggests correlation between nutritional status and clinical manifestations in children with cryptosporidiosis is no different than for other illnesses;[609] in nearby Jamaica, however, malnutrition is thought to predispose to severe *Cryptosporidium* infection.[501] Failure to thrive has been reported in many infants either as a result of or contributing to persistent illness.[318,369,435,710,796] In one nutritional center in Chile, 24.3% of children were infected with cryptosporidiosis, although only one third had diarrhea.[875]

Both severe diarrhea and respiratory symptoms attributed to *Cryptosporidium* have been associated with acute measles, when transient immune suppression occurs.[214,314,681] Chickenpox has occasionally been noted in conjunction with cryptosporidiosis, and in one day care center outbreak, children who recently had chickenpox had enteric cryptosporidiosis more often (p = 0.02) than those without chickenpox.[369,766,790]

In contrast to the above scenarios, in immunologically healthy individuals, diarrhea almost always lasts less than 30 d. Occasionally, however, there is an exception, such as the 33-year-old immunologically healthy man with an 18 kg weight loss due to 5 weeks of life-threatening *Cryptosporidium*-related diarrhea; this patient recovered and remained healthy for at least six months of followup.[230]

Other clinical symptoms besides diarrhea associated with cryptosporidiosis have been increasingly described, although the causal role for *Cryptosporidium* has not been as well established as for diarrheal disease. These include cholecystitis, hepatitis, pancreatitis, reactive arthritis, and a variety of respiratory problems.

Gallbladder disease, primarily acalculous cholecystitis, but sometimes sclerosing cholangitis, has been increasingly reported, especially in HIV-infected patients in whom chronic gallbladder carriage may be responsible for the inability to eradicate intestinal cryptosporidiosis.[111,163,208,299,300,348,396,401,514,639,724] The first case was actually described in 1981, in which an AIDS patient with diarrhea and intestinal cryptosporidiosis underwent a cholecystectomy because of cholestasis, a dilated common bile duct, and no visible stones; organisms were found postoperatively in the gallbladder mucosa,[300,639] Approximately a dozen other cases have been recorded since then. Usual symptoms include fever, right upper quadrant nonradiating pain, nausea, and vomiting. Although not discussed in all reported cases, some, but not all, patients had simultaneous diarrhea. Jaundice may also occur. Alkaline phosphatase and bilirubin have been elevated whenever measured. Noninvasive imaging (sonography or computed tomography) usually shows dilated extrahepatic biliary ducts associated with distal common bile duct (CBD) stenosis, although occasionally the CBD is normal and the gallbladder simply enlarged with a thick wall; ducts may have a beaded appearance.[111,208,348] Diagnosis has generally been made by histological examination of the gallbladder (sometimes retrospective rereview of specimens), by ampullary biopsy, or by direct examination of bile for oocysts; not all patients have oocysts detectable in fecal specimens. Cytomegalovirus has been found concurrently in at least six patients;[111,348,396,401] *Enterobacter cloacae* was associated in another instance.[396] Sphincterotomy proved beneficial in some instances,[724] although prognosis is usually poor, and pancreatitis and cholangitis are possible sequellae.[348,514,724] T-tube drainage and cholecystectomy have also been used to relieve cholestasis although biliary oocyst drainage continued after cholecystectomy in one patient.[514]

Cryptosporidium has been identified in liver biopsy specimens attached to epithelia of bile ducts in three cases, all coincident with cholecystitis.[299,396,514] Hepatitis with markedly elevated liver enzymes (GGT > 718 U/l) was noted in one patient.[396]

Symptomatic pancreatis and concurrent cryptosporidiosis have been described.[299,320,348,396,422] In one case, an immunologically healthy 14-year-old farmer's daughter had severe abdominal pain and a serum amylase of 14,400 U/l (nl <300) 1 week after diagnosed cryptosporidial enteritis. Extensive workup showed enlargement of the pancreas with ascites and no other etiology; symptoms resolved spontaneously over 6 weeks.[320] The other cases were all in immunologically compromised patients: an AIDS patient with right upper quadrant pain, elevated alkaline phosphatase and amylase, and dilated common and intrahepatic bile ducts also had an enlarged head of his pancreas with a dilated beaded main pancreatic duct; *Cryptosporidium* oocysts drained from a nasobiliary tube for at least 36 h.[299] Two other AIDS patients had pancreatitis accompanying cholecystitis.[348,396] In another case, *Cryptosporidium* oocysts were found at autopsy in the pancreatic duct of a child with severe combined immune deficiency.[422] *Cryptosporidium* has also been found in the intrahepatic bile ducts and the pancreatic ducts associated with squamous metaplasia at autopsy of an AIDS patient with chronic intestinal cryptosporidiosis.[603a]

Reactive arthritis has been attributed to *Cryptosporidium* three times. An HLA-B27-negative 27-year-old man had diffuse arthralgias and right knee effusion 7 d after *Cryptosporidium* enteritis. The effusion was sterile, and other intestinal pathogens sometimes associated with reactive arthritis were not detected; the patient improved over a 4-month period.[321] Two immunologically healthy children had joint involvement (feet, ankles, wrists, hands), in conjunction with *Cryptosporidium*-associated diarrhea;[733] clinical resolution was complete.

Cryptosporidium infection of the respiratory tract, particularly in immunologically impaired patients, has been the subject of a growing number of case reports.[119,252,285,314,356,408,422,497,509,547,874] The significance of this extraintestinal site is unclear, especially since pulmonary infections are common in immunodeficient individuals; respiratory symptoms have been noted with unusual frequency in persons with cryptosporidiosis identified in epidemiologic surveys.[114,183,324,501,609,728,729,742,796,797,823,863,869] While the species of *Cryptosporidium* infecting poultry does produce respiratory symptoms in poultry (see Chapter 9), it has not been shown to cause infection in humans. Whether *Cryptosporidium* of the respiratory tract simply represents organisms aspirated from another anatomical location or is a true pathogen remains conjectural.

Respiratory symptoms associated with *Cryptosporidium* infection have been nonspecific: shortness of breath, hoarseness, wheezing, croup, and, most often, cough are described. Diarrhea has not been noted in all patients, although stool examinations for *Cryptosporidium* have not always been performed. Chest X-ray findings have been unremarkable, generally limited to occasional infiltrates or increases in bronchial markings. *Cryptosporidium* oocysts have been identified in sputum, tracheal aspirates, bronchoalveolar lavage fluid, brush biopsy specimens and alveolar exudate on lung biopsy; organisms have rarely been found attached to the surface of bronchial mucosal cells[252,285,422] or in macrophages.[497] Concurrent pulmonary cytomegalovirus, *Pneumocystis carinii,* or mycobacterial infection have been noted in most cases, although *Cryptosporidium* has been the only pulmonary pathogen isolated in four HIV-seropositive cases, all of whom were alive at least 9 months after diagnosis.[356] Most immunodeficient patients with *Cryptosporidium* in their respiratory tracts do not recover. In one case of laryngotracheitis in which *Cryptosporidium* oocysts were identified in a sterile tracheal aspirate from a child in New Guinea with measles and cryptosporidial diarrhea, full recovery occurred within 45 d.[314] In another instance, in a hypogammaglobulinemic patient with recurrent sinusitis and chronic intestinal cryptosporidiosis, *Cryptosporidium* was actually isolated from aspiration of the maxillary sinus in the absence of any other pathogen;[209] the patient's sinus headache resolved after a course of spiramycin, although his intestinal infection did not.

There have been no published cases of disseminated cryptosporidiosis, although organisms were identified outside the colon within submucosal blood vessels in one cancer patient.[271] In an unusual situation, a 16-month-old boy with Trisomy 21, bloody diarrhea, and massive hemolysis leading to hemolytic-uremic syndrome, *Cryptosporidium* oocysts were identified on fecal examination; a causal association seems doubtful.[95]

Other than identification of *Cryptosporidium* organisms, there are no characteristic laboratory findings in the infected individual. Leukocytosis is not a feature of infection, although eosinophilia has been reported.[53,317,440,591,638,741] Smooth muscle antibodies (IgG) sometimes found in intracellular infections were identified in 11 of 24 (45.8%) patients in one series from Thailand.[115] Abnormal tests of malabsorption have also been noted.[29,55,166,272,527,549,636,741,751,769,870,871,881] Abdominal radiographic findings have been non-specific, including prominent mucosal folds, air-fluid levels, distended loops of intestine, and disordered motility.[64,91,101,591] Computed tomography has shown thickened mucosal folds and bowel walls.[296,761]

Asymptomatic infections have been described primarily during evaluation of possible outbreaks in day care centers or related to travel (see Section IV) and, less often, in the context of geographical surveys (Table 2). Accuracy with which asymptomatic infections are detected depends on the diagnostic proficiency and rigor of the laboratory, and since *Cryptosporidium* may be very difficult to diagnose,[180] asymptomatic infections may be underestimated. Only 90 of 6077 (1.5%) asymptomatic control subjects surveyed in geographically based studies (Table 2) excreted *Cryptosporidium* and, and 41 individuals were from a single study.[526] However, in Peru and Chile, as many as 30% of childhood infections have been reported to be asymptomatic.[840,875] Of 154 adult patients undergoing endoscopy for symptoms other than diarrhea, 18 (11.7%) had *Cryptosporidium* detected in duodenal aspirates; 7 of these patients also had organisms identified in stool examinations.[692b] A single case of asymptomatic cryptosporidiosis has been described in an HIV-infected patient in whom fecal examination was performed as part of an AIDS protocol evaluation.[912] Whether these all represent true asymptomatic infections or oocysts identified sometime after an undiagnosed perhaps brief, mild or atypical clinical illness remains speculative.

For details on diagnosis and treatment, see Chapters 1 and 2.

D. DURATION OF OOCYST SHEDDING

While oocysts may be shed intermittently in any patient regardless of immune status, immunologically intact persons with *Cryptosporidium*-associated diarrhea eventually stop excreting oocysts entirely (Table 1). Initially, oocyst shedding and clinical illness were thought to coincide,[41a,186] but as diagnostic experience increased and improved fecal staining and concentration techniques became available, oocyst excretion was discovered to exceed diarrhea and most other associated clinical symptoms. Several studies have examined the duration of oocysts detectable in feces after onset of symptoms in immunologically intact patients.[41a,75,186,317,364,386,390,731,767] Early reports suggested that oocysts were visible up to 2 weeks after diarrhea stopped,[75,317,364] although this interval varied considerably in different patients and varied according to diagnostic techniques. With a modified Ziehl-Neelsen stain, 44% of 43 cases shed oocysts for 14 d or more after beginning of illness;[364] using sucrose flotation, 82% of 33 patients had detectable oocysts for at least 2 weeks, 42% of 14 patients for more than 3 weeks, and 21% of 7 patients for more than 4 weeks.[75] In 50 immunologically healthy patients extensively followed, the range of illness was between 2 and 26 days; after symptoms ceased, however, oocysts were still detectable, sometimes intermittently, in 19 patients for between 1 and 15 d (mean 6.9 d), and an occasional positive specimen occurred as long as 75 to 85 d after infection.[386,390] In eight healthy patients infected during a day care center-related outbreak, five persons excreted oocysts for 1 to 5 weeks after symptoms stopped, and there was no relation between oocyst shedding and the duration or severity of diarrhea.[767] In another day care center-related outbreak, children excreted oocysts as long as 48 d after onset of symptoms, which lasted up to 28 d.[171] One study of 25 patients, however, suggested that oocyst shedding did correlate strongly with symptoms in 76% of patients and that in 16%, abdominal pain in fact continued after oocysts were no longer detectable; the median time from onset of symptoms to negative stool specimens in 49 patients was 18 d with a maximum of 50 d.[731] Oocyst detection after remission of symptoms is important in transmission of infection, particularly if the initial illness was mild and the affected individual was unaware of harboring an infectious agent. There has been no well-documented second symptomatic infection in an immunologically healthy host.

TABLE 2
Geographic Surveys Based on Fecal Examination for *Cryptosporidium* Oocyst Excretion

Geographic location	Population examined (date)	Principals (no. positive for Crypto./ no. examined)	Controls (no. positive for Crypto./ no. examined)	Ref.
		Africa		
Ghana				
Kumasi	Children w/diarrhea[a] (no date given)	61/474 (12.9%)	NR[b]	2
	Patients w/diarrhea (10/85—12/85)	114/1070 (10.7%)	16/300 patients without diarrhea (5.3%)	3
Guinea-Bissau				
Rural villages	Children <5 years w/diarrhea (3/84—4/84)	6/48[c] (12.5%)	4/222 children without diarrhea (1.8%)	138, 139
Liberia				
Urban slums and rural villages	Children w/diarrhea (no date given)	22/278[d] (7.9%)	NR	351
	Children <5 years (no date given)	29/374 (7.8%)	NR	353
	Children w/diarrhea (1/83—4/83)	20/237 (4.8%)	6/102 children without diarrhea (5.9%)	354
Nigeria				
Ogun State	Patients w/diarrhea (6/86)	11/420 (2.6%)	0/59 patients without diarrhea	681
Zaria	Patients w/diarrhea (8/86—2/87)	16/75[e] (21.3%)	NR	431
Rwanda				
Kigali	Patients w/diarrhea (10/83—1/84)	23/293[f] (7.8%)	0/94 patients without diarrhea	114
	Diarrheal specimens submitted to laboratory (10/83—1/86)	227/2717 children (8.4%); 35/1346 adults (2.6%)	NR	114
Butare	Children w/diarrhea ± measles; (3/84—4/84)	2/48 without measles; 6/24 w/measles[g] (11.1%)	1/55 without diarrhea ± measles (1.8%)	214
South Africa				
Durban	Children hospitalized w/diarrhea (1/85—3/85)	31/259 (11.9%)	0/103 children without diarrhea	742, 893
	Specimens from children <2 years w/diarrhea (8/85—7/86)	39/1222 (3.2%)	NR	548
Sudan				
Khartoum	Children hospitalized w/gastroenteritis (4/85—5/85)	5/82 (6.1%)	0/37 children without gastroenteritis	693
Uganda				
Kampala	Patients w/AIDS (10/85—11/85)	11/23 (47.8%)	NR	727a

TABLE 2 (continued)
Geographic Surveys Based on Fecal Examination for *Cryptosporidium* Oocyst Excretion

Geographic location	Population examined (date)	Principals (no. positive for Crypto./ no. examined)	Controls (no. positive for Crypto./ no. examined)	Ref.
Zaire				
Kinshasa	Adults w/diarrhea (no date given)	4/46 (8.7%)	NR	338
		Asia		
Bangladesh				
Dhaka	Dairy farm attendants and family members with diarrhea (12/82—11/83)[h]	7/88 attendants (7.9%); 7/77 family members (9.1%)	0/155 family members without diarrhea	664
	Patients w/diarrhea (1/84—5/84)	25/578 (4.3%)	NR	728
	Patients w/diarrhea (1/84—12/84)	71/2056[i] (3.5%)	NR	729
India				
Bombay	Patients w/acute diarrhea (6/86—12/86)	21/375 (5.6%)	0/65 patients without gastro-intestinal symptoms	717
Calcutta	Children w/acute diarrhea (1/85—9/86)	18/402 (4.5%)	NR	205
Chandigarh	Children w/diarrhea (no date given)	5/375 (1.3%)	NR	507
Kashmir	Patients w/diarrhea (7/86—10/87)	5/250[j] (2%)	NR	583
Varanasi	Children <3 years w/diarrhea (no date given)	3/100 (3%)	0/50 children without diarrhea	736a
Vellore	Children <3 years w/diarrhea <3 d (8/83—2/85)	89/682 (13.1%)	41/418 children without gastro-intestinal or infectious disease (9.8%)	526
Thailand				
Bangkok	Children w/diarrhea (1/85—6/85)	13/410 (3.2%)	1/410 without diarrhea (age matched) (0.2%)	792
		Caribbean		
Jamaica				
Kingston	Specimens submitted to enteric laboratory (4/85—9/85)	25/513 (4.9%)	NR	501
Haiti				
Port-au-Prince	Children <2 years w/diarrhea (10/82—9/84)	138/824[k] (16.7%)	NR	609
	Patients w/AIDS (7/83—12/84)	65/131 (49.6%)	NR	213

TABLE 2 (continued)
Geographic Surveys Based on Fecal Examination for *Cryptosporidium* Oocyst Excretion

Geographic location	Population examined (date)	Principals (no. positive for Crypto./ no. examined)	Controls (no. positive for Crypto./ no. examined)	Ref.
Trinidad/Tobago				
Port of Spain	Hospitalized patients w/diarrhea (no date given)	3/214 (1.4%)	NR	797
Middle East				
Israel				
Jerusalem	Hospitalized children w/diarrhea (6/85—10/85)	30/221 (13.5%)	NR	710
	Outpatient children w/diarrhea (6/85—10/85)	4/77 (5.2%)	NR	710
Kuwait				
Riadyh	Specimens submitted to hospital laboratory (1/85—12/85)	0/5614	NR	126
Saudi Arabia				
Damman/Alkhobar	Patients w/diarrhea (11/86—5/87)	2/321[l] (0.6%)	0/NG[m]	407
Pacific				
Australia				
Adelaide	Specimens submitted to laboratory (1981—1984)	11/9056[n] (0.1%)	NR	492
Alice Springs	Children w/diarrhea (7/84—unspecified)	9/94 (9.6%)	NR	492
Fairfield	Specimens submitted to hospital laboratory (11/84—10/86)	55/2248 (2.5%)	NR	97
Melbourne	Patients hospitalized w/gastroenteritis (2/81—6/82)	36/884[o] (4.1%)	0/320 patients without gastro-enteritis	823
New South Wales	Patients w/diarrhea or parasitic infection (6/85—7/85)	7/140 (5%)	NR	611
North Queensland	Diarrheal specimens submitted to laboratory (7/85—6/86)	36/780[p] (4.6%)	NR	183
New Zealand				
Auckland	Children w/diarrhea (12/84—3/85)	8/36 (22.2%)	NR	883
New Plymouth	Patients w/gastro-intestinal symptoms (6/84—5/85)	54/1273 (4.2%)	NR	140, 141
Philippines				
Manila	Patients w/diarrhea (8/83—12/84)	19/735 (2.6%)	NR	181
Palawan	Patients w/diarrhea (no date given)	1/19 (5.2%)	NR	443

TABLE 2 (continued)
Geographic Surveys Based on Fecal Examination for *Cryptosporidium* Oocyst Excretion

Geographic location	Population examined (date)	Principals (no. positive for Crypto./ no. examined)	Controls (no. positive for Crypto./ no. examined)	Ref.
		Europe		
Denmark				
Copenhagen	Patients w/gastro-intestinal symptoms (Spring, 1983)	16/800 (2%)	0/120 patients without gastro-intestinal symptoms	359
Finland				
Helsinki	Specimens from different patients submitted to parasitology laboratory (12/82—3/83)	14/154[q] (9.1%)	NR	387
	Continuation (3/83—4/84)	119/4545 (2.6%)	NR	389
	Travellers to Leningrad (no date given)	0/34 before trip; 9/34 after trip (26.5%)	0/17 nontravellers	388
Tohmajarvi	Patients at health center without acute medical problem (6/83—8/83)	7/233 (3.0%)	NR	648
France				
Paris	Patients w/AIDS (no date given)	28/132 (21.2%)	NR	682a
Rouen	Children w/gastro-enteritis (11/84—1/86)	11/260 (4.2%)	0/181 children without diarrhea	122
Tours	Children hospitalized w/diarrhea (4/84—3/85)	4/190 (2.1%)	0/37	55
Germany				
Berlin	Children w/diarrhea (11/84—10/85)	20/142 (14.1%)	0/166 children without diarrhea	323
Freiburg	Specimens submitted to laboratory (6/85—8/85)	14/1712[r] (0.6%)	NR	253
Gottingen	Patients w/urticaria (2/81—12/83)	18/178[s] (10.1%)	NR	662
Westphalia	Patients w/gastro-enteritis (6/86—10/86)	30/1600 (1.9%)	NR	764
Ulm	Immunocompromised patients without diarrhea (no date given)	0/160	0/60 immuno-logically healthy adults without diarrhea	406
	Patients w/diarrhea (no date given)	2/250 (0.1%)	As above	406
Italy				
Rome Orvieto	Children w/diarrhea (1/84—12/84)	9/124 (7.2%)	NR	135i

TABLE 2 (continued)
Geographic Surveys Based on Fecal Examination for *Cryptosporidium* Oocyst Excretion

Geographic location	Population examined (date)	Principals (no. positive for Crypto./ no. examined)	Controls (no. positive for Crypto./ no. examined)	Ref.
Portugal				
Lisbon	Day care center attendees[t] (9/87—10/87)	28/103 (27.1%)	NR	180c
Spain				
Barcelona	Patients with gastro-enteritis (no date given)	25/107[u] (23.3%)	2/103 patients without gastro-intestinal symptoms (2.0%)	657
Madrid	Specimens submitted to laboratory (no date given)	3/339[v] (0.9%)	NR	487
	Children w/diarrhea (no date given)	7/139 (5%)	7/593 patients without diarrhea (2 adults included) (1.2%)	486
Salamanca	Children <14 years evaluated at hospital (1/87—12/87)	8/699 (1.1%)	NR	264
Sweden				
Lund	Patients w/diarrhea (10/83—9/85)	20/698 (2.8%)	NR	60
Malmo	Patients w/diarrhea (3/84—9/85)	9/780 (1.2%)	0/519 routine employment physical exams	60
Switzerland				
Bern	Patients w/diarrhea (1/86—12/86)	31/2367[w] (1.4%)	NR	505
United Kingdom				
England				
Blackburn	Specimens submitted to laboratory (4/84—9/84)	24/2174[x] (1.1%)	NR	904
	Children admitted to hospitals (4/84—3/86)	48/742 (6.5%)	NR	796
Bristol	Patients w/diarrhea (10/83—1/84)	43/867 (4.9%)	NR	364
Liverpool	Children w/gastro-enteritis (6/83—4/84)	27/1967 (1.4%)	NR	317
	Continuation (no date given)	72/5242 (1.4%)	NR	318
	Patients w/diarrhea (6/83—5/85)	65/4028[y] (1.6%)	NR	73, 76
London	Children hospitalized w/diarrhea (9/83—2/84)	7/213 (3.2%)	1/112 children without diarrhea (0.9%)	369
	Patients submitting specimens to laboratory (6/85—10/85)	2/741 (2.7%)	NR	863

TABLE 2 (continued)
Geographic Surveys Based on Fecal Examination for *Cryptosporidium* Oocyst Excretion

Geographic location	Population examined (date)	Principals (no. positive for Crypto./no. examined)	Controls (no. positive for Crypto./no. examined)	Ref.
	Patients w/AIDS (1/81—6/87)	26/234 (11.1%)	NR	173
Manchester	Day care cnter attendees[t] (no date given)	13/74 (17.6%)	NR	313
Newcastle-upon-Tyne	Specimens submitted to hospital laboratory (1/84—12/84)	12/2197 (0.5%)	0 healthy volunteers	517
Sussex	Specimens submitted to laboratory (no date given)	14/800[z] (1.8%)	NR	590
Yorkshire	Specimens (diarrheal) submitted to laboratory (8/84—9/84)	12/166 (7.2%)	NR	906
Ireland				
Galway	Children w/diarrhea (2/85—2/86)	41/935 (4.3%)	NR	177
Scotland				
Paisley	Specimens submitted to laboratory (1/86—12/87)	83/NG[aa]	NR	732
Wales	Patients w/diarrhea (5/83—8/83)	7/500 (1.4%)	NR	142
	Continuation (no date given)	25/1500 (1.6%)	NR	144
	Central and South America			
Argentina				
Rosario	Children w/diarrhea (2/87—2/88)	16/210 (7.6%)	NR	912a
Brazil				
Belem[bb]	Children w/diarrhea (no date given)	3/94 (3.2%)	0/56 children without diarrhea	489
Fortaleza	Patients w/diarrhea (5/78—10/80)	9/117 (7.7%)	0/22 children without diarrhea	869
Sao Paulo	Patients w/AIDS (6/86—4/87)	19/157 (12.1%)	NR	218
Chile				
Arica/Antofagasta	Children <2 years w/diarrhea	44/263 (16.7%)	1/132 children without diarrhea (0.7%)	50
Santiago	Children hospitalized w/diarrhea (no date given)	4/100 (4%)	NR	872
	Nutritional child care center attendees[t] (no date given)	17/70 (24.3%)	3/54 children at an unaffected nutritional center (5.6%)	875
	Children hospitalized w/gastrointestinal disease (2/85—3/85)	11/180 (6.1%)	NR	790, 873

TABLE 2 (continued)
Geographic Surveys Based on Fecal Examination for *Cryptosporidium* Oocyst Excretion

Geographic location	Population examined (date)	Principals (no. positive for Crypto./ no. examined)	Controls (no. positive for Crypto./ no. examined)	Ref.
	Children hospitalized w/digestive symptoms (9/85—9/86)	48/750 (6.4%)	0/144 asymptomatic children	876
Costa Rica				
Urban slum and rural villages	Children w/diarrhea (1/82—12/82)	12/278[cc] (4.3%)	0/90 children without diarrhea	523, 524
Ecuador				
Quito	Children w/diarrhea (no date given)	19/169 (11.2%)	0/209 children without diarrhea	302
El Salvador				
San Salvador[bb]	Children w/diarrhea (3/87)	8/210 (3.8%)	NR	682
Guatemala				
Guatemala City[bb]	Children <1 year followed prospectively (7/85—6/86)	13/158 diarrheal[dd] episodes (8.3%)	7/1122 asymptomatic episodes (0.6%)	184
Mexico				
Soledad	Specimens from children submitted to laboratory (7/84—8/84)	18/57 (31.5%)	NR	755
Venezuela				
Caracas	Children <3 years w/diarrhea (3—4/83 and 5—9/84)	13/120 (10.8%)	NR	635
		North America		
Canada				
British Columbia	Patients w/diarrhea (10/83—10/84)	46/7300 (0.6%)	NR	553
	Patients w/diarrhea and all children submitting specimens to laboratory (10/83—6/85)	74/NG[aa]	NR	231
Manitoba	Diarrheal stool specimens submitted to laboratory (10/83—10/84)	39/3656[ee] (0.9%)	NR	510
	Diarrheal stool specimens submitted to laboratory (10/83—10/85)	78/6124[ff] (1.3%)	NR	512
Newfoundland	Specimens from different patients submitted to laboratory (5/84—11/84)	19/1621 (1.2%)	NR	670

TABLE 2 (continued)
Geographic Surveys Based on Fecal Examination for *Cryptosporidium* Oocyst Excretion

Geographic location	Population examined (date)	Principals (no. positive for Crypto./ no. examined)	Controls (no. positive for Crypto./ no. examined)	Ref.
United States				
California	Day care center attendees[t] (9/84)	7/11 (63.6%)	NR	43
Florida	Day care center attendees[t] (6/85)	28/84 (33.3%)	NR	767
Georgia	Day care center attendees[t] and staff (2/84—8/84)	12/90 (13.3%)	NR	43
Massachusetts	Specimens from different patients submitted to parasitology laboratory (2/83—1/84)	47/170 (2.8%)	NR	897
Michigan	Specimens submitted to laboratory (3/84—2/85)	63/1752[gg] (3.6%)	NR	116
	Day care center attendees[t] (8/84)	27/42 (64.3%)	NR	43, 171
New Mexico	Day care center attendees[t] (9/84)	12/31 (38.7%)	NR	43
New York	Specimens submitted to laboratory (6/88—7/88; 9/88—10/88)	2/400 (0.5%)	NR	67
	Asymptomatic day care center attendees (9/87)	6/22 (27.3%)	NR	179
Oklahoma	Day care center attendees[t] (8/84—11/84)	80/186 (43.0%)	NR	324, 536
Ohio	Pediatric stool specimens submitted to laboratory (no date given)	7/2780[hh] (0.3%)	NR	311
Oregon	Different patients submitting specimens to laboratory (1/84—12/84)	6/1710 (0.4%)	NR	739
Pennsylvania	Day care center attendees[t] (6/84)	14/45 (31.1%)	NR	11
South Carolina	Specimens from different patients submitted to parasitology laboratory (1/84—12/84)	25/582 (4.3%)	NR	358
Texas	Day care center attendees[t] (8/84)	18/50 (36.0%)	NR	791
	Specimens from children submitted to laboratory (7/83—10/83)	5/553 (0.9%)	NR	886
Washington	Patients at health center without acute medical problem (8/83—2/84)	1/89 (1.1%)	NR	779

TABLE 2 (continued)
Geographic Surveys Based on Fecal Examination for *Cryptosporidium* Oocyst Excretion

[a] The definition of diarrhea (number of stools per day and stool consistency) used in these studies varies.
[b] NR means none reported.
[c] The prevalence was 9.2% in the Balantas, the ethnic group surveyed with the most domestic animals and the only group with cattle. No *Cryptosporidium* was identified in the Muslim Mandinga group, who use pit latrines and keep fewer animals.
[d] 15/154 (9.7%) were from urban slum areas, and 7/124 (5.6%) were from rural villages.
[e] 29% of excretors were adults, and 8% were children less than 16 years of age.
[f] 20/193 (10.4%) were children, and 3/100 (3.0%) were adults.
[g] The association with measles is statistically significant ($p < 0.05$).
[h] 14% of 208 calves with diarrhea and 1% of 202 age-matched other calves also had *Cryptosporidium* sp. on fecal examination.
[i] 73% of infections were in children < 2 years of age.
[j] Prevalence was 15.8% in patients with malignant diseases and 0.85% in patients who were not immunocompromised.
[k] 533/824 (16.3%) were urban slum children, and 291/824 (17.5%) were rural children.
[l] Specimens from 209 children and 112 adults were examined and both positives were from children. All patients were from a Muslim community with good sanitary facilities and no domestic animals.
[m] No control specimen was positive, but neither denominator data nor details of the control group were given.
[n] 11 positives represent 10 patients.
[o] 33/697 (4.7%) were children, and 3/187 (1.6%) were adults ($p < 0.05$).
[p] 11 cases were from a day care center outbreak; 13/151 (8.6%) from an isolated aboriginal community in the wet tropics; and 12/629 (1.9%) from the dry tropics.
[q] Of specimens from 1422 patients submitted and concentrated, 154 were stained and examined for *Cryptosporidium;* 14/1422 positives gives only 0.98% prevalence.[232]
[r] 14 specimens represent 9 patients.
[s] 8 patients also had gastrointestinal symptoms.
[t] This study is outbreak associated.
[u] 18/22 (82%) children in a day care center and 7/85 (8.0%) hospitalized others had *Cryptosporidium.*
[v] 3 positive specimens represent 2 patients (brothers); 90 specimens (2 positives) were diarrheal, while 249 (1 positive) were not.
[w] 24 were children infected with *Cryptosporidium.*
[x] There was no difference between urban or rural locations; 14/153 (0.9%) individuals from hospital populations and 10/651 (1.5%) individuals from general practice populations were positive for *Cryptosporidium.*
[y] 61 children includes 2 with IgA deficiency and 1 leukemia. All adults were contacts of infected children.
[z] 14 positive specimens represent 10 patients.
[aa] NG means no denominator data given.
[bb] This is in an urban slum area.
[cc] 4/95 (4.2%) from rural areas and 8/183 (4.4%) from urban environments were positive for *Cryptosporidium.*
[dd] 20/130 (15.4%) children excreted *Cryptosporidium* at some time during the study period; the numbers recorded here represent specimens collected during diarrheal episodes, and the control represents specimens collected during healthy periods.
[ee] More persons from rural than urban areas were affected; 25% of 148 separate bovine specimens examined were positive for *Cryptosporidium.*
[ff] Infection rate was 31.7/100,000 in children under 5 years of age compared to 1.1/100,000 for adults.
[gg] 63 positives represent 53 patients.
[hh] 7 specimens represent 3 patients.

III. IMMUNE RESPONSE

There is scant information about the function of the human immune response to *Cryptosporidium.* Clearly both cellular and humoral immunity are important, as evidenced by development of chronic cryptosporidiosis both in HIV-infected patients with T cell abnormalities and in patients with gammaglobulin deficiencies; this contrasts with development of transient

Cryptosporidium infection in immunologically normal individuals. The role of different antibody classes as well as T helper cells (CD4+), T suppressor cells (CD8+), and T cell products, such as gamma interferon (a regulator of infection with other intracellular pathogens) or interleukin-2 (reported to have cured *Cryptosporidium* in two patients receiving it as part of an AIDS treatment protocol),[405] remains to be investigated. The importance of immunologic mediators in clearance of organisms during infection or subsequent protection, either from infection or from symptomatic expression of disease, is unknown.

Antibody response has been assessed both in individual cases and in larger series using either immunofluorescent assays (IFA) (with infected animal intestinal sections or fecal oocysts as antigen) or enzyme-linked immunosorbent assay (ELISA) tests[134,147,148,420,752,772,819,838,840,841] Specific serum IgM, IgG, and probably IgA, which persist for varying lengths of time after infection, have been identified. Serial evaluation of general, not class specific, antibody response of seven animal workers involved in the first documented *Cryptosporidium* outbreak in immunologically healthy persons, showed that 2 to 6 days after onset of infection, two had anti-*Cryptosporidium* antibodies, while five were seronegative.[134] Between 60 and 90 d after infection, however, all seven patients had elevated or increased antibody titers, and after a year, all five of the patients examined still had detectable titers.

In another investigation using ELISA technology, 39 of 41 patients with cryptosporidiosis for an undetermined period developed specific anti-*Cryptosporidium* IgG within 6 weeks of clinical diagnosis, including all of 26 HIV-infected patients; the two without IgG were early in infection and had detectable specific anti-*Cryptosporidium* IgM.[838] In three immunologically healthy patients followed longitudinally, IgM disappeared within 4 months and IgG within 12 months, although in another patient specific IgM antibodies persisted for at least 4 months. In HIV-infected patients, anti-*Cryptosporidium* IgG was detectable in ten cases for up to 10 months (the course of their illness until death), which included perimortem specimens; four of these patients had detectable specific IgM in at least one specimen much earlier in their HIV infection, consistent with residual ability to process antigen.[838] Persistence of IgG occurred in 77% (41 of 53) of infected Peruvians followed for 12 months; 7 of 31 persons seropositive for specific anti-*Cryptosporidium* IgM remained seropositive after 12 months, 6 with concomitant IgG.[840] Antibodies (IgG and/or IgM) persisted for at least 12 months in some U. S. Peace Corps volunteers during service in Africa.[841] Whether the persistent antibodies in all these studies represent continuous undetectable or latent infections (perhaps with production of separate antibodies to different life cycle stages), new infections (perhaps with antigenically different *Cryptosporidium* isolates), repeated exposure, or simply normal duration of antibody response remains conjectural.

Using Western blot technique, sera from 93% of 40 patients with cryptosporidiosis reacted with a single 23,000-Da calf-derived *Cryptosporidium* oocyst antigen separated by polyacrylamide gel electrophoresis;[839] sera from calves, humans, and horses obtained after *Cryptosporidium* infection reacted with a 20,000-Da *Cryptosporidium* antigen in a similar study.[538] Another group, again using similar technology but *Cryptosporidum* oocysts of goat and human origin as antigen, found that nine of ten acute or convalescent human sera and all goat sera tested recognized a 23,000- and a 32,000-Da. *Cryptosporidium* antigen.[494] The importance of these antigens in development of the human immune response needs further investigation.

In another series, ten paired serum samples drawn from patients at different times after *Cryptosporidium* infection were tested for specific anti-*Cryptosporidium* IgG, IgA, IgM and IgE using IFA technology with oocysts as antigen.[147,148] All showed at least a fourfold antibody titer rise, the greatest in IgE and the least in IgG. Uninfected rural control groups were predominantly positive for IgG alone and urban control groups for IgA or IgM. Only the oocyst wall and not the sporozoites fluoresced. The source of antigen made little difference in antibody titer, although in two instances, especially in a child whose cryptosporidiosis was likely derived from a scouring lamb, antibody titers were highest using homologous *Cryptosporidium* isolates as

TABLE 3
Transmission of *Cryptosporidium*

Person-to -person
Family and household members
Sexual partners
Health care workers
Day care center workers and attendees
Animal-to-person
Pets (cats, dogs)
Farm animals (cattle, sheep)
Laboratory animals
Environment
Nosocomial
Water

IFA antigen.[148] These studies represent the only data on the human specific anti-*Cryptosporidium* IgA response and are the only ones as well in which specific IgE responses were detected, unusual for a protozoan infection in humans.

IV. TRANSMISSION

Cryptosporidium oocysts need to be orally ingested to cause infection. The major sources for oocysts which infect humans are: other humans, animals, and the environment, particularly water, but possibly fomites, food or milk products as well (Table 3 and Chapter 1, Table 5).

Person-to-person transmission has been established between household and family members, sexual partners, children in day care centers and their caretakers, and health care workers and their patients. Family members may be detected in the course of screening after presentation of an index case. In each of two families who travelled in Africa, two children 3 years of age or younger had diarrhea associated with cryptosporidiosis; in one family, the mother but not the father became likewise infected, while in the other family, neither parent developed symptoms, although both had significant detectable antibodies to *Cryptosporidium* and may have been asymptomatically infected.[752] Family members have been especially diagnosed during evaluation of day care center outbreaks of cryptosporidiosis. In Michigan, 71% (10 of 14) of families with a symptomatic child had additional infected family members;[171] 32% (11 of 34) of family contacts in an outbreak in Florida had oocysts identified on stool examination, most frequently in persons known to have changed diapers;[767] 23% (6 of 26) of household members exposed to a child with cryptosporidiosis also had *Cryptosporidium* infection (median day of onset was 15 d after onset in the child) in an outbreak in Oklahoma, contrasted with 2% (1 of 53) household members exposed to a child without cryptosporidiosis ($p < 0.01$).[324] Other epidemiologic surveys and occasional case reports also describe increased cryptosporidiosis in household contacts frequently asymptomatic.[60,113,140,147,183,313,323,358,532,732,764,812,863,897]

Farming families may be especially susceptible to infection: 9.1% (7 of 77) of family members of infected farm attendents in Bangladesh shed oocysts.[664] Presumed secondary intrafamilial spread to a mother and sibling occurred 16 d after onset of initial infection in three family members exposed together to a presumed infected calf.[685] Immunocompromised family members are likely to be at particular risk; one bone marrow transplant recipient, for example, became infected after she cared for her two toddler-aged children during their episodes of cryptosporidiosis.[168] Spread of infection between sexual partners, both heterosexual and homosexual also occurs.[420,644,863,866]

The occurrence of cryptosporidiosis in child care centers has been described throughout the world, as well as at least eight separate outbreaks in the U.S. alone (see Table 2).[10,43,55,116,179,180c,183,231,313,323,324,358,364,492,512,536,657,739,767,856a,875,886,897] Some outbreaks may not be detected. Infected persons may not be bothered enough to seek medical attention: in one instance in Michigan, only three children saw a physician.[43] Some children remain asymptomatic: in a survey conducted in September in an upper middle class day care center in New York without a diarrheal outbreak, 27% (6 of 22) of asymptomatic children excreted oocysts; some of these children may have had unknown infections during the summer months.[179] In another survey of 30 randomly chosen children from 30 separate home day care centers in Grand Junction, Colorado, 9 (30%) excreted *Cryptosporidium* oocysts, and 6 of these were asymptomatic.[219]

Generally, in child care center-associated outbreaks, children less than 3 years old and their teachers are at greatest risk. Source for *Cryptosporidium* oocysts in these outbreaks is usually unclear. Asymptomatic carriers may excrete oocysts for days or weeks after resolution of clinical symptoms and are likely important reservoirs of infection. Frequently more than half of the symptomatic children in a day care setting can be documented to have cryptosporidiosis (Table 2). Also, typically, an outbreak lasts no longer than 2 months after identification of the index case; once an outbreak terminates it is unlikely that new sick children or oocysts will be found, although transmission to new settings remains possible. Teachers are often affected during outbreaks, and in one instance, a rotating staff worker became ill 2 weeks before the outbreak began in the children.[171] Of 18 caretakers, 4 (22%) in one center and 1 of 14 in another were infected with *Cryptosporidium,* detected during mass screening.[179,767] One outbreak may have been coincident with an outbreak of chickenpox and another occurred in a Chilean center caring for malnourished children, suggesting that transient or acquired immunosuppression may be a predisposing factor.[766,875] Numbers of infected individuals detected may vary with diagnostic technique used: in an outbreak in Oklahoma, 18.1% of fecal specimens containing oocysts were identified with Sheather's sugar flotation, and 18.1% using modified acid fast staining, but 22.9% of specimens were positive by one technique alone and only 13.5% positive by both.[536]

Individual cases and clusters have been associated with exposure in hospitals: one nurse became ill while caring for an infected 13-month-old child in Britain,[70] another while caring for an infected bone marrow transplant patient,[224] and another in the course of her work with infected renal transplant patients.[695] A medical intern with compatible symptoms was seropositive for anti-*Cryptosporidium* antibodies 2 weeks after caring for an infected AIDS patient; 8 of 26 other exposed hospital personnel in this study also had serologic evidence of infection compared to 3 of 18 unexposed employees.[420] It has also been suggested that individual patients may become infected during hospitalization.[70,357,512,537]

Three nosocomial outbreaks of cryptosporidiosis have now been recorded from Italy, South Africa, and Argentina.[518,695,894] In the first example, six patients on a bone marrow transplant unit were infected with *Cryptosporidium*, five symptomatically.[518] One of the six patients shared a room with a patient with cryptosporidiosis before being transferred to the transplant unit. No staff member or relative (0 of 22) was found to be infected; water samples from floor cleaning machines and toilet-cleaning rags were heavily contaminated with *Cryptosporidium*. In Durban, South Africa, during a randomized trial of spiramycin in hosptialized children, *Cryptosporidium* oocysts were found in stools of nine patients not participating in the study up to 2 weeks following an initial negative examination, consistent with hospital spread.[894] In a renal transplant unit in Buenos Aires, Argentina, 11 of 14 patients with acute diarrhea had cryptosporidiosis; hospital transmission was thought possible through shared toilet facilities.[695] Whether these are actually cases of person-to-person, fomite, or water-borne infection is not clear.

Animal-to-person (zoonotic) transmission has been described from household pets, laboratory animals, and farm animals. Infected cats or dogs have been likely sources;[419,464,512] however, in these instances, direction of spread is impossible to verify, and pets may have become infected from their owners as well.[416] At least 20 cases of infected laboratory workers have been reported,[17,41a,102,186,355,462,644,674] the majority in contact with infected calves. In one case a researcher became infected during oral inoculation of rabbits.[102] Another laboratory worker was thought to have become infected after inhaling *Cryptosporidium*-containing airborne droplets expelled from an infected calf.[355] Farm animals, again mainly cattle, have been implicated in a number of instances. Zoonotic exposure was likely important in the first case ever reported[591] and subsequent reports as well.[101,140,177,253,419,443,512,648,664,685,739,764] Asymptomatic oocyst excretion has been noted in 6 of 7 infected persons (total of 233 persons screened, 63 persons with direct contact to cattle) in one farming community in Finland.[648] Sheep, particularly bottle-fed orphan lambs, have also been recently described as a probable source[147,151] as have infected water buffalo in the Philippines.[443]

The environment, especially water, has been increasingly recognized as a source for *Cryptosporidium* oocysts. Oocysts are highly resilient organisms: they remain viable at 4°C in the laboratory for up to 12 months,[189,556,824] and must be kept at 45°C for 5 to 20 min[21] before they lose infectivity. Theoretically, therefore, consumption of raw or improperly pasteurized milk, or improperly preserved food products, or those washed with contaminated water, may be potential sources of infection, although, there is no conclusive evidence of this.[147,253,311,313,510,685,772,796]

Water has been implicated as a source of *Cryptosporidium* oocysts from case clusters connected with travel and from clearly documented water borne outbreaks, including one associated with swimming in a pool containing oocyst-contaminated water.[257] *Cryptosporidium* has been isolated from drinking water (raw and city finished), river, stream, or reservoir water in North America, South America, and Europe and is likely ubiquitous in water throughout the world.[46,47,147,200,258,322,368,502,544,602,699,704,743,840] It is not surprising that prevalence of cryptosporidiosis is higher in less developed regions of the world where water supplies may be unprocessed and that travel, particularly to areas with poorer water supplies, may be a risk factor for cryptosporidiosis.

The first association of cryptosporidiosis with travel was suggested in 1983, when 12 of 14 Finnish patients found to be excreting *Cryptosporidium* oocysts were retrospectively discovered to have become ill within 12 d of visiting Leningrad, U.S.S.R., and to have had no contact with each other.[387] In a subsequent prospective study, three groups of Finnish students were screened for *Cryptosporidium* before and after a visit to Leningrad and 9 of 34 students became infected.[389] In St. Lucia in the Caribbean, four North American tourists traveling together developed symptomatic cryptosporidiosis within 9 d of their visit; although no *Cryptosporidium* were found in the river water supplying hotel drinking water, a statistically significant correlation with tap water consumption was noted for these tourists compared to asymptomatic tourists.[499] Additional travel-related cases have been reported in visitors to Egypt, the Central African Republic, Mexico, Pakistan and New Guinea.[183,231,249,752,772] See Chapter 3 for discussion of waterborne outbreaks.

V. EPIDEMIOLOGY

Human infection with *Cryptosporidium* has been identified on all six inhabited continents (Table 4). Large-scale surveys have generally examined fecal specimens from selective, and not necessarily comparable, populations: in more developed countries, other than outbreak situations, specimens are most often from adults or children who have submitted fecal samples to a specific diagnostic laboratory or who sought medical attention for a gastrointestinal complaint; generally little information is available about these patients. In less developed countries, numerous field surveys have also been performed. Any survey which identifies oocysts obviously gives prevalence data on active infection, although accuracy of such data relies on number of stool specimens examined from single individuals, number and kinds of diagnostic techniques used, and proficiency with these. Serologic prevalence surveys, based on detection of specific anti-*Cryptosporidium* IgG, IgM, or IgA have also been performed in some areas of the world using IFA or ELISA technology. These correlate with infection sometime in life, but not necessarily active infection.

TABLE 4
Geographic Distribution of Reported Human Cases of Cryptosporidiosis[a]

Africa	Central/South America	Middle East
Ghana	Argentina	Egypt (549)
Guinea-Bissau	Brazil	Israel
Liberia	Chile	Kuwait
Nigeria	Costa Rica	Saudi Arabia
Rwanda	Ecuador	
South Africa	El Salvador	**North America**
Sudan	Guatemala	
Uganda	Mexico	Canada
Zaire (338)	Peru (840)	United States
	Venezuela	
Asia	**Europe**	**Pacific**
Bangladesh	Belgium(851)	Australia
China (240)	Czechoslovakia (385)	New Guinea (314)
India	Denmark	New Zealand
Pakistan (249)	Finland	Philippines
Thailand	France	Tasmania (532)
	Germany	
Caribbean	Greece (397)	
	Italy	
Haiti	Portugal	
Jamaica	Spain	
St. Lucia (499)	Sweden	
Tobago	Switzerland	
Trinidad	United Kingdom (England, Ireland, Scotland, Wales)	

[a] One reference for each country not included in Table 2 is listed in parentheses.

As shown in Table 2, there have now been more than 100 geographically based surveys for *Cryptosporidium* oocyst excretion from at least 40 different countries around the globe. Although variations in population examined (numbers of specimens vs. numbers of patients) render these studies not comparable in the strictest sense, Table 5 nevertheless summarizes the range and mean prevalence rates of active *Cryptosporidium* infection in various land areas. In the more industrialized countries of Europe and North America, prevalence rate, excluding documented outbreaks, is generally between 1 and 3%. In contrast, mean prevalence rates from elsewhere in the world are between 4.9% in Asia and 10.4% in Africa. The higher prevalence of cryptosporidiosis in some less developed countries may relate to relative lack of clean water and sanitary facilities, crowding of households, and large number of animals in close proximity to residences. In general, urban areas in lesser developed countries have more cryptosporidiosis than less crowded rural areas (although in one survey in Ghana,[3] the prevalence in rural populations was higher than in urban populations), while the reverse appears to be true in more developed countries (Table 2)

Seroprevalence studies (Table 6) support the observations of increased cryptosporidiosis in urban slum areas of the less developed world, although seroprevalence rates in Europe and North America were usually between 25 and 35%, suggesting *Cryptosporidium* infections, perhaps asymptomatic, may be more common in these regions than fecal oocyst excretion surveys have suggested. In one longitudinal serosurvey of U.S. Peace Corps volunteers, after 1 year of service

TABLE 5
Geographic Prevalence of *Cryptosporidium* Oocysts in Fecal Specimens[a]

	Range (%)	Mean (%)	Number of references included[b]
Africa	2.6—21.3	10.4	9
Asia	1.3—13.1	4.9	7
Caribbean	1.4—16.7	7.6	3
Central/South America	3.2—31.5	9.7	9
Pacific	0.1—22.2	6.0	10
Europe	0.1—14.1	3.5[c]	23
North America	0.3—4.3	1.6	10

[a] Only the latest data prevalence rate was included if there were multiple studies from the same location; no data was included from outbreaks, if known, or from specialized groups, such as AIDS patients.

[b] See Table 2 for precise references; References 3, 50, 113, 135i, 736a, 779, 912a were included in Table 2 after Table 5 was prepared and are therefore excluded here.

[c] If two references (323, 662) with unusually high prevalence rates (10.1 and 14.1%) are excluded, the mean becomes 2.6%.

TABLE 6
Seroprevalence of *Cryptosporidium*

	% seropositive (no. of individuals)	Ref.
U. S. Peace Corps volunteers before overseas service	32 (24/75)[a]	841
Healthy U. S. Centers for Disease Control employees	35 (7/20)[a]	322
Personnel in one U. S. hospital	25 (11/44)[b]	420
Random sera from two laboratories in Great Britain	26 (26/100)[b]	148
Random blood donors in Edinburgh	86 (18/21)[b]	819
Peruvian children and adults	64 (267/415)[a]	840
Venezuelan children	64 (54/84)[a]	840

[a] IgG alone.

[b] Multiple antibody classes.

in Africa, 14% (8 of 56) developed anti-*Cryptosporidium* IgG (15 of 56 were already seropositive). After the second year, a similar percentage became newly seropositive.[841] These data suggest increased infection for previously unexposed workers in a probable endemic area.

Seasonal differences in infection rates have been proposed by some with more infections in warmer or more humid months.[116,526,553,664,728,823,897] Also proposed by some but not substantiated is a synergistic association between *Giardia lamblia* or another enteropathogen and *Cryptosporidium*.[6,236,368,388,389,536,896]

Epidemiologic data on the relationship between cryptosporidiosis and HIV-infected patients show that in 1986 in the U.S., an estimated 3.6% of late-stage HIV-infected patients had cryptosporidiosis at time of report to the CDC.[587] However, in patients with AIDS and diarrhea, 15% of those evaluated at the National Institutes of Health, Bethesda, MD, and 16% of those evaluated at the Johns Hopkins Hospital, Baltimore, had *Cryptosporidium* infection, the most common pathogen in the latter series.[442,744] In a retrospective review of 100 charts from patients treated at Montefiore Medical Center or North Central Bronx Hospital, New York,

Cryptosporidium was identified in 5 of 36 (14%) AIDS patients with diarrhea.[48] In two hospitals in Great Britain and Brazil, 11 and 12% of AIDS patients, respectively, had cryptosporidiosis, and 19% of those in Great Britain were thought to have died as a direct result.[173,218] In a series from Paris, 21.1% of 132 prospectively studied AIDS patients were infected with *Cryptosporidium*, the most frequent enteropathogen identified.[682a] Finally, 48.7% of 23 AIDS patients with enteropathy in one hospital study from Uganda had cryptosporidiosis.[727a] As the number of late-stage HIV-infected persons increases, there will likely be more cryptosporidiosis in the overall HIV-infected population.

BIBLIOGRAPHY

Information cited in this chapter is found in References 2, 3, 6, 10, 11, 17, 19, 21, 29, 30, 41 to 44, 46 to 48, 50, 53 to 55, 57, 59, 60, 64, 65, 67, 70 to 76, 79, 85, 91, 92, 95 to 97, 101, 102, 106, 111, 113 to 116, 119, 120, 122, 126, 134, 135i, 138 to 142, 144, 146 to 148, 151, 153, 159, 163 to 166, 168, 171 to 173, 175 to 177, 179, 180, 180c, 181, 183, 184, 186, 189, 200, 201, 205, 208, 209, 213, 214, 218, 219, 224, 230 to 232, 236, 240, 249, 250, 252, 253, 257 to 260, 262, 264, 268, 269, 271, 272, 285, 296, 299 to 302, 311, 313, 314, 316 to 318, 320 to 324, 334, 338, 348, 351, 353 to 359, 361, 364, 368, 369, 385 to 391, 395 to 397, 399, 401, 402, 404 to 408, 416, 419, 420, 422, 431, 435, 440, 442, 443, 450, 462, 464, 486 to 489, 492, 494, 495, 497 to 499, 501, 502, 505, 507, 509, 510, 512, 514, 516 to 518, 523 to 527, 532, 535 to 538, 540, 544 to 549, 553, 556, 567, 583, 587, 591, 595, 600, 602, 609, 611, 635 to 639, 642 to 644, 646, 648, 656, 657, 660, 662, 664, 665, 670, 674, 681 to 682a, 685, 692b, 693, 695, 699, 704, 705, 708 to 710, 717, 721, 724, 727a to 729, 731 to 733, 736a, 739, 741 to 744, 751, 752, 755, 761, 764, 766, 767, 769, 772, 773, 776, 777, 779, 790 to 792, 796, 797, 812, 819, 823, 824, 830, 831, 833, 838 to 842, 851, 852, 856a, 863, 866, 869 to 876, 881, 883, 886, 893 to 897, 904, 906, and 911 to 913.

Chapter 5

CRYPTOSPORIDIOSIS IN RUMINANTS

Kenneth W. Angus

TABLE OF CONTENTS

I. RUMINANT HUSBANDRY: EVOLUTION AND GLOBAL SIGNIFICANCE

The domestication of ruminants dates back to the end of the Old Stone Age period of prehistory.[686] Sheep and goats probably became domesticated around the beginning of the Neolithic period, that is, approximately 9000 B.C., while domestication of cattle came much later, around 6000 B.C.[432] Primitive man almost certainly killed cattle initially as a source of food; the use of oxen as draft animals came later, and for thousands of years, cows' milk was used solely for rearing the young calves. Sheep and goats were important to Neolithic man not only for food, but also for the protection against the bitter cold afforded by their skins.[686] From these primitive times, evolving farming methods gradually led to the exploitation of these species as sources of animal protein, milk and milk products, leather, wool, fiber for cashmere and mohair production, and for the clearing of brush land. Very recently, the red deer and some other deer species have been added to the growing list of farmed ruminant species that help to feed and clothe the ever-increasing world population.

Much of the energy reaching our planet from the sun becomes stored through photosynthesis in forms which man cannot readily use directly as a source of food energy. We depend to a great extent on ruminant species to convert a whole variety of materials into nutritious and palatable foodstuffs. Livestock farming helps to conserve the topsoil and serves to return numerous essential elements back into the ground for recycling.[686] With over one third of the world population in the grip of chronic malnutrition,[445] improvements in livestock production and efficiency in the developing countries are urgent priorities.

In the highly developed industrial countries, methods of livestock farming have become increasingly intensive, with the accent on bigger and more elaborate units. In many European countries at least, this trend has led to over-production and a recession in livestock production. The application of sophisticated technology to reduce labor costs requires a high fuel-energy input, which paradoxically has the effect of eroding profit margins.[445] However, Third World countries have been quick to adopt Western ideas, which are gradually superseding the traditional flock and herd management systems of thousands of years of static village-based economies. A threat to all such enterprises, both in the affluent industrial societies and in less favored regions of the world, is the impact of neonatal disease, particularly enteric infections. Cryptosporidiosis is only one of several which need to be controlled.

II. SPECIES OF *CRYPTOSPORIDIUM* THAT INFECT RUMINANTS

Only two species of *Cryptosporidium* have been identified in ruminants. The intestinal species, *C. parvum,* is the one which most commonly causes ruminant cryptosporidiosis.[198] However, an abomasal species, *C. muris,* has been found in cattle in the U.S.[25] and in mountain gazelles in the Federal Republic of Germany.[658]

III. CRYPTOSPORIDIOSIS IN CATTLE (*BOS TAURUS*)

A. HISTORICAL BACKGROUND

The first report of *Cryptosporidium* sp. infection in cattle appeared in 1971.[607] Endogenous cryptosporidial stages were detected on histological examination of the jejunum of an 8-month-old heifer which had chronic diarrhea. This report was soon followed by others[66, 543] in which similar infections were found in diarrheic 2-week-old calves. In 1976, Canadian workers reported the occurrence of *Cryptosporidium* sp. infection in calves aged less than 2 weeks with acute diarrhea, from beef and dairy herds.[564] The parasite was found in histological sections of the lower jejunum and ileum, associated with slight villus stunting. A variety of other enteropathogens, both viral and bacterial, were associated with the presence of the parasite in

TABLE 1
Prevalence of *Cryptosporidium* sp. Oocyst Excretion by Diarrheic Calves

Locality	Total herds surveyed[a]	Total calves examined	Percent excreting oocysts	Ref.
Hungary	110	497	(27)	578
Federal Republic of Germany	*	222	(40)	326
U.S. (Idaho)	41	171	(64)	11
Canada	26	161	(26)	712
Italy	47	997	(40)	270a
U.S. (Maryland)	12	136	(27)	449
Netherlands	11	375	(55)	211
Federal Republic of Germany	70	284	(44)	243
Denmark	*	1927	(17)	335
Canada	148	182	(25)	510
Finland	8	68	(76)	645
U.K. (South)	45	465	(23)	684
U.K. (North)	32	309	(25)	745
U.S. (Calfornia)	5	95	(21)	757
U.S.S.R. (Leningrad)	—	518	(26)	822b

[a] * = laboratory submissions.

most instances, but in five calves, no other common pathogens were detected. A similar study in Iowa in 1978[649] confirmed the Canadian findings, and the authors concluded that cryptosporidia probably were common enteropathogens of calves. The results of numerous investigations in many countries over the ensuing decade have amply supported this conclusion[34,117,158,191,414,822] (Table 1). However, the role of *Cryptosporidium* spp. as primary enteropathogens was uncertain until 1980, when evidence from field studies showed that the parasite was capable of causing clinical diarrhea in calves in the absence of concurrent infection with the common viral or bacterial agents normally incriminated in outbreaks of neonatal calf diarrhea.[810] Experimental confirmation followed when methods to free the infective oocysts from other contaminating pathogens became available.[329,825] Judging from the site of infection observed in histological studies, namely, the distal small intestine, all of these early outbreaks appear to have been associated with *C. parvum* infections. However, an abomasal form of cryptosporidiosis was found in cattle in the west central U.S. in 1985,[25] caused by a species of *Cryptosporidium* apparently indistinguishable from *C. muris*, the species originally found in a mouse stomach by Tyzzer in 1907.[806]

B. CLINICAL FEATURES OF BOVINE *C. PARVUM* INFECTIONS

Cryptosporidium spp. have direct life cycles, and in calves the life cycle of *C. parvum* is complete in about 4 d.[190] Infection is by the fecal-oral route. The main presenting sign of cryptosporidiosis in calves is diarrhea associated with profuse shedding of infective oocysts (in the order of 10^6 to 10^7/g of feces).[110,190] At least one large survey of the agents causing diarrhea in neonatal calves[684] found a highly significant association between *Cryptosporidium* sp. infection and diarrhea. Varying degrees of dullness, anorexia, fever, and loss of condition can occur, but monoinfections with *C. parvum* seldom cause the acute dehydration, collapse, and high mortality experienced in enterotoxigenic *Escherichia coli* (ETEC-K99+) or rotavirus infections, for example. These observations are in line with experimental transmission studies using oocyst preparations rendered free of live bacteria or viruses by chemical treatment.[329,825]

In field investigations, the true role of the parasite as a cause of diarrhea can be determined only if exclusion of other known enteropathogenic agents from feces samples can be demonstrated by recognized diagnostic methods. The most common agents likely to be found include ETEC-K99+, *Salmonella* spp., rotavirus, coronavirus, and other enteric viruses detectable by transmission electron microscopy (TEM) after negative staining.[684] Several studies which excluded these and some other possible candidate enteropathogens have demonstrated that diarrhea in *C. parvum* infections is usually mild,[13,645,810] though occasionally severe but transient,[212] resulting in varying degrees of morbidity, but generally low mortality.

Infection has been confirmed in calves as young as 4 d[745] and as old as 4 weeks,[89,810] but analysis of numerous reports indicates that oocyst excretion and diarrhea seldom occur in calves under 7 d old and tend to peak at approximately 2 weeks of age.[13,89,211,645] Because the life cycle of *C. parvum* in calves is only about 4 d,[190] clinical infection in a calf as young as 4 d would indicate heavy environmental infection in the calving area. The possible relationship between management practices and the age at which calves first become exposed to threshold doses of oocysts is discussed more fully under Section III.H (see below). However, delayed exposure to *C. parvum* is clearly important if there is a risk of concurrent ETEC-K99+ infection, because adherence of these coliforms to enterocytes diminishes rapidly after the first few days of life.[818] In contrast, calves between 1 and 2 weeks of age are highly vulnerable to rotavirus and coronavirus infections,[818] and investigations of diarrhea associated with *C. parvum* in this age group should always take into account the possibility of multiple infections.

In outbreaks where *C. parvum* is the dominant pathogen, the duration of diarrhea depends on various factors, such as the level of environmental infection, which determines the intake of infective oocysts, the pathogenicity of the infecting isolate, host susceptibility, and age at the time of primary infection. In transmission experiments, oocyst numbers of the order of 10^5 to 10^6 given to conventionally reared calves that had at least one feed of their dams' colostrum, and which were known to be free of viral and bacterial enteropathogens before oocyst dosing, caused acute diarrhea commencing on day 5 postinoculation and lasting for a mean of about 8 d.[110] At the peak of the diarrhea, oocyst numbers of the order of 10^6 to 10^7/g of feces were excreted, which is in line with natural infections.[190,198] Similar clinical findings and excretion patterns were also demonstrated in calves that had recovered from experimentally induced combined ETEC-K99+ and rotavirus infections.[810] There are several published accounts of severe diarrhea outbreaks, sometimes prolonged and intractable[40] or associated with high mortality,[712] in which *C. parvum* was the only enteropathogenic agent found. Such outbreaks could be the result of infection with highly pathogenic isolates of *C. parvum,* undue susceptibility of the host population, or delay in sampling for diagnosis beyond the normal time course of complicating infections.

C. CLINICAL FEATURES OF BOVINE *C. MURIS* INFECTION

The limited reports relating to the abomasal infection[25,27] indicate that affected animals do not develop diarrhea, though they may excrete oocysts for several months. *C. muris* also infects and causes prolonged oocyst excretion in older calves, steers, and even adult animals,[27] unlike *C. parvum* infections. In one study, infected animals had significantly elevated plasma pepsinogen levels and showed a significantly lower weight increment than animals of comparable age that did not shed oocysts.[27]

D. DIAGNOSIS OF BOVINE CRYPTOSPORIDIOSIS

Recognition of the presence of oocysts in the feces of diarrheic animals is indicative of infection. The oocysts of *C. parvum* are 4.8×5.0 µm, while those of *C. muris* are 8.0×6.0 µm.[158] The simplest method of confirming their presence in feces involves preparing and staining smears. The literature abounds with staining recipes for *Cryptosporidium* sp. oocysts, but most laboratories use either modifications of the Kinyoun acid-fast method[495] tailored to suit their own preferences, or auramine O staining followed by fluorescence microscopy. However, it would be naive to diagnose clinical cryptosporidiosis on the basis of recognizing oocysts in stained smears. Healthy, nondiarrheic calves can also shed sufficient oocysts to be detected in this way, presumably as a result of subclinical infections. What is required is some measure of the level of excretion. Although this can be approximated by numerical scoring of mean numbers of oocysts seen in representative fields of stained smears at specified magnifications,[430] or by counting absolute numbers in aliquots of feces suspensions,[825] true excretion data can be obtained only by flotation and enumeration of oocysts using a hemocytometer[110,235] and phase-contrast microscopy. Significant numbers indicating clinical infection are in the order of 10^6 to 10^8/g feces.[110, 191] Second it is dangerous to base the diagnosis on a few positive samples only. Realistic numbers of samples should be collected from nondiarrheic as well as diarrheic calves. If at least 25% of diarrheic calves are excreting significant numbers of oocysts, compared with nondiarrheic contacts, then cryptosporidiosis is probably a major factor in the diarrhea outbreak.[684] Third it is essential that feces samples are examined for other common enteropathogens. At the very least, tests for bovine rotavirus, coronavirus, and ETEC-K99+, cultural methods for *Salmonella* spp., and, if an electron microscope is available, TEM examination for negatively-stained enteric virus particles should be undertaken. The following general approach to major outbreaks of calf diarrhea has been recommended:[684]

1. Samples of feces should be examined from affected calves at various stages of the disease. (Some viruses, e.g., calici-like agents, are more easily identified at the onset, while *C. parvum* might be found later.)
2. Samples should be examined for rotavirus, coronavirus, *Cryptosporidium* spp., and *Salmonella* spp. routinely. Examination for ETEC-K99+ adhesins need be performed only on calves less than 1 week of age that have acute, watery diarrhea.
3. Quantitative as well as qualitative examinations for *Salmonella* spp. will distinguish between clinical cases or "active carriers" and "passive carriers".
4. Samples from healthy calves should also be examined.

Necropsy findings in bovine cryptosporidiosis are nonspecific. Although descriptions of congested mucosae and alterations in the color, odor, and consistency of gut contents have been reported, these have no pathognomonic significance. However, cryptosporidial endogenous stages and oocysts can be detected in Giemsa-stained impression smears made from the mucosa of the ileum of cadavers[649] and on histological examination of intestinal tissue fixed soon after death.[34] To delay fixation for more than about 20 min after death will result in loss of the surface epithelium containing the endogenous stages as a result of autolytic processes.[52] Best results are obtained by euthanasia of a moribund calf and immediately fixing and/or freezing gut samples from various locations for histologic and immunocytochemical examinations.[308] This approach may identify agents undetected in feces.[308] Serum antibodies can be detected in calves infected by *C. parvum*,[191,819] but because calves acquire antibody from their dams and because virtually all calves become infected with the organism, serological findings are not diagnostic for bovine cryptosporidiosis.[38]

E. PATHOLOGY OF BOVINE CRYPTOSPORIDIOSIS

1. *C. parvum* Infection

The results of some early investigations of calf diarrhea outbreaks[20,245a,629,649,712] indicated that infections were concentrated in the distal small intestine and were associated with mild to moderate villous atrophy, villous fusion, and changes in the surface epithelium, including metaplasia to a low columnar or cuboidal form, with some degeneration and sloughing of individual enterocytes and shortening of microvilli.[649] Infiltrates of mononuclear cells and neutrophils were seen in the lamina propria.[712] Infections were also seen in the cecum and colon[649,712] and occasionally in the duodenum.[712] Organisms were not seen in the crypts of Lieberkuhn in the small intestine, but could be found in crypts in large intestine sites.[649] Some crypts in all these sites were dilated and contained necrotic debris or dead leucocytes.[712] The preferential concentration of parasite stages in the distal small intestine was confirmed by orally infecting calves that had conventional gut flora using as inocula either fresh diarrheic feces[556] or potassium dichromate-stored feces[556] containing high concentrations of *Cryptosporidium* sp. oocysts.

The results of all these investigations and transmission experiments were always open to the criticism that either agents other than those excluded by the microbiological and other techniques used could have been acting synergistically with the parasite to produce the changes observed or, that the changes resulted from some form of interaction with the normal gut flora. These suspicions were to some extent heightened by the elegant studies of Hall and colleagues,[308] who used a battery of morphologic and immunochemical techniques to examine in detail the intestines of moribund diarrheic calves from farm outbreaks. Their studies showed that *Cryptosporidium* sp. could be associated with other enteropathogens identifiable by immunological methods in the tissues, but not by microbiological methods in the feces. To be valid, experimental transmission of the organism required the chemical sterilization of oocyst suspensions with 60% alcohol[825] or 3.2% peracetic acid.[329] In one series of experiments,[825] conventional, specific pathogen-free and gnotobiotic calves were given infected feces containing oocysts but no other common enteropathogens, similar feces that had been passaged once through specific pathogen-free mice, or feces containing oocysts treated with 60% alcohol after passage through gnotobiotic piglets. Most of the calves developed anorexia, depression and diarrhea by 2 to 7 d postinoculation (DPI). By 2 to 4 d, all calves had high concentrations of parasite stages in the lower jejunum and ileum as determined by histological examination, but few had cecal or colonic infections. Moderate infections were seen in the duodenum and upper jejunum. Calves euthanized between 5 and 9 DPI had no parasites in the duodenum or anterior jejunum, but the ileum, cecum and colon were heavily infected. Lesions were stunting and fusion of villi, which were covered by a poorly differentiated, cuboidal epithelium. Macrophage and neutrophil infiltrates were seen in the lamina propria, with inflammatory cells in some distended lacteals. The mucosa was hyperemic and edematous, with serum protein leakage at the base. Crypts were infected only in large bowel sites. No significant changes were seen in cecum or colon. Membrane-bound lactase concentrations declined during the clinical phase in the distal small intestine, then recovered to concentrations in excess of normal when the diarrhea stopped. In a separate study, gnotobiotic calves given oocyst suspensions sterilized with peracetic acid[329] became diarrheic at 2 or 3 DPI. By 5 d, the distribution of parasites was similar to that described by other workers. Measured villi in the distal small intestine were significantly shorter than control villi and crypts significantly longer. Some villous fusion was seen. Surface epithelium was altered to low columnar or cuboidal, while basophilic, degenerating cells were frequently seen. Numerous mitotic figures were seen in the crypts, and the mucosa was infiltrated by mononuclear cells and neutrophils. Stringent microbiological tests failed to detect any extraneous agents.

The mechanism(s) by which cryptosporidia damage the gut and produce their clinical effects is still fairly obscure. The morphologic changes described above largely represent a standard response of the intestinal mucosa to a certain grade of insult, which differs very little from that produced by a whole range of different enteropathogens.[308] Villous atrophy probably results from accelerated loss of damaged enterocytes from the surface, with compensatory crypt hypertrophy. By incorporating into its endogenous stages large numbers of enterocyte microvilli, the parasite might interfere with microvillous disaccharidase activity by sheer weight of numbers. This effect is probably exaggerated by an absolute reduction in available mature enterocytes due to cell sloughing, coupled with "re-clothing" of damaged villi with immature crypt cells. Sugars, particularly lactose, could enter the large bowel in an undegraded state, allowing bacterial overgrowth with formation of volatile free fatty acids, which promote a change in osmotic pressure across the gut wall.[38,198,356a] Hypersecretion of fluids and electrolytes in the proximal small intestine has led to the proposal that *C. parvum* can produce a cholera-like toxin capable of stimulating cyclic AMP production through the adenyl-cyclase system in the enterocytes.[198] It seems logical that the parasite must derive its energy in some way at the expense of the host, either by adapting the host cell for its own purposes, or by utilizing preferentially substances absorbed from the gut lumen. Further study in this direction seems justified.

2. *C. muris* Infection

This species colonizes the digestive glands of the abomasum of cattle.[25] Individual glands are infected with endogenous stages from the apex to the orifice of the gland. The centers of the glands are often dilated and lined by a flattish epithelium. The microvilli of the peptic glands are destroyed by the parasite,[25] which probably accounts for the elevated plasma pepsinogen concentrations found in infected animals. Accumulations of lymphocytes and plasmacytes can be found between infected glands.

F. INTERACTIONS WITH OTHER ENTEROPATHOGENS

As stated, *C. parvum* infections are most commonly linked to infections with rotavirus, coronavirus, ETEC-K99+, or *Salmonella* spp.[430,745] However, concurrent infections with ETEC-K99+ are confined to calves less than 1 week old, and, indeed, many infections with these coliforms tend to be found only in 1- to 2-d-old calves;[430] thus, even calves which ingested *C. parvum* oocysts shortly after birth would be too old for ETEC-K99+ infections by the time the parasite completed its life cycle. Both rotavirus[235,745] and coronavirus[430] infections would tend to coincide with *C. parvum* infections. An additive effect between the various agents in these multiple infections is highly likely. Indeed, *C. parvum* may precipitate subclinical rotavirus infection into clinical disease.[308] Reynolds and colleagues[683] showed that asymptomatic rotavirus infection injured the upper jejunum in calves, but suggested that the undamaged ileum was compensating for any loss of jejunal digestive or absorptive capacity. The possibility that *C. parvum* could negate the compensatory function of the distal small bowel by colonizing and damaging the ileum, thus promoting a clinical effect in asymptomatic rotavirus infection, should not be overlooked. Similarly, coronavirus infections in the large bowel might be complicated if the ileum were badly damaged by the parasite.

Concurrent infections with *C. parvum* and *Salmonella* spp. have been reported in calves up to at least 35 d old.[430] Species identified include *S. dublin, S. typhimurium, S. enteritidis,* and *S. seftenberg.*[430] It appears to be taken as read that *Salmonella* spp. and *C. parvum* are synergistic, though this has not been verified experimentally. Non-enterotoxigenic strains of *E. coli* that adhere to enterocytes mainly in the large bowel were detected with *C. parvum* in a recent survey.[53] Such *E. coli,* which can elaborate a shiga-like toxin, have been described in a dysentery-like syndrome of young calves;[307] thus, an additive effect with *C. parvum* is possible. *Campylobacter* spp., which are not regarded as pathogens in calves,[794] were found in greater numbers in *C. parvum*-infected diarrheic calves than in asymptomatic contact calves that were not excreting cryptosporidial oocysts. A similar possible interaction between *C. parvum* and *Campylobacter* spp. appears to occur in humans.[147] Cultural procedures for anaerobic bacteria are time consuming and are not routinely carried out in microbiological surveys of calf diarrhea outbreaks. Many anaerobic species form part of the normal gut flora.[235] A possible interaction between *Clostridium perfringens* and *C. parvum* has been suggested, based on experiments in which *Cl. perfringens* was cultured from feces of calves experimentally infected with oocysts and from alcohol-treated inocula of oocysts given to calves that subsequently died.[235] However, although *Cl. perfringens* serotypes are ubiquitous, mortality in *C. parvum*-associated diarrhea is usually low, suggesting that interaction between the parasite and the toxins of these bacteria is rare.

An association has been demonstrated between a calici-like virus (Newbury agent) and *C. parvum* infections in calves.[308] Newbury agent can cause diarrhea in very young calves,[125] but infections with the virus and *C. parvum* could overlap. Other viruses found in *C. parvum*-infected calves include bovine virus diarrhea virus[430] and toga-like viruses,[512] but the significance of these combined infections is unknown. Coinfections with *Giardia* spp. and *C. parvum* have been reported in calves.[247] However, the data presented showed that *Giardia* excretion peaked much later than *C. parvum* and continued beyond the age when calves had recovered from the cryptosporidial infection. Coexistence between *C. parvum* and *Eimeria* spp. of coccidia has been reported,[430] but the clinical significance of these coinfections is uncertain.

G. EPIDEMIOLOGICAL FACTORS IN BOVINE CRYPTOSPORIDIOSIS

1. The Oocyst

The epidemiology of cryptosporidiosis centers upon the availability and concentration of oocysts in the immediate environment. The oocyst has a tough, environmentally resistant outer shell which encloses four naked sporozoites.[198] Various factors influence the length of time that oocysts can survive; they are destroyed by freezing and thawing[734] and, to a certain extent, by desiccation,[23] but if they are not exposed to extremes of temperature, they probably remain viable for several months. Oocysts also are resistant to many commonly used farm disinfectants.[133] Cryptosporidial oocysts sporulate in the host and do not require a critical combination of environmental temperature and humidity for development of infectivity as do most other coccidial species; they are shed in the feces already fully developed and capable of infecting new hosts immediately. When this capability is combined with a short life cycle and a profuse excretion rate in diarrheic feces, it is easy to understand how environmental contamination with infective stages can be generated within a very short time in enclosed areas such as calf pens or houses.

2. Sources of Infection

a. Other Calves

The most dangerous source of infection for a calf is its neighbor, particularly where animals are communally housed and overcrowded. The same is equally true of calves reared outdoors in permanent paddocks; here, spread of infection might be more rapid because the cows' teats and udders become contaminated with infected calf feces in their lying area. This might explain why outbreaks of cryptosporidiosis are more severe in single or multiple suckling herds than in dairy herds,[684] where the calves are likely to have little or no contact with their dams. Infection can also be introduced onto premises by introducing bought-in infected calves purchased through markets.[757]

b. Animal Attendants

The clothing, footwear, and hands of attendants can spread infection to young calves, their feeding utensils, and bedding. The various paraphernalia used around calf houses, e.g., brushes, shovels, the wheels of barrows or tractors, can all contribute to the dissemination of infective oocysts.

c. Vermin

Both mice and rats carry *C. parvum*, and experimental cross-infection of mice and calves with calf isolates and mouse isolates[418] has led to the conclusion that such isolates are freely interchangable in the farm environment. Feed contaminated with mouse feces is a possible source of infection. Flies have been suggested as possible vectors, and other hypothetical sources of infection suggested by Anderson[23] include barn cats, dogs that eat calf feces, and wild birds.

d. Adult Animals

Subclinical infections have been confirmed in adult cattle,[155,512] so it is feasible, though not proven, that calving cows could excrete sufficient numbers of oocysts to infect their offspring at birth. While the genital tract of mice has been infected experimentally,[466] any suggestion that bovine uterine discharges could be a further source of infection for the newborn animal will require verification.

H. TREATMENT AND CONTROL OF BOVINE CRYPTOSPORIDIOSIS

Presently (at the time of writing), no specific anti-cryptosporidial drug has been identified, and few of the anticoccidial or antiprotozoal drugs licensed for animal use have demonstrable action upon the parasite. Claims for the efficacy of individual compounds such as decoquinate[712] or sulfaquinoxaline[578] in calf diarrhea outbreaks involving *C. parvum* were not borne out when they were tested in controlled experiments. Recently, claims for the efficacy of halofuginone lactate in calf cryptosporidiosis have been made by French scientists.[908a] In an uncontrolled field trial, 67 diarrheic dairy calves under 4 weeks old, which were excreting *Cryptosporidium* sp. oocysts but no other detectable enteropathogen, were treated on 3 consecutive days with 0.5 mg/kg halofuginone lactate in aqueous solution. At 5 d after the start of treatment, 97% had stopped excreting oocysts, though only 68% had ceased to scour, while altogether 22% of the treated calves died despite additional rehydration and other nonspecific supportive treatment. The authors concluded that while halofuginone lactate appeared to act against the parasite, the margin of tolerance was low, although mild depression of appetite was the only sign of possible toxicity. The drug would need to be given in a very dilute form with careful regulation of the dose. Similarly, a recent report claiming the successful use of aggressive dosing with amprolium, both for treatment and as a means of prophylaxis in bovine cryptosporidiosis in Ireland,[248] must be treated with caution until the recommendations can be tested under controlled conditions. This is not to say that chemotherapeutic agents are not helpful in the clinical situation in their capacity to control concurrent bacterial infections. Thus, the use of sulfaquinoxaline, either alone or combined with other sulfonamides or vitamins,[174,578] might be a useful measure in some outbreaks. It must be emphasized that the first priority, which should usually be undertaken before the pathogen status in any outbreak can be established, is to employ oral or parenteral rehydration measures based on glycine/electrolyte solutions, perhaps coupled with the use of kaolin-based absorbent and demulcent suspensions and/or antispasmodic drugs and to institute sensible isolation and hygiene measures to combat the spread of infection. A specific anti-cryptosporidial drug would be useful, but the cost of treating large numbers of animals and the difficulties involved in catching and dosing suckled calves, for example, has to be considered in the context of a disease which, in calves at least, appears to be self-limiting. Control measures based on disinfection and hygiene might be all that is required.

Because *C. parvum* is the cause of clinical diarrhea in calves, while *C. muris* is associated with very mild disease in older animals, control measures are mostly aimed at the former species. Control is obviously hampered to some extent by the lack of any specific drug therapy, and no satisfactory form of prophylaxis, either chemical or immunostimulatory, has been discovered as yet. Passive protection by normal or hyperimmune colostrum[239] is of very dubious value and probably does little to influence the scale or course of naturally occurring outbreaks of cryptosporidiosis in calves.[191,315a] Thus, consideration of the housing and management arrangements in any given situation is important.

The main target for any control regime is the oocyst. The primary aim is to kill it; failing this, effort should be concentrated upon delaying exposure of the vulnerable animal to any infected area for as long as possible. Stringent disinfection is the usual means adopted to free premises from pathogens. However, with *Cryptosporidium* spp. this is by no means straightforward, as the oocyst is remarkably impervious to chemical agents. Until very recently, the only rational methods available were steam heat, 10% formalin, or various concentrations of household ammonia.[34,133] A spray disinfectant that generates release of ammonia on working surfaces has been developed commercially (OO-CIDE, Antec International Ltd., Chilton Industrial Estate, Sudbury, Suffolk, England) and will destroy the infectivity of coccidial and cryptosporidial oocysts.[109] *In vitro* experiments have shown that high concentrations of hydrogen peroxide also can kill cryptosporidial oocysts.[109] Elimination of infection from calf pens and houses is an important primary strategy in controlling clinical cryptosporidiosis.

Available evidence suggests that it is very difficult, if not impossible, to keep calving areas free of cryptosporidial oocysts. In one report, newborn calves taken in plastic bags into brand new wooden boxes still excreted oocysts within a matter of days. Excretion by the dams with contamination of the calf during the birth process cannot be ruled out; thus, control at this stage might be difficult. Where calves are to suckle their dams, the best that can be achieved is to clean out the rearing area daily and to limit the amount of potentially infected feces in the vicinity. However, in dairy units, calves are removed from their dams often after only one feed of colostrum. There is evidence from reports of cryptosporidiosis in very well-managed rearing units[110,247,768] that oocyst excretion is delayed when calves are individually penned for a period after they are born. In two instances,[110,765] calves did not develop clinical disease until they were moved to communal housing. If calves can be kept fairly clear of infection for the first 2 to 3 weeks of their lives, the clinical effects of *C. parvum* infections are likely to be innocuous and transient. Furthermore, the structure of calf pens can have a bearing on the availability of infective oocysts. An epidemiological study in Brazil[605a] found that the use of raised slatted, floors reduced *C. parvum* infection significantly when compared with floors made of concrete or bare earth.

Where disinfection is not feasible, other measures have to be adopted. Efforts should be made to reduce exposure to infection by redistributing the animals. For example, where an explosive outbreak is under way in an outdoor calving paddock, uncalved cows and cows with recovered, nondiarrheic calves should be moved away to another field, while acutely diarrheic calves and their dams should be housed if possible. Later, the indoor premises could be completely disinfected, while consideration might be given to ploughing and reseeding the contaminated ground. Similarly, an outbreak on permanent pastures close to farm premises could be combatted by putting cows and calves on to hill grazings and transporting their feed out to them. This could be labor-intensive, but only in the short term. The possible influence of inclement weather must be considered where calves are being reared outdoors. Calves which contract *C. parvum* infection require a greater energy intake to maintain their core temperature than noninfected calves; thus, in severe winter weather, infected calves simply die of chilling and starvation.[190] It might be necessary to provide windbreaks or some other form of shelter in such conditions.

Measures designed for controlling outbreaks of cryptosporidiosis on the farm are equally applicable to other enteric infections. This is important because multiple agent outbreaks including *Cryptosporidium* spp. are far more common than monoinfections with the parasite. Aided by the knowledge that oocysts can be destroyed and that clinical disease can be deferred by individual penning from birth in sterilized pens or hutches, the following principles based on the experiences of experienced veterinary scientists[560] should be adopted:

1. Calf rearing should be on an "all-in, all-out" basis.
2. Individual pens should be stringently disinfected between batches of calves.
3. Calves should be born and raised in a clean, dry environment.
4. Ideally, newborn calves should be penned individually for 2 to 3 weeks.
5. Sick calves should be removed immediately from the company of healthy calves.
6. Healthy calves should have different attendants than those of sick calves.
7. Attendants should keep boots, protective clothing, etc. as free of feces as possible.
8. Utensils should be heat sterilized, if possible, daily.
9. Vermin, farm dogs, and cats should be controlled.
10. Colostrum management and nutrition should be satisfactory.
11. Appropriate prophylactic measures against other agents, such as rotavirus or ETEC-K99+ vaccines, should be employed.

BIBLIOGRAPHY

Information pertaining to this subject may be found in References 13, 20, 23, 25, 27, 34, 38, 40, 66, 89, 109, 110, 117, 125, 133, 147, 155, 158, 174, 190, 198, 212, 235, 245a, 247, 308, 329,

414, 418, 430, 466, 495, 543, 560, 564, 578, 605a, 607, 629, 645, 649, 683, 712, 734, 745, 757, 768, 794, 806, 810, 818, 819, 822, and 825.

IV. CRYPTOSPORIDIOSIS IN SHEEP (*OVIS ARIES*)

A. HISTORICAL BACKGROUND

Ovine cryptosporidiosis was first diagnosed in Australia in 1974[66] in diarrheic lambs aged 1 to 3 weeks on a farm that produced ovine cheese and other milk products. After 3 d with their dams, these lambs were penned in groups and artificially reared. Cryptosporidia, probably *C. parvum,* were found in the ileum of several lambs, and the presence of the full range of endogenous stages was confirmed by electron microscopy. The lambs were also infected with *Salmonella typhimurium,* and the role of the parasite in the outbreak, if any, was not clear. Cryptosporidiosis was again identified in 1978 in two young diarrheic lambs submitted to a diagnostic laboratory,[87] but other enteropathogens were apparently not sought and once more the role of the parasite as a possible pathogen was not established. A more comprehensive investigation was reported by Tzipori and colleagues in 1981,[815] involving a group of 48 lambs that were being artificially reared on cows' colostrum with a milk replacer; 40 lambs had acute diarrhea commencing between 5 and 12 d after birth and lasting for 7 to 16 d, and 16 lambs died. None of the lambs had ETEC-K99+ but one was excreting rotavirus. Other enteric virus particles were sought by TEM, but none were found. Numerous cryptosporidia were found in the lower jejunum, ileum, cecum, and colon in moribund lambs that were euthanized with intravenous barbiturate. Fecal homogenates containing *Cryptosporidium* sp. oocysts from these lambs were given orally to two specific pathogen-free lambs and caused diarrhea, depression, and anorexia commencing at about 3 DPI. No diarrhea occurred in a group of suckled lambs moved later into the same premises; thus, it was proposed that lack of colostrum caused young lambs to be susceptible to *Cryptosporidium* sp. infections.

Several diarrhea outbreaks in lambs in Idaho were associated with excretion of *Cryptosporidium* sp. oocysts.[15] Lambs 1 to 2 weeks old in a small 30-ewe flock had transient diarrhea and oocyst excretion with no other clinical signs. Lambs never stopped feeding throughout the period of diarrhea. Losses did occur in an orphan lamb-rearing unit attached to a shed-lambed range flock, but multiple infections including *Cryptosporidium* sp., an encapsulated *E. coli,* coronavirus-like particles, and bovine virus diarrhea-like particles were identified in several of these lambs. Finally, *Clostridium perfringens, E. coli,* and *Cryptosporidium* sp. were excreted by a hospitalized lamb which recovered from a bout of diarrhea. It was concluded by the Idaho workers[70] that uncomplicated cryptosporidiosis in lambs was mild, but that environmentally related stress, and/or concurrent infections with other agents, might cause some mortality. Quite the opposite experience was described in Scotland[31] on premises where cryptosporidial infections were suspected to be present. Here various groups of ewes were brought indoors sequentially to lamb during March to April. The earliest lambs were healthy, but diarrhea with excretion of *Cryptosporidium* sp. oocysts commenced in the late lambs of the first group. Subsequently most lambs had diarrhea commencing at about 10 d of age and lasting for about 7 d. Of 37 feces samples examined for *Cryptosporidium* sp. oocysts, 29 (78%) were positive; only a single sample contained rotavirus, while none contained ETEC-K99+ or other virus particles as judged by TEM. Widespread cryptosporidial infections were found in 12 lambs at necropsy, the greatest concentration being in the ileum. It was concluded that *Cryptosporidium* sp. was the predominant pathogen in the outbreak, though mortality seemed to be somewhat less in the offspring of ewes that had lambed in the same premises in earlier years.

Meanwhile, transmission experiments using calf feces containing *Cryptosporidium* sp. oocysts had shown that bovine isolates of the parasite could infect 6 other species, including lambs.[811] Further experiments were carried out in which young specific pathogen-free lambs were infected with bovine fecal homogenates containing *Cryptosporidium* sp. oocysts, but no other recognized enteropathogens, or with homogenates of suckling rat intestines containing cryptosporidia derived from bovine feces.[816] These inocula induced diarrhea as early as 2 DPI, and the pathologic lesions described[33,816] closely resembled those found in the natural disease.[31,66,815] Finally, inocula containing oocysts, but from which viruses and bacterial contaminants had been excluded by very stringent methods, induced diarrhea and pathologic changes in germ-free lambs, confirming that *Cryptosporidium* sp. was pathogenic for neonatal lambs.[746]

B. CLINICAL FEATURES OF OVINE CRYPTOSPORIDIOSIS

As in calves, the cardinal clinical sign is diarrhea, with excretion of numerous oocysts. Lambs experimentally infected with 10^6 *C. parvum* oocysts will excrete up to 10^7 oocysts per gram of feces for about a week after the start of diarrhea, though firm data on the excretion rate in natural outbreaks of cryptosporidiosis are lacking. Affected lambs often are dull, depressed, and anorexic and may move stiffly with tremor of voluntary muscles.[31,815] The life cycle of the parasite in neonatal lambs is 3 to 4 d,[746] and diarrhea usually coincides with the development of sexual stages in the gut,[746] but is seldom seen before lambs are 3 d old. However, the author has received reports from the field of diarrhea with profuse oocyst excretion in 2-d-old lambs. The few published reports of natural outbreaks indicate that the usual pattern is acute diarrhea commencing at 7 to 10 d, lasting from a few days only to up to 2 weeks. Intermittent diarrhea and constipation have been reported in lambs,[816] and cessation of diarrhea might be followed by relapse after a few days.[31] Although firm data are lacking, indications from unpublished as well as published accounts are that considerable mortality can occur in lambs infected by *C. parvum.* This has been verified by the results of experiments, in which some of the infected lambs died or were very ill, with severe lesions in their intestines.[746,816] As always in neonatal diarrheas, the situation can be complicated by concurrent infections with other enteric pathogens. However, as shall be shown below, the timing of the most common of these infections is far more precise than in calves; thus, the opportunity for *C. parvum* to act on its own is much greater.

C. DIAGNOSIS OF OVINE CRYPTOSPORIDIOSIS

Cryptosporidiosis should always be suspected where acute diarrhea occurs in lambs over 1 week old. The presence of typical oocysts in feces or gut contents can be confirmed by examining stained smears; both oocysts and endogenous stages can be demonstrated in impression smears of the ileal mucosa stained by Giemsa's method. Histologic confirmation is useful only if a moribund lamb can be euthanized with barbiturate and tissues immediately placed in fixative. Dead lambs become autolytic so quickly that their intestines are rarely suitable for histologic evaluation. Necropsy findings are vague and nonspecific; patchy congestion associated with very liquid, abnormally colored, and malodorous gut contents in emaciated carcasses may be found. Other agents must not be neglected; feces samples from very young lambs, i.e., under 6 d, should be tested for the presence of ETEC-K99+ and rotavirus, while all samples should be cultured for *Salmonella* spp. It is important to enquire if the ewes were vaccinated with a multicomponent clostridial vaccine, to assess the possibility of lamb dysentery as a complicating factor.

D. PATHOLOGY OF OVINE CRYPTOSPORIDIOSIS

Early reports described villus stunting, cryptitis, and increased crypt mitosis in the distal small bowel.[66,87] Additionally, villous fusion[31,815] and clothing of deformed villi with immature, cuboidal epithelial cells[31] was usually associated with the presence of numerous endogenous stages of the parasite. Experimental studies, first with conventional lambs,[31,815,816] then with virus- and bacteria-free oocyst inocula in germ-free lambs,[746] have provided more precise information on the lesions and pathogenesis. Germ-free lambs given approximately 10^5 oocysts became anorexic and developed profuse watery diarrhea on day 3 postinoculation.[746] Lambs killed at 12, 24, 48, and 72 h postinoculation were used to trace the progress of infection. At 12

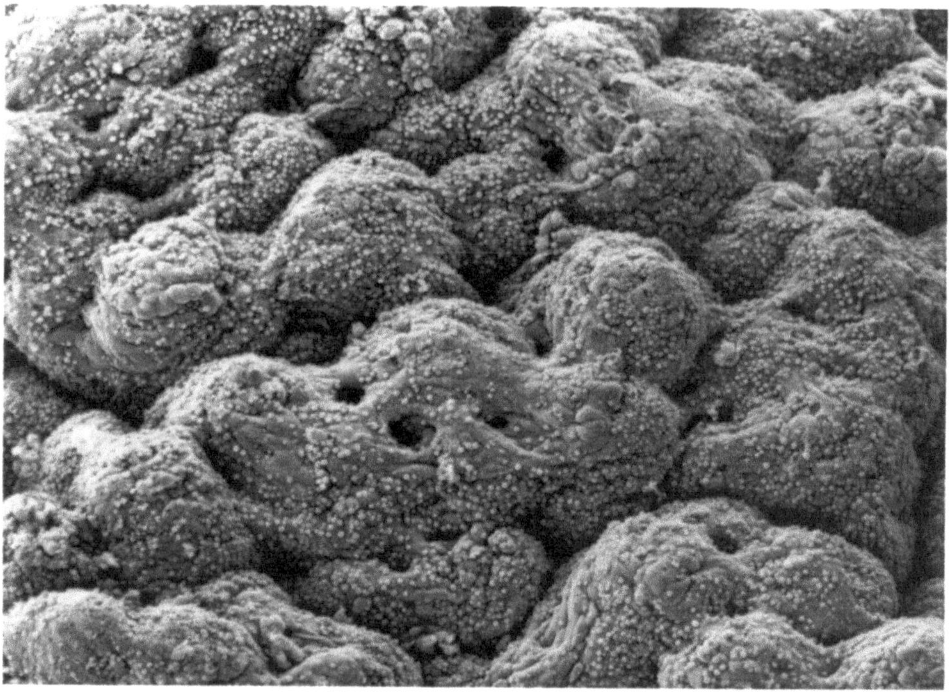

FIGURE 1. Stunting and fusion of intestinal villi in experimental ovine cryptosporidiosis.

h a few trophozoites were seen in a mid-jejunal site. By 24 h, trophozoites and some meronts were scattered throughout the midgut and ileum; concentrations of these stages in the distal small intestine increased during the next 24 h some 30-fold, and some were seen in the duodenum, though not in the large bowel, by 48 h. From then on, all endogenous stages were found in all gut sites, though the greatest numbers were in the distal jejunum and ileum. No lesions were seen up to 48 h after inoculation, but by 72 h, extensive mucosal lesions were apparent in the distal small bowel (Figure 1). These were severe villous stunting and fusion, replacement of the normal columnar and mucus-secreting epithelial cells clothing the villi by cuboidal cells, and copious infiltrates of mononuclear cells in the lamina propria. Large intestinal sites had dilated or atrophic crypts and similar mononuclear infiltrates. There was no difference in the lesions found in germ-free lambs from those seen in lambs that had been given "normal" calf gut bacteria.

E. INTERACTIONS WITH OTHER ENTEROPATHOGENS

In this context, an important consideration is the age at which diarrhea first appears. Once a lamb is 2 to 3 d old, it is no longer susceptible to ETEC-K99+ infection, and by 7 d it has an age immunity to rotavirus.[817] However, there is evidence that dual infections with these agents prolong the clinical course.[818] Thus, in lambs under a week old, multiple infections involving *C. parvum* and these other agents must be considered possible. Lamb dystentery caused by *Clostridium perfringens* type B is worth considering only if the ewes are unvaccinated or if colostrum from vaccinated ewes is thought to be inadequate. An important first step is to ascertain whether lambs are infected with *Salmonella* spp.; concurrent infection with cryptosporidia and *S. typhimurium* has been reported in lambs.[66] Nonenterotoxigenic *E. coli* have been identified in lambs infected with *C. parvum*[15, 211] and might play a synergistic role with the parasite in some outbreaks. The significance of such common gut inhabitants as *Neisseria* spp., *Proteus* spp. and *Streptococcus faecalis*,[31] or virus particles resembling bovine virus diarrhea virus or coronavirus,[15] all of which have been found in feces samples from diarrheic lambs as well as *C. parvum* oocysts, is unknown.

F. EPIDEMIOLOGICAL FACTORS IN OVINE CRYPTOSPORIDIOSIS

The close proximity of cattle, or the use of grazing land or housing recently occupied by calves infected with *C. parvum*, probably acts as the main source of infection for many outbreaks of lamb cryptosporidiosis.[38] The author has experience of one outbreak in which grazing land used as a calving area in January became a lambing area in April, with disastrous consequences. The single-suckled calves experienced a mild episode of cryptosporidiosis, with no losses, but there was a high mortality in the lambs. Infection also can be carried from contaminated calf-houses to young lambs by stockmen in the same way that they can spread bovine infections amongst calves, and many of the other potential sources mentioned in the section on bovine cryptosporidiosis apply to lambs. Lambs are often penned with their mothers for a day or so until a good bond is established and the lamb is sucking well. If these pens become contaminated with oocysts, they might act as a rich source of infection for subsequent lambs by fecal contamination of the ewes' udders. Adult sheep may be subclinically infected and might therefore infect their own lambs shortly after birth.[156,608a]

G. TREATMENT AND CONTROL OF OVINE CRYPTOSPORIDIOSIS

Treatment is mainly supportive and based on oral or parenteral rehydration. Specific anticoccidial drugs are of no value, apart possibly from halofuginone lactate, for which claims of efficacy in experimental *C. parvum* infections of lambs have been made.[576a]

The principles of control are essentially similar to those offered for bovine cryptosporidiosis. However, sheep management systems do not lend themselves quite so well to direct attack upon the oocyst as intensive calf rearing systems. Outdoor paddocks, lambing fields, or corrals once heavily contaminated are likely to remain so for months. However, removal of in-lamb ewes or ewes with recovered lambs to other parts of the farm, allied to housing and care of sick lambs, might be helpful. The following measures should be immediate and vigorous. Permanent lambing pens should be disinfected, while temporary pens for outside use should be dismantled and erected elsewhere. All bedding from such pens should be burned, all partitions or battens disinfected, and the original area fenced off until after lambing. Shelter should be provided to protect sick lambs from hypothermia. It is essential that lambs are hygienically handled at birth and are born into clean, dry surroundings. Particularly important is that they must receive an adequate amount of colostrum. Although specific passive protection against *Cryptosporidium* sp. infection via colostrum has not been established, it certainly will protect against other agents and, most important, acts as a vital source of energy and nutrients. Supplementing the colostrum of multiparous ewes with amounts drawn from ewes with single lambs or from colostrum banks held at –20°C should be considered, and if the flock's colostral status is poor overall, advice on proper nutrition during pregnancy should be sought. Finally, the trace element status of the ewes should be investigated, in case some imbalance is causing poor lamb viability or low resistance to infection.

BIBLIOGRAPHY

Information pertaining to this subject may be found in References 15, 31, 33, 38, 66, 87, 211, 576a, 608a, 746, 811, and 815 to 817.

V. CRYPTOSPORIDIOSIS IN GOATS (*CAPRA HIRCUS*)

A. HISTORICAL BACKGROUND

Caprine *Cryptosporidium* sp. infection was first reported in Tasmania, Australia, in 1981.[520]

A 2-week-old Angora kid died after a short episode of acute diarrhea. No bacterial or viral pathogens were recognized in gut contents, but cryptosporidia were found on histological and TEM examination of the jejunum and ileum. Also in Australia, an outbreak of acute diarrhea in young goats was investigated the following year.[822] Of 29 kids between 5 and 21 d old, 21 had acute diarrhea lasting for 3 to 7 d. Many became dehydrated and three died. Feces samples from nine diarrheic kids contained *Cryptosporidium* sp. oocysts, but none contained ETEC-K99+, rotavirus, or other virus particles as judged by TEM. Cryptosporidia were found throughout the whole length of the small intestine in two kids. In 1983, Current and colleagues in the U.S.[186] induced diarrhea in three young goats with *Cryptosporidium* sp. oocysts from a human patient with AIDS. Oocysts were shed from days 5 to 12 after inoculation, and at necropsy the ileum, cecum, and colon were heavily infected. In 1983, caprine cryptosporidiosis was reported for the first time in Europe by French workers,[653] who considered the parasite to be a common cause of diarrhea in goats. In the same year, Nagy and colleagues in Hungary[580] reported multiple infections in young goats, including *Cryptosporidium* sp., rotavirus, coronavirus, and ETEC-K99+. Certainly in France, where diarrhea is the most common cause of death in young goats,[908] *C. parvum* is thought to be the dominant pathogen, either with or without these other organisms.[292]

B. CLINICAL FEATURES OF CAPRINE CRYPTOSPORIDIOSIS

Acute diarrhea with dehydration are the main clinical signs in young goat kids (1 to 3 weeks old) with cryptosporidiosis.[822] The most vulnerable age group is between 5 and 15 d, and the period of oocyst excretion may last for 1 to 2 weeks.[292,822,908] However, data on the concentrations of oocysts in the feces of diarrheic kids are lacking. Mortality in neonatal goats might reach 50%,[654] and morbidity is invariably high. A second form of cryptosporidiosis has been reported in which progressive emaciation developed in kids 5 to 8 d old, without any associated diarrhea.[654] Histological detection of swollen, blunt villi, heavily parasitised with cryptosporidial stages, in the small intestine, confirmed the diagnosis in two affected herds,[654] and mortality was high. Persistent diarrhea with progressive loss of condition was also found in goats aged about 6 weeks, associated with oocyst shedding in the absence of other enteropathogens.[654] The source of infection for these animals was believed to be a group of neonatal goats with cryptosporidiosis and diarrhea, which was sharing the location with the older group. All the older kids recovered spontaneously. There is, in addition, a single report of cryptosporidiosis and intermittent diarrhea in a 6 month-old goat,[226] in the absence of viral or bacterial enteropathogens. Subclinical infection has also been confirmed by ileal biopsy in 6 week-old goats.[528]

C. DIAGNOSIS OF CAPRINE CRYPTOSPORIDIOSIS

Demonstration of oocysts by examination of stained feces smears will confirm *Cryptosporidium* sp. infection. Histological confirmation of the degree and distribution of infection is useful if a terminally ill animal can be euthanized with intravenous barbiturate, or if a recently dead animal can be necropsied within 20 min or so of death. It must be remembered that many pedigree goats are highly valued; thus, the owner is unlikely to accede lightly to a request to kill a sick kid. It is essential that feces samples from ill and healthy kids be submitted for laboratory examination for other infectious agents. Multiple infections with two or more enteropathogens are very common in diarrhea outbreaks in goats under 4 weeks old.[580] In view of the public health hazard of *Salmonella* spp., cultural methods for these bacteria should be carried out routinely.

D. PATHOLOGY OF CAPRINE CRYPTOSPORIDIOSIS

Very little information is available on the lesions of cryptosporidiosis in young goats. Necropsy findings are nonspecific, though distension of the small intestine by clear, yellow fluid contents has been reported.[654] Shortening of villi, congestion, focal leucocytic infiltration, minor surface erosion and metaplasia of the surface epithelium to an immature form was reported in a 16-d-old diarrheic kid,[520] and villous atrophy with some fusion and cross-bridging was seen in a kid in another diarrhea outbreak.[822] In the form of cryptosporidiosis associated with emaciation rather than diarrhea,[654] heavily parasitized, swollen, and blunt villi have been described. Villous atrophy and crypt hyperplasia were also reported in asymptomatic, 6-week-old goats[528] and in ileal mucosal scrapings from an infected 6-month-old goat.[226]

E. INTERACTIONS WITH OTHER ENTEROPATHOGENS

Several agents, including *C. parvum,* have been incriminated in neonatal goat diarrhea.[292,580] Agents commonly found include ETEC-K99+, rotavirus and coronavirus. Nonenterotoxigenic *E. coli* and *Clostridium perfringens* have also been found in association with the parasite in goats.[136] Adenovirus infection causing concurrent pneumonia is a serious complication of multiple enteric infections in kids under 4 weeks old and may be diagnosed by demonstration of the viral particles in diarrheic feces by TEM.[580]

F. EPIDEMIOLOGICAL FACTORS IN CAPRINE CRYPTOSPORIDIOSIS

Although several surveys indicate that cryptosporidiosis is widespread in goat herds,[292,654] the salient epidemiological factors influencing infection are largely a matter for conjecture. However, poor hygiene is the most important predisposing factor in neonatal enteritis in goats[908] and must have a bearing on the severity of outbreaks of cryptosporidiosis in neonates. The possibility that infection can persist for long periods in pens or boxes, or in paddocks and permanent grazing or browsing areas where young goats are nursed by their dams, should be the starting point for future epidemiological studies, while contact with cattle or other ruminants, and/or spread by attendants, must be considered as factors causing the introduction or spread of infection. Purchase of infected kids which are shedding *C. parvum* oocysts is another means by which the disease can be introduced to a herd. The possibility that adult goats become subclinically infected with *C. parvum,* with a low but significant excretion of oocysts, must be considered.

G. TREATMENT AND CONTROL OF CAPRINE CRYPTOSPORIDIOSIS

Treatment of cryptosporidiosis in neonatal goats is symptomatic, and aggressive rehydration therapy is a matter of some urgency.[654] Treatment with antibiotics is justified only where coinfections with bacterial pathogens have been identified and the sensitivity to specific antibiotics of the bacterial species involved has been determined. Otherwise, antibiotic therapy is not only wasteful, but may be detrimental on the grounds that it might interfere with the normal gut flora.[654] No treatment for the progressive emaciation, nondiarrheic form of the disease has been offered, but it might be worth allowing affected artificially reared kids to resume suckling their own dams or to suckle surrogate females in order to increase their energy intake. As an alternative, milk replacer could be fed regularly by stomach tube and multivitamin injections used to try to stimulate appetite.

Control should be based, where practical, on stringent disinfection of pens or boxes where affected kids have been housed, with removal and destruction of all contaminated bedding between successive batches of animals.[654] Contact with calving cows or lambing ewes should be avoided, and sick kids should be isolated away from recovered animals or pregnant females. Animal attendants should take steps to prevent spread of infection by wearing easily washed protective clothing. Adequate colostral intake is important for passive protection against other enteropathogens, and it may be necessary to obtain blood samples from randomly selected kids to check antibody status. The nutritional and trace element status of the breeding stock may require examination, in case deficiencies are contributing to poor viability in the neonates.

BIBLIOGRAPHY

Information pertaining to this subject may be found in References 136, 186, 226, 292, 520, 580, 653, 654, 822, and 908.

VI. CRYPTOSPORIDIOSIS IN DEER (*CERVUS* SP.)

A. HISTORICAL BACKGROUND: FARMED DEER

Deer farming is a recent innovation, dating from the early 1970s in Scotland and New Zealand.[310] Commercial deer farming has developed considerably since these early beginnings, and venison production has steadily gained ground in the last decade in the UK and other northern European countries.[309] Red deer (*Cervus elephas*) and fallow deer (*Dama dama*) are the most commonly farmed species, but much of the available knowledge relating to deer diseases centers upon the red deer.

The earliest reports of cryptosporidiosis in red deer emanated from Scotland[814] and New Zealand.[605] These both described diarrhea outbreaks in artificially reared deer calves with high mortality associated with *C. parvum* infections. Diarrhea developed 1 to 2 weeks after the calves had been housed for orphan rearing, and in one outbreak[814] lasted from 2 to 14 d. Diligent search for other common enteropathogens by both groups of researchers proved negative, though astrovirus particles were found in feces in one outbreak.[814] Investigation of the cause of death in naturally reared deer calves in the same outbreak revealed that they, too, were infected with cryptosporidia.[814] Field investigations have recently confirmed that cryptosporidiosis can be a hazard in naturally reared red deer calves,[39,40] and might cause high mortality in the absence of other recognized enteropathogens.

B. CLINICAL CRYPTOSPORIDIOSIS IN FARMED RED DEER

Published information on clinical signs of *C. parvum* infections in deer calves is extremely fragmentary. A report by Tzipori and colleagues[814] on an outbreak of cryptosporidiosis in artificially reared deer calves recorded the duration of diarrhea and found a strong association between diarrhea and oocyst shedding. However, clinical details were confined to an account of a single calf that scoured intermittently for 2 weeks then became constipated, bloated, and emaciated. New Zealand workers[605] reported that affected calves had no initial loss of appetite, but developed a pasty, yellow scour and died within a few days. A prospective longitudinal study of the spread of infection in an experimental deer unit was carried out in 1986.[39,108] During this study, 7 out of 30 artificially reared calves died of the disease, and 92% of the calves eventually excreted oocysts. Affected calves became reluctant to feed and either became diarrheic or constipated, and some had recurrent bloat. Unpublished field reports from diagnostic laboratories or veterinary practitioners[39] have described outbreaks of cryptosporidiosis in which infected deer calves under 1 week old died suddenly in adverse weather conditions. Deer calves that shed *C. parvum* oocysts and have acute diarrhea can die overnight if the weather turns wet and chilly.

C. DIAGNOSIS OF CERVINE CRYPTOSPORIDIOSIS

If faced with an outbreak of acute, unexplained sudden deaths in deer calves under 1 month old, cryptosporidiosis should always be considered as a possibility. Oocysts can be demonstrated readily in stained smears of feces or gut contents, and their presence confirms infection. However, as in all other ruminant species, diagnosis depends on stringent search for other recognized enteropathogens. If these are absent or confined to very few samples, and if *C. parvum* infection as judged by quantitative or semiquantitative estimation of oocyst numbers per gram of feces, backed up if possible by histological examination of the intestines of recently dead calves, is clearly severe and widespread, then the parasite probably is the dominant pathogen. Samples from healthy as well as sick calves should be examined and compared. In the presence of a background of infection, deaths may be due to triggering factors such as inclement weather or inadequate colostral intake.

D. PATHOLOGY OF CERVINE CRYPTOSPORIDIOSIS

Necropsy findings in cervine cryptosporidiosis include patchy reddening or congestion of the small intestinal mucosa and the presence of fluid, yellow contents in the large bowel. In calves which die after a prolonged illness, the carcass is usually emaciated. Typically, *C. parvum* colonizes the lower jejunum, ileum, cecum, and spiral colon of deer. Stunted, fused, and swollen villi can be found in the distal small intestine, associated with accumulations of mononuclear cells and neutrophils in the lamina propria.[605] In prolonged infections, total villous atrophy with marked crypt hypertrophy have been described.[814] The large bowel might be very heavily infected and infiltrated with leukocytes, while crypts might be distended or atrophic, and there might be a deficiency of mucus-secreting cells.[605,814]

E. INTERACTIONS WITH OTHER ENTEROPATHOGENS

Both rotavirus and strains of *E. coli* are known to infect deer in the first week or so of life.[131] Rotavirus infection is widespread in wild deer[68] and infections cause anorexia, depression, and profuse watery diarrhea with dehydration in deer less than 10 d old. Interaction between rotavirus and *C. parvum* is certainly possible, as the normal prepatent period of the parasite in deer is around 8 d.[39] However, rotavirus particles have seldom been found in deer calves with cryptosporidiosis.[97,99] Coliform infections can cause an acute yellow, foamy diarrhea leading to death in untreated animals in under 3 d;[424] external factors such as chilling or poor hygiene contribute to outbreaks. The strains of coliforms found in deer can be either ETEC-K99+ or nonenterotoxigenic strains; the latter is more likely to interact with *C. parvum*. Enterotoxemia caused by various serotypes of *Clostridium perfringens* must be considered in any outbreak of sudden deaths in young deer,[132] though there is no published evidence of any interaction with *C. parvum*. Astroviruses, which are not regarded as pathogens of deer,[814] have been seen in deer feces as well as *C. parvum*. Small, round virus particles were seen in the feces of one deer calf with cryptosporidiosis, but the significance of these or coronavirus-like particles seen occasionally, is unknown.

F. EPIDEMIOLOGY, TREATMENT, AND CONTROL OF CRYPTOSPORIDIOSIS IN FARMED RED DEER

Sources of *C. parvum* infection include association with infected bovine calves or with accommodation previously used by cattle, while animal attendants working simultaneously with young deer and scouring bovine calves may spread the infection. In one study[108] it was clear that deer calves introduced into rearing pens acquired the infection from the pens, as the prepatent period was always longer than the life cycle of the parasite. It appears likely that outbreaks in naturally reared deer calves arise from exposure to contaminated grazings. However, there is a certain amount of circumstantial evidence implicating asymptomatic infection and oocyst shedding by hinds at the time of parturition.[40] In this eventuality, deer calves could become infected very shortly after birth from their dams.

Treatment is symptomatic and is based on the rehydration measures used to combat dehydration in bovine calves. New Zealand workers described the successful therapeutic use of adult deer serum given at the rate of 20 ml per feed, later replaced by 30 ml bovine colostrum per feed, until recovery,[605] though the study was not controlled. Similarly, the prophylactic use of these amounts of deer serum and bovine colostrum, given four times daily with milk replacer to 750 ml for 3 weeks, proved to be a successful method of rearing deer calves, though, again, the study was not controlled.[605] Where licensed slaughter premises are permitted on deer farms, blood could feasibly be collected when adult deer are butchered, and serum separated and frozen away for subsequent use.

Control measures should include avoidance of contact with cattle or other ruminants; thus, attendants who work among bovine calves should not mingle with or attend to deer calves. Pens required for artificial rearing of deer calves should be stringently disinfected. Where cryptosporidiosis occurs outdoors, sick calves should be isolated if possible, and hinds that have still to calve should be moved to other grazings. Young deer like to make themselves inconspicuous by hiding in patches of tall grass or scrub, and these might become heavily contaminated with oocysts. Such areas might require fencing off. Ideally, shelter should be available in the event of driving rain or chill winds, to reduce hypothermia. The nutritional status and endoparasite control measures used in the adult stock might need to be modified if colostral production appears to be inadequate. Finally, if there is evidence of concurrent infection with *C. parvum* and *Clostridium perfringens*, vaccination using a double dose of a multicomponent sheep anticlostridial vaccine should be considered,[132] though not until the calves are at least 8 weeks old, in case of interference with immunity by colostral antibodies.

G. CRYPTOSPORIDIOSIS IN WILD DEER

Cryptosporidial infections have been confirmed in the roe deer (*Capreolus capreolus*),[425] fallow deer (*Dama dama*), sika deer (*Cervus nippon*), mule deer (*Odocoileus hermionus*), Eld's deer (*Cervus eldi thamin*), axis deer (*Cervus axis*), and barasingha deer (*Cervus duvauceli*).[340] These were virtually all diarrheic neonatal deer, hand-raised in a wildlife park, and diagnosis was made by finding oocysts of *Cryptosporidium* sp. in feces smears stained by auramine O. Fluid and electrolyte replacement, antidiarrheals, antibiotics, and other supportive measures were recommended.[340] Investigations for other possible pathogens were not carried out, and the significance of these infections is not known.

BIBLIOGRAPHY

Information on this subject may be found in References 39, 40, 108, 340, 425, 605, and 814.

VII. CRYPTOSPORIDIOSIS IN BUFFALO (*BUBALUS BUBALIS*)

Cryptosporidiosis in buffalo was first recognized in Italy when 17 of 229 (7.4%) of 5- to 15-d-old diarrheic water buffalo calves were found to be excreting typical oocysts.[135a] The following year, an investigation of the intestinal protozoal burdens in 264 dairy water buffalo from 6 herds in the provinces of Caserta and Salerno was carried out using reliable methods for identifying the various organisms in fresh or stained feces preparations. *Cryptosporidium* sp. oocysts were found in 17 of 264 (6.4%) animals.[135b] No mortality was reported, but there was implicit in the findings some concern over a possible human health hazard, since in these provinces of Italy water buffalo milk is used for the manufacture of mozzarella cheese.

Cryptosporidiosis has been reported also in Egypt.[373] Of 87 diarrheic buffalo calves aged 2 to 12 d, 18 (21%) had *Cryptosporidium* sp. oocysts in feces or gut contents. No mention was made of other enteropthogens. In four dead calves, histological findings included villous atrophy and fusion, while cryptitis was associated with decreased goblet cell numbers. Both villi and crypts were infected by numerous endogenous stages of the parasite. The lamina propria contained infiltrates of lymphocytes, plasma cells, neutrophils, and eosinophils.

These limited studies give little indication of the extent of the problem in buffalo. Control would clearly depend on the system of management under which buffalo are kept; thus, in the semiintensive dairies in Italy, control measures could be similar to those recommended for bovine calves.

Information on this subject may be found in References 135a and b and 373.

VIII. CRYPTOSPORIDIOSIS IN OTHER RUMINANT SPECIES

Death associated with severe cryptosporidial infection occurred in a 1-d-old Persian gazelle (*Gazella subgutterosa*).[241] This animal also was infected with an *E. coli* strain and was deemed to be immunologically immature. The authors pointed out the dangers to neonatal animals or animals in wildlife exhibits where interspecies exchange of *Cryptosporidium* sp. is possible.

Diarrhea with oocyst excretion has also been diagnosed in Persian gazelles.[340] A particularly interesting report from the Munich Zoo[658] described *C. muris* infection in the abomasum of mountain gazelles (*Gazella gazella cuvieri*). Mucosal glands were hypertrophied, and the mucosa was hyperplastic and infiltrated with small lymphocytes.

Two surveys of cryptosporidial infection in neonatal exotic ruminants have been carried out recently. *C. parvum* infection was confirmed either histologically or by auramine O staining of feces smears[340] in blackbuck (*Antelope cervicapra*), sable antelope (*Hippotragus niger*), scimitar-horned oryx (*Oryx gazella dammah*), and fringe-eared oryx (*Oryx gazella callotis*). Additionally, the infection was confirmed by histologic examination in addax (*Addax nasomaculatus*).[849] In 2 blackbuck, infection was widespread and concentrated in the small intestine, but sparse in the cecum and colon.[849] Concurrent infection with *Salmonella typhimurium* complicated the pathology. The policy of bringing weak, possibly colostrum-deprived neonates of various species into communal wooden pens from outdoor compounds was believed to be an important factor in the epidemiology,[849] as diarrhea only occurred approximately 1 week after penning.

Cryptosporidium sp. oocysts were found in the feces of diarrheic neonatal impala (*Aepyceros melampus*), springbok (*Antidorcas marsupialis*), nilgai (*Boselaphus tragocamelus*), Turkomen markhor (*Capra falconeri*), Addra gazelle (*Gazella dama ruficollis*), slender-horned gazelle (*Gazella leptoceros*), eland (*Taurotragus oryx*) and Armenian mouflon (*Ovis orientalis gmelini*),[340] but other possible pathogens were not sought, and the significance of *Cryptosporidium* sp. infections in these exotic ruminants must await further study.

Information on this subject matter may be found in References 241, 340, 658, and 849.

IX. CONCLUSIONS

Cryptosporidium sp. is widespread in ruminant species throughout the world. Where ruminants are grouped together in livestock units or in wildlife parks, the parasite becomes concentrated in the environment and disease outbreaks can result. The mortality rate in bovine cryptosporidiosis becomes critical only when other enteropathogens are present concurrently, but in lambs, goats, and young deer, the parasite can cause serious losses without the collusion of other agents, though stress factors probably play an important part in the fatality rate in these species. No therapeutic or prophylactic measures are available against the organism, and control depends on disinfection, hygiene, and attention to husbandry. Cryptosporidiosis is a zoonosis, and infective oocysts can be acquired directly or indirectly by humans from infected animals; thus, ruminant species may be potential reservoirs of infection. Particularly at risk are immunocompromised persons, such as those with AIDS, in whom cryptosporidiosis is one of the least tolerable and manageable of all the infections to which they are susceptible.

Chapter 6

CRYPTOSPORIDIOSIS IN PIGS AND HORSES

C. W. Kim

TABLE OF CONTENTS

I. CRYPTOSPORIDIOSIS IN PIGS (*SUS SCROFA*)

The pig has been shown to be a natural as well as an experimental host for *Cryptosporidium*. Infection does not appear to be a serious cause of enteritis in either neonatal or postweaning pigs.[828] However, the pig has served as a suitable experimental host for studying the pathogen icity and species specificity, as well as for testing the efficacy of chemotherapeutic drugs against the parasite.

A. HISTORICAL BACKGROUND

Initial accounts of natural infection in pigs were reported in the U.S. in 1977 in a pig with necrotic enteritis in South Dakota[88] and in three purebred Duroc pigs in Kansas.[403] In 1982, piglets in New South Wales,[481] as well as suckling and fattening pigs in the Federal Republic of Germany,[722] and a diarrheic nursing piglet in South Dakota[90] were reported to be infected with *Cryptosporidium*. *Cryptosporidium* was also detected in mucosal scrapings, as well as in histologic sections of intestines, in neonatal pigs submitted for routine diagnostic evaluation during a 3-year period from 1980 to 1982 in Canada.[713] Cryptosporidiosis, based on shedding of oocysts in feces, was reported in piglets in Vietnam.[426] These reports showed that porcine cryptosporidiosis was found worldwide, but the pathogenic or clinical significance of the infection was not clear because of the limited number of observations.

A series of reports in 1987 from Canada,[714] Czechoslovakia,[860] Bulgaria,[83] and the U.S.[787] of cryptosporidiosis in piglets further emphasized the wide geographic distribution of the parasite, although the infection was not very intense.

Early experimental infections were produced per orally by inoculating piglets with calf feces containing the oocysts.[556,813] These reports established the pig as a suitable experimental host for infection with *Cryptosporidium* oocysts from calves, as evidenced by pathological changes, primarily in the ileum, shedding of oocysts in the stool, and diarrhea. The first experimental infection in which the inoculum was believed to contain only *Cryptosporidium* oocysts in the calf feces involved gnotobiotic piglets.[821] The capability of establishing cryptosporidiosis in sites other than the intestine was shown by inoculating the trachea and conjuctiva of piglets.[325]

B. SPECIES OF *CRYPTOSPORIDIUM*

From studies to date, a single species of *Cryptosporidium* is thought to infect pigs. Isolates from infected calves have been found to be infective for piglets.[426a,821] Passage of calf and human *Cryptosporidium* isolates experimentally through different animals does not appear to alter the pathogenicity for piglets.[824] These cross-transmission studies demonstrate that *Cryptosporidium* is not host specific.

C. NATURAL INFECTIONS

The prevalence of cryptosporidiosis in pigs under natural conditions is still unknown,[562] although infections have been reported worldwide from the U.S.,[88,90,403,787] New South Wales,[481] the Federal Republic of Germany,[722] Canada,[713, 714] Italy,[135c,] Vietnam,[426] Czechoslovakia,[860] Japan,[907a] Chile,[6a] and Bulgaria.[83]

Most infections were detected in piglets approximately 1 to 12 weeks old. In a retrospective study conducted between 1981 and 1985 in Ontario, Canada, *Cryptosporidium* was detected histologically in 184 of 3491 pigs (5.3%).[714] Infected pigs ranged in age from 1 to 30 weeks, with 60% being 6 to 12 weeks of age. The infection was not found in pigs less than 1 week old. The infection rate diminished in pigs over 12 weeks of age. The 5.3% infection rate was considered an underestimation of the true infection rate in pigs, in view of the fact that the rate was based on histological rather than fecal examination. In a survey of 200 market pigs in southern California, 10 (5%) were excreting *Cryptosporidium* oocysts when examined by rectal swabs;[787] 7 were feeder pigs and 3 were butcher hogs. The presence of the parasite in market swine should be a warning to pork producers, abattoir workers, and veterinarians to take precautionary measures in handling these animals.

D. PATHOGENESIS AND CLINICAL MANIFESTATIONS

The parasite has been observed in the small[481,714] and large[403,714] intestine and the cecum.[860] While some investigators find parasites in the brush border of the villous epithelial cells in the jejunum and ileum,[481] others find them embedded in the microvillous border of crypt epithelial cells of only the large intestine, with a few phagocytosed by leukocytes within the crypts.[403] Both investigators report finding a few parasites in the lumen. In another study,[714] the distribution of *Cryptosporidium* is more extensive, including the jejunum, ileum, cecum, and colon, as well as the lumen, with most parasites located in enterocytes in the dome epithelium covering Peyer's patches in the terminal portion of the ileum. The villi appear mildly to moderately atrophied, especially over the ileal domes, with infiltration of the lamina propria by large numbers of mononuclear cells and a few eosinophils. Endogenous stages identified by electron microscopy include trophozoites and merozoites, with rare macrogametes and oocysts.

Diarrhea, the most prominent clinical manifestation, was observed in 26% (48 of 184) of infected pigs submitted from 133 farms in Canada.[714] Of these, 20 pigs had one or more enteric pathogens, such as adenovirus and *Salmonella* sp. Therefore, *Cryptosporidium* could not be established as the sole etiologic agent of diarrhea, especially because diarrhea in farm animals can be a multifactoral problem. Similarly, the diarrhea in market pigs in southern California could not be attributed solely to *Cryptosporidium* without simultaneously conducting bacteriological and virological studies.[787]

The shedding of oocysts in pigs has not been consistently observed by all investigators. Evidently, oocysts can be detected from both clinically healthy young pigs as well as those with diarrhea, suggesting that a carrier state can exist in which adult swine develop resistance to the parasite, but in this particular circumstance the stress of transporting or handling the market swine might have induced the shedding of oocysts.[787]

Because a subclinical infection may be the norm in conventionally farm-raised pigs,[714] whether cryptosporidiosis occurs as a primary disease, a subclinical infection, or whether the parasite acts as a predisposing agent to other enteropathogens needs further investigation. In any event, the shedding of oocysts in apparently healthy mature swine creates the possibility of these animals serving as reservoirs of infection to their young.

E. EXPERIMENTAL INFECTIONS

One study found that suckling piglets are susceptible to an oral inoculation of *Cryptosporidium* oocysts originally derived from diarrheic calves.[813] The piglets, exposed in the first week of life, developed diarrhea accompanied by shedding of oocysts in the feces 4 days postinoculation (DPI). The parasite was distributed throughout the small and large intestines with extensive mucosal damage, particularly in their terminal ileum, with stunting and fusion of villi and cross-bridging between adjacent villi, as well as cellular infiltration of the lamina propria and sloughing of the villous tips. In another study, artificially reared 1-d-old piglets inoculated with *Cryptosporidium* from feces of calves resulted in pathological changes in both small and large intestines, with shedding of oocysts in the feces beginning 3 1/2 DPI.[556] All six infected piglets had atrophy of the ileal villi accompanied by infiltration of the lamina propria by mononuclear cells and neutrophils. The diarrhea ceased within 1 d of the cessation of shedding of oocysts. These two reports established the pig as a susceptible host for *Cryptosporidium* from calves.

The first experimental infection in which the inoculum was believed to contain only *Cryptosporidium* oocysts in the calf feces involved 12 1- to 15-d-old gnotobiotic piglets.[821] The mucosal changes in gnotobiotic piglets were consistent with those seen in experimental infection of conventional piglets,[813] but in piglets of 3 d or less in age, the entire intestine was infected and the mucosa severely damaged with greater involvement of the upper small intestine. All 12 piglets excreted oocysts in their feces. Diarrhea was present in all the piglets, except in the 15-d-old piglets. The absence of diarrhea in older piglets was confirmed later,[824] in which infection of specific pathogen-free (SPF) 4-week-old piglets during weaning with calf isolates failed to induce postweaning diarrhea. Clinical manifestations were also absent in 20- and 24-d-old piglets that were inoculated with 2.5×10^7 and 5×10^7 bovine *Cryptosporidium* oocysts, respectively. Although oocysts were detected for 6 to 10 and 5 to 11 d, respectively, developmental stages of the parasite were observed in histological sections of the intestine only from the piglet inoculated with 5×10^7 oocysts.[135c]

The establishment of cryptosporidiosis outside the intestinal tract was accomplished by inoculating ten hysterectomy-derived 3- to 6-d-old piglets with oocysts originally isolated from a calf and concentrated by sugar flotation into the trachea and onto the conjunctival sacs.[325] Focal changes were observed in the trachea, where the epithelium was irregular, low, stratified, and composed of ovoid to squamous cells with numerous intraepithelial lymphocytes. There were extensive perivascular infiltrates of lymphocytes, monocytes, and macrophages in the lamina propria below affected foci, when examined by electron microscopy. Changes in the conjunctiva consisted of somewhat flattened surface epithelial layer, goblet cell depletion, and lymphocyte and monocyte infiltrations in affected foci. The fact that *Cryptosporidium* from feces can excyst, complete its life cycle, and cause morphologic damage in extraintestinal mucosae under experimental conditions raises the question as to whether the parasite can do the same in nature.

F. CHEMOTHERAPY

Chemotherapy of cryptosporidiosis has been uniformly unsuccessful, although a variety of antimicrobial agents has been tested in various animal hosts, including piglets.[558,828] In an attempt to test for the efficacy of DL-α-difluoromethylornithine (RMI 71782), which inhibits ornithine decarboxylase and is known to be active against a closely related coccidium, *Eimeria tenella,* six 1-d-old hysterectomy-derived, colostrum-deprived piglets were infected with *Cryptosporidium.*[558] The dialyzed inoculum was prepared from the feces of a man with immunodeficiency due to dysgammaglobulinemia type I. All six inoculated piglets shed oocysts in the feces, starting 4 or 5 DPI, and all had diarrhea for several days during the period of oocyst shedding. As reported for other infected pigs, the ileal villi were atrophic and covered with squamous to cuboidal, basophilic epithelial cells. The parasite was also observed on surface and crypt epithelium in the cecum and colon, where the epithelium was squamous to cuboidal. Unfortunately, the difluoromethylornithine had no demonstrable therapeutic effect on the infection, although altered polyamine levels in the intestinal mucosa was evidence that the treatment was adequate to maintain inhibition of ornithine decarboxylase activity in host tissue. Chemotherapeutic study was extended to test additional drugs in gnotobiotic pigs.[828] After an incubation of 2 to 3 d and just before clinical diarrhea was expected to begin in control piglets, the treatment was started and maintained for 4 to 6 d. Clinical as well as histological assessments proved none of the drugs tested to be effective against *Cryptosporidium.*

BIBLIOGRAPHY

Information pertaining to this subject matter may be found in References 6a, 83, 88, 90, 135c, 325, 403, 426, 481, 556, 558, 562, 713, 714, 722, 787, 813, 821, 824, 828, 860, and 907a.

II. CRYPTOSPORIDIOSIS IN HORSES (*EQUUS CABALLUS*)

Data on cryptosporidiosis in horses are sparse. Most are based on natural infection.

A. HISTORICAL BACKGROUND

Clinical cryptosporidiosis in horses was first reported in immunodeficient Arabian foals in the U.S. in 1978.[750] Cryptosporidiosis was also reported in two Arabian foals with combined immunodeficiency in Australia.[274] These early findings suggested that perhaps a state of immunodeficiency was almost necessary before the infection could be established in horses. However, in a serologic survey of 22 horses, a high percentage (91%) was shown to have antibodies to *Cryptosporidium*,[819] which suggested that cryptosporidiosis might be more widespread in immunocompetent horses than was formerly believed. Indeed, cryptosporidiosis in immunocompetent horses was confirmed by other investigators,[167,242,256,760] and immunodeficiency does not appear to be a prerequisite for cryptosporidiosis in horses.

B. SPECIES OF *CRYPTOSPORIDIUM*

The limited number of experimental infections to date in foals has involved inoculum of bovine *Cryptosporidium*. Whether the same species is involved in natural infections in horses is not definitive, but the bovine species of *Cryptosporidium* appears to be the etiologic agent in most of the infections. The most recent report[167,417] of the size range of the oocysts from horses indicates size ranges consistent with that of *C. parvum*. On the other hand, the first report of cryptosporidiosis in foals in Italy indicated that the 18-, 21-, and 64-d-old foals were supposedly infected with *C. muris*, although the diameter of the oocysts ranged from 3.9 to 5.5 μm, with an average of 4.5 μm.[135d]

C. NATURAL INFECTIONS

In 1981 it was reported that 91% of 22 Scottish horses examined by the indirect immunofluorescent test, using cryostat sections of the ileum of a 1-d-old lamb inoculated with the parasite that was originally isolated from calves, had antibodies to *Cryptosporidium*.[819] Although the parasite was initially reported only in foals with combined immunodeficiency in the U.S.[750] and Australia,[274] reports from France,[158a,760] Canada,[256] Spain,[242] and the U.S.[167,417] showed that immunocompetent foals could also be infected with *Cryptosporidium*. However, natural infections in immunocompetent horses appeared to be incidental findings, particularly since the parasite was not detected after examining 14 Quarterhorse neonatal foals suffering from foal heat diarrhea in one study,[680] and 52 foals with diarrhea in another study.[834] Although a recent report[167] showed that all pony foals reared in a helminth-free environment, as well as several Thoroughbred, Quarterhorse, and pony foals reared on pastures with their dams, were infected with *Cryptosporidium*, the fact that no oocysts were detected in a similar survey conducted a year later, using the same procedures in examining 300 samples from 58 Quarterhorses, Arabian, and pony foals, the prevalence of the infection still seemed uncertain until a recent report from Normandy which showed that *Cryptosporidium* was highly infectious for both foals and mares. In fact, a higher infection rate was observed in the mares as evidenced by oocysts excreted in the feces, although diarrhea was absent.[158a] The pregnant mares also seemed quite active in spreading the oocysts, suggesting that the mares can be the source of infection for their foals.

D. PATHOGENESIS AND CLINICAL MANIFESTATIONS

In the first report of cryptosporidiosis in Arabian foals with inherited combined immunodeficiency, which consisted of lymphopenia (<1000/mm^3), absence of IgM, and depletion of lymphocytes in various lymphoid organs, all died between 21 and 56 d, despite therapy and special care.[750] Five of six foals were infected with *Cryptosporidium* as evidenced by the presence of different developmental stages (trophozoites and schizonts) of the parasite in the microvillous border of epithelial cells of the intestine and a wide variety of tissues, including the stomach, and along the luminal surface of the epithelium lining the common bile duct, gall bladder, and major pancreatic ducts in two foals. Although the foals had diarrhea, it was not possible to separate the effects of *Cryptosporidium* from those of the concomitant adenoviral enteritis. Because *Cryptosporidium* was not found in the sections of intestines and other organs in any of the 23 foals with normal immune systems, the findings suggested that an immunodeficient state was conducive to the establishment of the infection. Two purebred Arabian foals from Australia with combined immunodeficiency were also infected with *Cryptosporidium*.[274] One foal, by a sire heterozygous for severe combined immunodeficiency, had numerous *Cryptosporidium* in histological sections of the small intestine, but none in the stomach, large intestine, gall bladder, and bile or pancreatic ducts as had been reported earlier from the U.S.[750] Although diarrhea in these two foals could not be attributed solely to *Cryptosporidium*, the absence of adenoviral inclusions and other pathogens and the characteristic lesions of cryptosporidiosis in the small intestine suggested a causal relationship. A recent report of two immunodeficient Thoroughbred male foals that developed diarrhea between days 5 and 11 and died shortly thereafter showed numerous organisms in the histological sections of the small intestine with lesions consistent with cryptosporidiosis.[655a] This report again stressed the contributory role of immunodeficiency to death, despite therapy and nursing care.

Of 82 immunocompetent foals examined, 13 foals between the ages of 3 weeks and 3 1/2 months were infected with *Cryptosporidium*.[760] None had diarrhea, and examination for rotavirus was negative. Cryptosporidiosis in immunocompetent horses was confirmed in two foals, a 1-week-old Percheron and a 6-week-old Arabian.[256] In the Percheron, numerous oocysts were detected by sugar flotation of fecal samples, as well as various stages of the parasite throughout the length of the small intestine and lumen in histologic sections. Despite continued therapy and nursing care, the foal died 2 d after admission. In the older Arabian foal, oocysts were observed in sugar flotation fecal samples, but 1 month later, it stopped shedding oocysts. The source of infection was not known, inasmuch as the feces of the mares did not reveal oocysts. Recently, 22 immunocompetent pony foals reared in a helminth-free environment were found infected with *Cryptosporidium*.[167] Oocysts first appeared in the feces between 9 and 28 d after birth, and the period of oocyst shedding varied from 2 to 18 d. A high percentage (64%) of the foals had diarrhea that began before the shedding of oocysts. Of the foals with intractable diarrhea, three died during the period of oocyst shedding. Other enteropathogens were rarely detected in the surviving foals; coronavirus was demonstrated by electron microscopy in two that had diarrhea, and *Salmonella poona* was found in a third pony that shed oocysts after a period of diarrhea. The same study reported *Cryptosporidium* oocysts in feces of 2 of 17 (11.7%) Thoroughbred foals, 3 of 14 (21.4%) Quarterhorse foals, and 3 of 24 (12.5%) pony foals reared in pastures with their mares (dams). In a recent study involving 22 infected pony foals with diarrhea, oocysts were detected intermittently between 5 and 56 d of age, although episodes of diarrhea did not necessarily correlate with oocyst shedding.[417] Seven of these foals died. An earlier report from Spain showed even a higher mortality rate of 84.2% (16 of 19) in infected foals with diarrhea.[242] As reported for mature swine that were found to be shedding oocysts and, thus, could possibly serve as reservoirs of infection to their young, 2 of 4 mares were found to be shedding oocysts after their foals were detected passing oocysts, although the mares did not have diarrhea.[417] The high infection rate in mares (29 of 36, 20 of 21, 19 of 22) over a 3-year period of observation in Normandy leaves little doubt that the mares can serve as a source of infection for their foals.[158a] However, the clinical significance of natural cryptosporidiosis in immunocompetent foals remains unclear.

Because enteritis due to *Cryptosporidium* and foal heat diarrhea share temporal and clinical similarities, a possible association between these syndromes was investigated by examining 14 Quarterhorse foals for oocysts of *Cryptosporidium* in the feces.[680] Although all 14 foals exhibited diarrhea between 7 and 22 d of age, they were consistently negative for oocysts of *Cryptosporidium*, which suggested that the parasite was not an etiological factor in foal heat diarrhea.

E. EXPERIMENTAL INFECTIONS

Transmission experiments of bovine *Cryptosporidium* to newborn, colostrum-fed,[834] or colostrum-deprived[829,834] foals induced only subclinical infections. In a review in 1983, Tzipori[824] referred to an unpublished study in which colostrum-deprived or suckling foals up to 10 d old had asymptomatic infections after inoculation with calf *Cryptosporidium*. The infection produced no apparent mucosal lesions, and the shedding of oocysts in the feces was very brief. This limited number of experimental studies suggests that the foal may not be a very susceptible host to experimental cryptosporidiosis, at least not to *Cryptosporidium* of bovine origin.

F. CHEMOTHERAPY

There has been no chemotherapeutic study directed specifically against cryptosporidiosis in horses.

BIBLIOGRAPHY

Information pertaining to this subject matter may be found in References 135d, 158a, 167, 242, 256, 274, 417, 655a, 680, 750, 760, 819, 824, 829, and 834.

Chapter 7

CRYPTOSPORIDIOSIS IN CATS, DOGS, FERRETS, RACCOONS, OPOSSUMS, RABBITS, AND NON-HUMAN PRIMATES

Michael W. Riggs

TABLE OF CONTENTS

I. CRYPTOSPORIDIOSIS IN CATS

A. SPECIES OF *CRYPTOSPORIDIUM* INFECTING CATS

Cryptosporidium infection in the domestic cat (*Felis catus*) was first reported in 1979 when oocysts were detected in feces from 5 of 13 cats examined.[370] Developmental stages were limited to the small intestine. Adult cats, but not mice (ICR strain) or guinea pigs (Hartley strain), were infected by oral challenge with oocysts collected in feces or small intestine mucosal scrapings from naturally infected cats. Based on these cross-transmission studies, the name *Cryptosporidium felis* was proposed to designate the species infecting cats. However, the organism described is morphologically and developmentally indistinguishable from *Cryptosporidium parvum*.[808,844] Lack of transmission to mice in the single transmission trial conducted may have been age related, because 7-week-old mice, which may be resistant to *Cryptosporidium* infection,[328,734] were used. Furthermore, *C. parvum* isolates from calves and human beings are infectious to cats.[61,186,619] Therefore, until distinct differences can be shown, *C. felis* should be considered *C. parvum*.

A second species of *Cryptosporidium*, originally isolated from the house rat (*Rattus norvegicus*), infects the stomach but not the intestinal tract of cats.[372] Its site specificity and oocyst morphology resemble *C. muris*,[806,844] although oocysts are slightly larger. The organism has been designated *C. muris* strain RN 66.[372]

Cryptosporidium curyi was proposed as a third *Cryptosporidium* infecting domestic cats.[598] Oocysts measuring 32.3 × 31.1 μm and containing four sporozoites were detected in the feces of eight cats. Attempts to infect domestic cats with the oocysts were unsuccessful. Additional morphologic, developmental, and transmission studies are required before this proposed genus and species can be considered valid.

B. CLINICAL AND PATHOLOGIC FEATURES

Spontaneous or experimental intestinal cryptosporidiosis was subclinical in nine cats ranging in age from 1 month to adult.[370] Stages of *Cryptosporidium* were observed within villous tip epithelial cells throughout the small intestine, but were most numerous in the ileum. Organisms were not observed in the small intestinal crypts, stomach, cecum, or colon. Lesions at the light microscope level were not described, but parasitized cells showed degenerative changes ultrastructurally. Electron micrographs delineating the life cycle and structure of all endogenous stages were presented. Intraintestinal release of sporozoites from oocysts was observed, extending the concept of autoinfection, first suggested by Tyzzer,[808] to the infection in cats. The prepatent and patent periods in experimentally infected cats were 5 to 6 d and 7 to 10 d, respectively. One naturally infected cat shed oocysts for at least 5 months, but the relationship between this persistent infection and immunocompetency was not determined. Additional cases of subclinical intestinal cryptosporidiosis have been reported in cats.[61,86,186,837] Oocysts were detected in the feces of a 4-month-old cat without clinical signs after diagnosis of cryptosporidiosis in its immunosuppressed owner.[86] In another report, subclinical infection was produced in three 6-week-old kittens by oral inoculation with oocysts isolated from the feces of an AIDS patient.[186] The kittens shed oocysts 5 to 12 days postinoculation (DPI). On 6 DPI, a few cryptosporidia were observed in sections of ileum and cecum from one kitten, but lesions were not described.

Several reports have associated intestinal *Cryptosporidium* infection with clinical disease.[86,290e,554,619,655] The infection was diagnosed in a feline leukemia virus (FeLV)-negative adult domestic longhair cat with anorexia, weight loss, and diarrhea lasting 2 months.[655] Thickening of the small intestinal and cecal walls and mesenteric lymph node enlargement were observed at necropsy. Histologic lesions included villous fusion, goblet cell and crypt hyperplasia, occasional distension of crypts with necrotic debris, and a mild lymphocytic-neutrophilic enteritis. Numerous stages of *Cryptosporidium* were observed within the intestinal lumen and epithelial cells of crypts and villi. Unlike subclinical cases, cryptosporidia were concentrated in crypts and the lower half of villi in the small intestine rather than in villous tips. The cecum was also infected. Ultrastructurally, there was microvillar loss on infected cells. Fecal flotation and aerobic bacterial culture were negative for other intestinal pathogens. Viral particles were not seen in intestinal epithelium by electron microscopy, but a viral enteritis could not be excluded as a cause for the clinical signs and lesions. In another report, a 3-week-old kitten inoculated with *C. parvum* oocysts obtained from a calf shed oocysts 5 to 12, and 34 to 35 DPI.[619] On 5 to 9 DPI, the cat's feces were described as almost diarrheic, but concurrent intestinal infections were not excluded. Intestinal cryptosporidiosis was subsequently reported in two symptomatic cats.[86] The first case was in a 6-month-old FeLV-negative cat with diarrhea lasting 8 weeks, inappetance, and weight loss. *Cryptosporidium* oocysts were detected in feces by flotation and safranin staining. Stages of *Cryptosporidium* were observed in the jejunum, but lesions were not described. Oral ulceration and a positive complement fixation titer suggested concurrent calicivirus infection. The extent of testing for other pathogens was not specified. The second case was in a 6-month-old FeLV-negative cat with persistent diarrhea. An untyped *Campylobacter* was isolated from feces. Cryptosporidia were detected in feces and a small intestinal biopsy, and oocyst shedding persisted for 6 weeks after initial diagnosis. Stools became firmer after oocyst shedding ceased. The role of *Cryptosporidium* infection in the diarrhea in these two cats was unclear. Most recently, intestinal cryptosporidiosis was reported in a 4-year-old domestic longhair cat with intermittent diarrhea lasting 1 year.[554] At the time of initial examination the cat was FeLV-positive by ELISA. It was infected with *Toxocara* sp. and intestinal coronavirus and had histologic evidence of colitis in a biopsy specimen. Aerobic fecal culture was negative for bacterial pathogens. The cat was treated for *Toxocara* infection with pyrantel pamoate and inflammatory bowel disease with prednisone. Diarrhea decreased in frequency to periodic episodes which were responsive to prednisone. At 18 months after the initial examination, the cat was reexamined for anorexia, dehydration, and weight loss. *Isospora* sp. oocysts were identified in feces. At 5 months later, blood dyscrasias and polyclonal gammopathy developed and the cat was euthanatized and necropsied. *Cryptosporidium* infection had not been identified up to this point. Histologically, cryptosporidia were observed in mucosal epithelium of the duodenum, jejunum, and ileum. Infection was heaviest in the jejunum, and organisms were concentrated near villous tips. Cryptosporidia were not seen in the stomach or colon. Small intestinal lesions included mild goblet cell hyperplasia and diffuse lymphoplasmacytic enteritis. The lesions were attributed to *Cryptosporidium* infection secondary to FeLV-associated immmunodeficiency. The duration of *Cryptosporidium* infection and its relationship to the protracted course of diarrhea could not be determined.

Zoonotic transmission of *Cryptosporidium* from cats has been suggested.[86,419,887] In two cases, cats and immunodeficient human beings living in the same household were infected with *Cryptosporidium*. However, it could not be determined if the cats transmitted the infection to their owners or vice versa or if another source of exposure existed.

In the one report on gastric cryptosporidiosis in cats, three 1- to 2-month-old cats inoculated orally with *C. muris* (strain RN 66) oocysts developed heavy gastric infections but were asymptomatic.[372] The prepatent period was 15 to 21 d. Peak oocyst shedding (10^7 oocysts per cat per day) occurred around 30 DPI. Oocyst shedding decreased between days 40 to 50, but returned to peak levels 100 and 140 DPI. The relationship between the cats' immunologic status and persistent infection with *C. muris* RN 66 was not determined, although the cats were ELISA negative for FeLV infection.

C. PREVALENCE AND GEOGRAPHIC DISTRIBUTION

The prevalence of cryptosporidiosis in domestic cats is unknown. Antibodies against *Cryptosporidium* were detected in 20 of 23 cats examined in Scotland.[819] Oocysts were detected in feces from 4 of 300 cats surveyed in the Federal Republic of Germany[61] and 20 of 507 juvenile and adult cats surveyed in Japan.[837] Oocysts were not detected in feces from ten cats in France.[158b] Intestinal cryptosporidiosis has been reported in domestic cats from the U.S., Japan, Czechoslovakia, the Federal Republic of Germany, and England.[61,86,370,554,619,655,837] Gastric cryptosporidiosis (experimental) has been reported in cats from Japan.[372]

BIBLIOGRAPHY

Information pertinent to the subject matter of this section may be found in References 61, 86, 158b, 186, 290e, 370, 372, 419, 554, 598, 619, 655, 819, 837, and 887.

II. CRYPTOSPORIDIOSIS IN DOGS

A. SPECIES OF *CRYPTOSPORIDIUM* INFECTING DOGS

Intestinal cryptosporidiosis in the domestic dog (*Canis familiaris*) was first reported in 1983.[892] This and subsequent reports[186,254,805] involved a *Cryptosporidium* sp. morphologically indistinguishable from *C. parvum*. Furthermore, *C. parvum* isolates from calves and humans are infectious to dogs.[61,186]

Two 1- to 2-month-old pups inoculated orally with *C. muris* (strain RN 66) oocysts passed a few oocysts in feces, but evidence of infection was not clearly established.[372]

B. CLINICAL AND PATHOLOGIC FEATURES

Most reports of intestinal cryptosporidiosis in the domestic dog have involved the young or immunosuppressed. The initial report was in a 1-week-old Pomeranian puppy.[892] The pup was presented for acute diarrhea after litter mates died from undetermined causes. Histologic examination was limited to the small intestine of the diarrheic pup. Numerous cryptosporidia were observed in the epithelium of villous tips and crypts. Villi were mildly blunted and the lamina propria had a mild increase in mononuclear cells. Concurrent infections could not be excluded. In another report, cryptosporidiosis was diagnosed in a 6-month-old Siberian husky with canine distemper virus infection.[805] The pup had diarrhea and seizures. An immunodeficient state secondary to distemper virus infection was indicated by (1) neutropenia and bone marrow myeloid hypoplasia, (2) IgG hypoglobulinemia, (3) subnormal lymphocyte mitogenic responses, and (4) lymphoid depletion in thymus, spleen, and lymph nodes. The dog had also been treated with corticosteroids (dexamethasone and prednisone) at 19 weeks of age. *Trichomonas, Giardia,* and *Cryptosporidium* were identified in feces. Cryptosporidia were seen in small intestinal villous epithelial cells but lesions were not observed. Due to concurrent infections, the role of *Cryptosporidium* infection in the diarrhea could not be determined. In six cases of cryptosporidiosis in puppies, the infection was not associated with vomition or diarrhea.[186,254,738] The first case was in a 6-week-old mixed breed pup presented in a semicomatose condition.[738] The pup subsequently died, but the cause of death could not be determined. Cryptosporidia were observed within villous and crypt epithelial cells in the ileum. Ultrastructurally there was microvillar effacement in parasitized enterocytes, but other intestinal lesions were not observed. The second case was in a 6-week-old mixed breed pup presented for seizures.[738] The clinical signs and subsequent death were attributed to toxaphene intoxication. Sections of small and large intestine contained cellular debris in scattered crypts and increased numbers of neutrophils and mononuclear cells in the lamina propria. A few cryptosporidia were observed in ileal enterocytes. Parvovirus infection was excluded. The third subclinical case of cryptosporidiosis was reported in a 3-month-old Cockapoo with canine distemper virus infection.[254] Cryptosporidia were observed in villous and crypt epithelium of the jejunum. Ultrastructurally, microvilli surrounding developmental stages were blunted. No other intestinal lesions were described and the infection was considered an incidental finding. In three 8-week-old puppies which were inoculated orally with *Cryptosporidium* oocysts from feces of an

AIDS patient, the prepatent and patent periods were 5 and 7 d, respectively.[186] Clinical signs were not observed. In puppies inoculated orally with oocysts from calves, the prepatent and patent periods were 2 to 14 d and 3 to 33 d, respectively.[61]

C. PREVALENCE AND GEOGRAPHIC DISTRIBUTION

The prevalence of intestinal cryptosporidiosis in domestic dogs is unknown. Oocysts were not detected in feces from 57 juvenile and adult healthy dogs in Finland,[641] 200 dogs in the Federal Republic of Germany,[61] or 101 healthy pet dogs in an urban area of Scotland.[736] However, in another study, antibodies against *Cryptosporidium* were detected in 16 of 20 dogs examined in Scotland.[819] In Japan, oocysts were detected in feces from 3 of 213 asymptomatic juvenile dogs examined.[837] In France, oocysts were detected in feces from 13 of 29 dogs.[158b] Most reports of spontaneous intestinal cryptosporidiosis in domestic dogs have been from the U.S.[254,738,805,892]

BIBLIOGRAPHY

Information pertinent to the subject matter of this section may be found in References 61, 158b, 186, 254, 372, 641, 736, 805, 819, 837, 892.

III. CRYPTOSPORIDIOSIS IN FERRETS

A. SPECIES OF *CRYPTOSPORIDIUM* INFECTING FERRETS

There has been only one report of intestinal cryptosporidiosis in the ferret (*Mustela putorius*).[677] The *Cryptosporidium* sp. involved was morphologically indistinguishable from *C. parvum*. Gastric cryptosporidiosis has not been reported in ferrets.

B. CLINICAL AND PATHOLOGIC FEATURES

Intestinal cryptosporidiosis was first reported in two dexamethasone-treated laboratory ferrets, aged 4 and 8 months.[677] Numerous cryptosporidia were observed in villous epithelial cells of the small intestine. Organisms were not seen in the crypts or colon. The diagnosis of cryptosporidiosis, previously unrecognized in ferrets, prompted a survey of colony ferrets for infection. Feces from 22 4- to 18-month-old ferrets were examined by an acid-fast staining method. Nine 4- to 6-month-old ferrets had *Cryptosporidium* oocysts in feces. Oocysts were subsequently detected in feces from 31 of 44 3-month-old newly arriving ferrets; 19 ferrets shed oocysts for up to 4 weeks, and small intestinal infection was confirmed histologically in 15; 24 infected ferrets treated with dexamethasone for 2 to 20 weeks had persistent small intestinal infection and shed oocysts for the duration of treatment. Cryptosporidia were observed at all levels of the small intestine, but the ileum was the most heavily infected site. Mild villous atrophy was seen in a few animals. Gastric, colonic, pulmonary, or gall bladder infections were not observed. Interestingly, pancreatic duct infection was observed in one dexamethasone-treated ferret. *Cryptosporidium* infection in this site has been previously reported only in human beings, non-human primates, and horses.[236] *Cryptosporidium* infection was subclinical in all ferrets examined, including those treated with dexamethasone.

C. PREVALENCE AND GEOGRAPHIC DISTRIBUTION

The prevalence of *Cryptosporidium* infection in ferrets is unknown. Recent recognition of *Cryptosporidium* infection in a large percentage of colony ferrets suggests that the infection may be common.[677] The infection has been reported in the U.S.[677]

BIBLIOGRAPHY

Information pertinent to the subject matter of this section may be found in Reference 677.

IV. CRYPTOSPORIDIOSIS IN RACCOONS

A. SPECIES OF *CRYPTOSPORIDIUM* INFECTING RACCOONS

The *Cryptosporidium* sp. infecting the intestinal tract of raccoons (*Procyon lotor*) is morphologically indistinguishable from *C. parvum*.[749] Gastric cryptosporidiosis in raccoons has not been reported.

B. CLINICAL AND PATHOLOGIC FEATURES

Cryptosporidiosis in the raccoon was first reported in a young wild-caught animal.[137] Stages of *Cryptosporidium* were observed in small intestinal villous epithelial cells and were most numerous on villous tips. Cryptosporidia were not seen in crypt epithelium. The infection was associated with villous atrophy and enteritis, but there was concurrent intestinal infection with another suspected coccidian parasite. Clinical signs were not observed.

C. PREVALENCE AND GEOGRAPHIC DISTRIBUTION

In a recent survey of 100 wild raccoons, *Cryptosporidium* oocysts were detected by an immunofluorescence assay in feces from 13 juveniles.[749] Surveyed animals had been trapped or shot, precluding evaluation for clinical disease. Cryptosporidiosis in raccoons has been reported in the U.S.[137,749]

BIBLIOGRAPHY

Information pertinent to the subject matter of this section may be found in References 137 and 749.

V. CRYPTOSPORIDIOSIS IN OPOSSUMS

A. SPECIES OF *CRYPTOSPORIDIUM* INFECTING OPOSSUMS

Spontaneous cryptosporidiosis has not been reported in opossums (*Didelphis virginiana*). Experimental intestinal infection was produced in 11-week-old nursing opossums by oral inoculation with *C. parvum* oocysts isolated from calves.[476]

B. CLINICAL AND PATHOLOGIC FEATURES

In experimentally infected opossums, oocysts were shed in feces and stages observed in epithelial cells of the ileum, cecum, and colon 6 to 10 DPI.[476] Mild ileal villous atrophy was observed in opossums examined 7 and 8 DPI. Clinical disease was not observed, but four of seven infected opossums had semiformed feces in the rectum.

C. PREVALENCE AND GEOGRAPHIC DISTRIBUTION

Experimental cryptosporidiosis in opossums has been reported in the U.S.[476]

BIBLIOGRAPHY

Information pertinent to the subject matter of this section may be found in Reference 476.

VI. CRYPTOSPORIDIOSIS IN RABBITS

A. SPECIES OF *CRYPTOSPORIDIUM* INFECTING RABBITS

Cryptosporidium cuniculus was proposed to designate the species infecting the intestinal tract of the rabbit (*Oryctolagus cuniculus*).[367] The organism described is morphologically and developmentally indistinguishable from *C. parvum*. This observation and subsequent transmission of *C. parvum* from calves to rabbits warrants designation of the species infecting the intestine of rabbits as *C. parvum*.[236]

Cryptosporidium muris (strain RN 66) infects the stomach of orally inoculated rabbits.[372]

B. CLINICAL AND PATHOLOGIC FEATURES

Intestinal cryptosporidiosis reported in two New Zealand white rabbits was not associated with clinical disease.[675] Stages of the organism were observed by light and electron microscopy in the villous epithelium of the jejunum and ileum. Crypt epithelial cells and the colon were not infected. The only lesions described were a slight decrease in the villous to crypt ratio and disruption of microvilli. In another report, subclinical intestinal cryptosporidiosis was described in an adult colony rabbit.[367] Organisms were observed in villous epithelial cells of the ileum. Lesions included villous atrophy, degeneration of parasitized epithelial cells, and elongation of microvilli adjacent to organisms. Intestinal *Cryptosporidium* infection was considered an incidental finding in a wild caught juvenile cottontail rabbit (*Sylvilagus floridanus*).[707] Unlike the reports involving domestic rabbits, cryptosporidia were observed in epithelial cells of the cecum and colon, but not the small intestine. Intestinal lesions were not observed. *Cryptosporidium* infection in weanling 4- to 5-week-old rabbits from a commercial rabbitry was associated with diarrhea.[633] However, the role of *Cryptosporidium* in the diarrhea could not be determined because the rabbits had other intestinal infections.

Three 6-week-old New Zealand white rabbits inoculated orally with *C. muris* strain RN 66 oocysts developed gastric infection and passed oocysts for 11 to 43 d, but did not develop clinical signs.[372]

C. PREVALENCE AND GEOGRAPHIC DISTRIBUTION

There are few reports on the prevalence of intestinal cryptosporidiosis in rabbits. The infection was detected in 12 of 149 weanling rabbits from 29 commercial rabbitries by an acid-fast staining method for fecal oocysts and by histologic examination.[633] In another report, the infection was considered common.[367] Naturally occurring or experimental intestinal cryptosporidiosis has been reported in rabbits from the U.S., Belgium, and the Federal Republic of Germany.[236,367,633,633b,675,707] Experimental gastric cryptosporidiosis has been reported in rabbits from Japan.[372]

BIBLIOGRAPHY

Information pertinent to the subject matter of this section may be found in References 236, 367, 372, 633, 633b, 675, and 707.

VII. CRYPTOSPORIDIOSIS IN NON-HUMAN PRIMATES

A. SPECIES OF *CRYPTOSPORIDIUM* INFECTING NON-HUMAN PRIMATES

Cryptosporidium rhesi was proposed to designate the species infecting rhesus macaques (*Macaca mulatta*).[457] However, reports of intestinal, hepatopancreatic, and pulmonary cryptosporidiosis in rhesus and other macaques have involved an organism morphologically indistinguishable from *C. parvum*.[162,427,776,890] Gastric infection with *C. muris* has not been reported in non-human primates.

B. CLINICAL AND PATHOLOGIC FEATURES

1. Macaques — Immunocompetent

The initial report of intestinal cryptosporidiosis in non-human primates involved two juvenile rhesus monkeys (*Macaca mulatta*).[427] One monkey developed pneumonia and persistent diarrhea following oral inoculation with staphylococcal enterotoxin and was euthanatized. Cryptosporidia were observed in epithelial cells of the intrahepatic bile ducts, common bile duct, and gallbladder. Mucosal hyperplasia, cholangitis/pericholangitis with neutrophilic exudate in occasional crypts, and periductal fibrosis were associated with the infection. A few cryptosporidia were seen in epithelial cells of the pancreatic ducts, associated with mucosal hyperplasia, lymphoplasmacytic inflammation, neutrophilic exudate, and perilobular fibrosis. There was no evidence of cytomegalovirus or adenovirus infection, either of which may produce lesions in the hepatopancreatic system similar to those observed. However, these and other agents could not be ruled out conclusively. Enteritis and bacterial pneumonia were present, but cryptosporidia were not observed in the duodenum or lung. The second monkey was found comatose and emaciated and was subsequently euthanatized. Many cryptosporidia were observed in villous and crypt epithelial cells of the jejunum and ileum, and fewer in the duodenum, cecum, and colon. Cryptosporidia were not observed in the bile or pancreatic ducts or gall bladder. Small intestinal lesions included hyperplasia of submucosal lymphoid tissue and crypt epithelium, lymphoplasmacytic enteritis, crypt epithelial necrosis, and villous atrophy. Mild inflammation was noted in the cecum, but no lesions were described in the colon. Concurrent infections could not be excluded. Of diagnostic importance, it was noted that bulbous apical cytoplasmic extrusions on epithelial cells in the gall bladder and bile and pancreatic ducts of uninfected healthy monkeys can be confused with cryptosporidia at the light microscope level. Giemsa staining and electron microscopy are useful in differentiating these host structures from cryptosporidia.[427] Similar host-derived apical cytoplasmic blebs have been observed on intestinal epithelial cells in mice, calves, and pigs.[690a] In another report, *Cryptosporidium* was considered the cause of persistent diarrhea in four 3- to 10-month-old colony-born macaques.[890] Multiple fecal cultures for *Salmonella, Shigella,* and *Yersinia* were negative. *Trichomonas* and *Chilomastix* were present in feces from one monkey. Testing for other enteropathogens was not reported. Two monkeys died and two were euthanatized. Stages of *Cryptosporidium* were observed in villous and crypt epithelial cells in the small intestine and were especially numerous in the ileum. None were seen in the stomach or colon. Lesions included villous atrophy and fusion, enterocyte necrosis, and crypt hyperplasia. Significant inflammation was not described. Ultrastructurally, infected cells showed degenerative changes and elongation of microvilli adjacent to parasites. Ultrastructural evidence for intraintestinal excystation of oocysts was noted, extending the concept of autoinfection previously suggested by others to occur in mice[808] and cats[370] to the infection in primates. The immunologic status of the monkeys was not determined, although two monkeys had thymic lymphoid depletion. Lymphoid hyperplasia and sinus histiocytosis were described in mesenteric lymph nodes of three monkeys. In other reports focusing on diagnostic methods for cryptosporidiosis, shedding of *Cryptosporidium* oocysts was temporally associated with diarrhea in simians.[127,776]

2. Macaques — Immunodeficient

Of comparative medical interest are reports of disseminated cryptosporidiosis in macaques with immunodeficiency syndromes caused by infection with simian immunodeficiency virus (SIDV) or a type D retrovirus (simian retrovirus).[337,412,413,453,455,531,606] SIV is a lentivirus related to HIV-1, the etiologic agent of human AIDS, and more closely related to HIV-2. The immunodeficiency syndromes in macaques have clinical and pathologic features in common with human AIDS, including the development of opportunistic infections and tumors. In one of the initial reports involving an adult rhesus monkey with lymphoma and probable simian AIDS, cryptosporidia were observed in epithelial cells of the intrahepatic bile ducts.[365] Organisms were associated with mild cholangitis and periductal fibrosis, but the relationship of these lesions to *Cryptosporidium* infection was unclear because of a concurrent viral enteritis. A healthy 8-month-old colony-born rhesus monkey inoculated with tumor homogenates from this monkey developed lymphoma. The inoculated monkey also developed a hematologic profile similar to that described for simian AIDS, cytomegalovirus hepatitis, viral enteritis, and small intestinal cryptosporidiosis, suggesting an acquired immunodeficiency state. Transmission of simian AIDS by inoculation of healthy adult macaques with a cell-free lymphoma filtrate was subsequently confirmed,[454] and the etiologic viral agent was later identified.[413] Inoculated monkeys developed lymphocyte dysfunction as indicated by impaired proliferative responses to lectins and reduced T4 (helper/inducer):T8 (suppressor/cytotoxic) cell ratios and multiple opportunistic infections including hepatopancreatic cryptosporidiosis. Two immunodeficient rhesus monkeys had viral enteritis and hepatitis and cryptosporidia in epithelial cells of the gallbladder, major intrahepatic bile ducts, and pancreatic ducts. Hyperplasia, inflammation, and fibrosis of the bile and pancreatic ducts were described.[454] There was no evidence of intestinal cryptosporidiosis in either monkey. In an epizootic of simian AIDS involving 29 rhesus macaques, the causes of death or euthanasia were usually septicemia and persistent diarrhea.[606] Seven juvenile and young adult monkeys involved in the epizootic had cryptosporidia in epithelial cells of the ileum and colon. In three of these monkeys, cryptosporidia were also present in the gall bladder and jejunum. All monkeys had a variety of concurrent intestinal and extraintestinal infections. In recent reports on simian AIDS, persistent diarrhea was the immediate cause of death in most monkeys, and cytomegalovirus, adenovirus, and *Cryptosporidium* were considered the most important agents involved.[69,107] Of 19 SIV-infected monkeys, 6 had intestinal cryptosporidiosis.[69] In two, cryptosporidia were disseminated in epithelial cells of the trachea, bronchioles, peribronchiolar glands, bile ducts, gall bladder, pancreatic ducts, and small intestine. Tracheal and bronchial infections were associated with epithelial sloughing and hyperplasia and loss of respiratory cilia. One of these monkeys had a concurrent mild *Pneumocystis* infection. Respiratory cryptosporidiosis had previously been reported only in birds and human beings with AIDS.[236] One monkey had hyperplasia of pancreatic duct epithelium associated with severe adenovirus infection and occasional cryptosporidia. This monkey also had pericholangitis and severe epithelial hyperplasia in bile ducts and the gall bladder associated with cryptosporidia. In two monkeys, *Cryptosporidium* infection was limited to the small intestine and heaviest in villous epithelium, with only occasional crypt involvement. Severe villous atrophy and fusion, epithelial degeneration and sloughing, and lymphoplasmacytic enteritis were described. Concurrent viral (cytomegalovirus, adenovirus), bacterial (*Campylobacter*), and helminthic (*Strongyloides*) intestinal infections were present. In another report, intestinal *Cryptosporidium* infection was identified in seven colony-born 4- to 10-week-old rhesus monkeys which died from multisystem bacterial and viral diseases.[162] In six monkeys, cryptosporidia were seen in villous and crypt epithelial cells in the small intestine, but no light microscopic lesions were described and the infection was considered nonpathogenic. Ultrastructurally, there was loss of microvilli on parasitized cells. In one monkey, a colony-born infant with mild diarrhea, a heavy *Cryptosporidium* infection in the jejunum and ileum was associated with villous atrophy. A few cryptosporidia were also observed in the colon. Concurrent intestinal infections could not be excluded. Cryptosporidia were not observed in epithelium of the gallbladder, pancreatic ducts, bile ducts, or respiratory tract of any monkey. Thymic lymphoid depletion was observed in four monkeys. Prior to the onset of clinical disease (diarrhea, pneumonia), six monkeys had been inoculated with Mason-Pfizer monkey virus, a type D retrovirus (SRV-3) subsequently shown to induce an immunodeficiency state in newborn macaques.[244] Retrospectively, the thymic lymphoid depletion observed in four of these monkeys and multisystem/multiagent infections in all monkeys suggest that immunodeficiency may have existed.

3. Prosimians and New World Monkeys

In addition to macaques, intestinal cryptosporidiosis has been reported in a 3-month-old squirrel monkey (*Saimiri sciureus*),[129] a red-ruffed lemur (*Varecia variegata rubra*),[340] and a cotton-topped tamarin (*Saguinus oedipus*).[340] In the squirrel monkey, cryptosporidia were associated with villous atrophy, epithelial sloughing, and neutrophilic enteritis in the distal jejunum and ileum. Clinical signs and evaluation for other enteropathogens were not reported. In the lemur and tamarin, *Cryptosporidium* oocysts were detected in feces by the auramine O staining method. Both monkeys had diarrhea but further characterization was not reported.

C. PREVALENCE AND GEOGRAPHIC DISTRIBUTION

In an epidemiologic survey of colony-born pig-tailed macaques (*Macaca nemestrina*) at the Regional Primate Research Center in Washington State, *Cryptosporidium* was identified in approximately 4% of the diarrhea episodes.[706] *Cryptosporidium* was usually recognized as a component of multiple-agent intestinal infections and seen in monkeys less than 6 months of age. *Cryptosporidium* infection may have been more common than observed because all diarrheic monkeys were not evaluated for this agent. During an outbreak of cryptosporidiosis at this facility, oocysts were detected in feces from 69 of 160 diarrheic simians, most of which were infants.[127] In contrast, no cases of cryptosporidiosis were diagnosed in several hundred adult rhesus monkeys at another facility.[162] Collectively, these reports suggest that the infection is most common in the young. Cryptosporidiosis in non-human primates has been reported in the U.S.[69,107,127,129,162,337,340,365,412,413,427,453-455,531,606,706,776,890]

BIBLIOGRAPHY

Information pertinent to the subject matter of this section may be found in References 69, 107, 127, 129, 162, 337, 340, 365, 412, 413, 427, 453 to 455, 531, 606, 706, 776, and 890.

VIII. SUMMARY AND CONCLUSIONS

The relatively recent nature of reports on intestinal cryptosporidiosis in cats, dogs, ferrets, raccoons, opossums, rabbits, and non-human primates suggests that the infection is more common than is currently recognized. The agent involved is morphologically indistinguishable from *C. parvum*, but other species infecting the intestine may exist. Intestinal *Cryptosporidium* infection appears to be most common in the young of these animals and may be asymptomatic. Clinical disease has sometimes been associated with the infection in cats, dogs, rabbits, and non-human primates. Additional studies are needed to clarify its role as a primary pathogen in these species. In macaques, particularly those with SIV infection, *C. parvum* may be an important cause of intestinal, hepatopancreatic, and respiratory disease. Naturally occurring gastric infection with *C. muris* has not been reported in cats, dogs, ferrets, raccoons, opossums, rabbits, or non-human primates. Additional studies are needed to assess the zoonotic potential of *C. parvum* in the group of animals discussed in this chapter.

Chapter 8

CRYPTOSPORIDIOSIS IN RODENTS

Lance E. Perryman

TABLE OF CONTENTS

I. HISTORICAL BACKGROUND

Organisms of *Cryptosporidium* sp. were first recognized and described in rodent species. Tyzzer noted the appearance of organisms in the gastric glands of common mice in 1907 and suggested the name *C. muris*.[806] In 1910, he provided a complete description of the organism and demonstrated transmissibility through feeding trials.[807] *C. parvum* was identified by Tyzzer in 1912.[808] He described the morphologic characteristics of *C. parvum* and showed it was distinct from *C. muris*. He also characterized the life cycle of the organism and suggested that autoinfection by newly formed sporozoites was possible, a point now accepted as contributing to the persistence of infection in immunocompromised hosts. Furthermore, he observed similar organisms in the intestinal tracts of rabbits. Tyzzer's observations were particularly insightful in that they predicted the species of *Cryptosporidium* infectious for mammals, suggested that more than one mammalian species was susceptible to *C. parvum*, and provided a framework for understanding the difference in the pathogenesis of disease in immunocompetent vs. immunodeficient hosts.

II. CRYPTOSPORIDIOSIS IN MICE

A. *CRYPTOSPORIDIUM PARVUM*

1. Prevalence

Tyzzer observed natural infections of house mouse (*Mus musculus*) by *C. parvum*.[808] Of 115 wild mice trapped at the calving site of a dairy operation 35 (30%) shed oocysts during 14 to 21 d captivity.[418] Wild mice, as well as at least eight laboratory strains of mice (including BALB/c, CBA, C57 Black, Swiss Webster, BALB/c nude, and CBA nude), have been infected with *C. parvum*.[195,233,236,280,312,315,328,418,466,734,808,811] *C. parvum* oocysts obtained from bovine feces are infectious for mice, and oocysts from mice will infect calves.[21,35,418,561,674,689,808] These observations suggest that in commercial cattle operations, infection could be maintained, in part, by a mouse-calf cycle.

2. Diagnosis

Euthymic mice infected with *C. parvum* do not usually develop diarrhea or exhibit other detectable signs of disease.[328,808] Infection is diagnosed by demonstration of oocysts in feces or detection of the organism in intestinal epithelial cells by histological examination of mucosal smears or fixed sections.[328,689,808] Indirect immunofluorescence microscopic examination has also been used to demonstrate endogenous stages of *Cryptosporidium* in the ileum of infected mice.[134]

3. Lesions

The first report of *C. parvum* infection in mice described organisms in the apical borders of intestinal epithelial cells distributed from the proximal duodenum throughout the small intestine, without significant tissue injury or host inflammatory cell response.[808] Subsequently, organisms were observed in the cecum and colon as well.[233,238,689,734] Transient villus atrophy, crypt hyperplasia, and inflammatory cell accumulations in the distal small intestines of euthymic BALB/c mice have been found in infected mice. In contrast, infection of athymic (nude) BALB/c mice resulted in persistent infection and diarrhea.[328] Villous atrophy, crypt hyperplasia, and inflammation resembled or exceeded that observed in euthymic mice. However, cystic hyperplasia with abscessation of crypts was observed in the large intestines of infected BALB/c nude mice. Organisms were more numerous in the cecum and colon of infected nude mice, as compared to euthymic mice.[328]

4. Experimental Observations

Susceptibility to infection — Wild mice; outbred Porton, Schneider Swiss White, and Swiss Webster mice; and inbred CBA, C57 Black, BALB/c, Porton, Hairless [HR/HR-ADR], CBA nude, and BALB/c nude mice are susceptible to infection when orally challenged with *C. parvum* oocysts.[21,35,189,190,195,233,280,328,418,561,597,734,808,842a] Susceptibility appears to be age related, as extensive infections occur in mice 1 to 4 d of age, but only a transient infection could be established in mice at 21 d of age or older.[734] However, the intestinal tracts of adult CD1 and BALB/c strain mice maintained in gnotobiotic environments were more readily colonized by *C. parvum* than were the intestinal tracts of age matched mice of both strains maintained in conventional environments.[315] Treatment with antibiotics to reduce intestinal microflora did not render conventionally reared adult mice as susceptible to infection as their gnotobiotic age-matched counterparts. Therefore, undefined factors in addition to age and intestinal microflora appear to influence susceptibility to infection.[315]

Routes of infection — Young mice are readily infected when orally inoculated with oocysts. In one study, ID_{50} in neonatal mice was between 500 and 1000 oocysts.[233] Nontraditional routes of infection with oocysts have also been evaluated. Intrauterine infections were established in adult BALB/c mice using oocysts isolated from bovine feces by sucrose density centrifugation and treated with 5.25% sodium hypochlorite to facilitate excystation.[466] Oocysts were injected through the uterine wall after abdominal surgery or were infused intracervically. The life cycle was completed in the uterine mucosa, and infectious oocysts were released into the vagina. Intestinal infection was established in suckling BALB/c mice by rectal inoculation with oocysts. The ID_{50} in mice inoculated intrarectally was approximately 1000.[689]

To test activity of agents directly against sporozoites, a mouse model was developed in which free sporozoites were inoculated. For this, oocysts were isolated from bovine feces using ether extraction, sieving, and hypochlorite digestion, and then were excysted *in vitro*.[689] Infective sporozoites were separated from intact and excysted oocyst debris by anion-exchange chromatography. Sporozoites were not infective when administered orally, but readily infected the ileum, cecum, and colon of suckling BALB/c mice when inoculated directly into the intestinal tract via rectum. The ID_{50} for isolated sporozoites inoculated by this route was approximately 8000.[689] Infection of the cecum of a mouse inoculated via rectum with 2×10^5 sporozoites is shown in Figure 1.

Distribution of organisms in the host — Infection of susceptible mice by oral administration of oocysts or intraintestinal injection of sporozoites results in colonization of the small intestine, cecum, and colon.[233,238,280,312,315,689,734,808] In contrast to infection by *C. muris*, the stomach is not parasitized by *C. parvum*. Although stages of *C. parvum* in the liver, lungs, and heart, as well as the intestinal tract, of experimentally infected mice were reported,[622] *C. parvum* in extra-intestinal sites in mice is uncommon, unless oocysts are injected by a nontraditional route.[466]

Role of host immune response — Infection of immunocompetent mice results in subclinical infection of short duration accompanied by minimal tissue injury. However, exposure of nude mice to *C. parvum* oocysts results in persistent infection, diarrhea, and significant alterations of intestinal morphology.[328] Because nude mice have significantly reduced T cell function, these results show that T lymphocytes are important for clearing cryptosporidial infection in mice. However, the mechanisms by which T lymphocytes contribute to clearance of infection are unresolved. The increased duration of infection in immunodeficient nude mice parallels the observations of persistent cryptosporidiosis in immunodeficient humans and horses.[236,750]

Mice naturally exposed to *C. parvum* oocysts develop antibodies to cryptosporidial antigens.[561] Following immunization, mice produce antibodies reactive with oocyst,[57,772] sporozoite,[538,690] or mixtures of oocyst and sporozoite antigens.[490] Many of the antigens recognized by mouse antibodies are complex glycoconjugates of lipids or proteins, with apparent molecular weights ranging from 70,000 to greater than 100,000.[57,490,690] A major

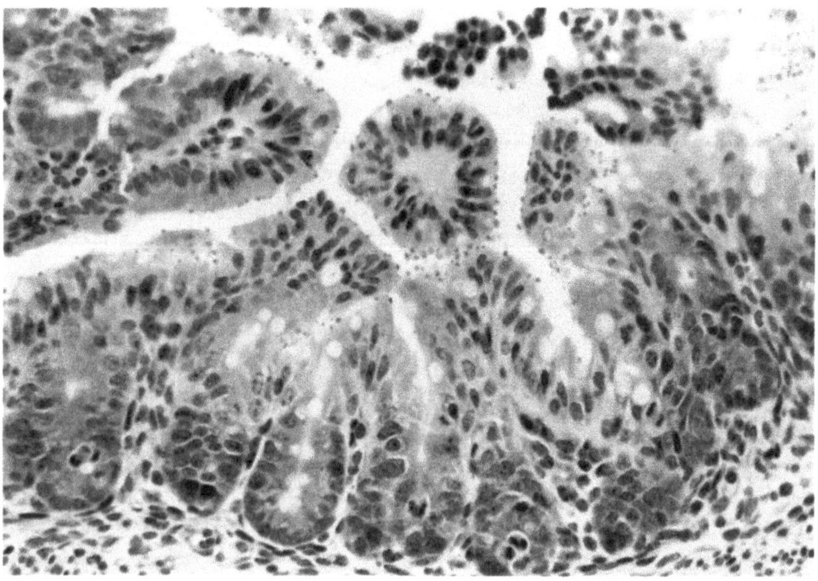

FIGURE 1. Photomicrograph of cecum from a neonatal BALB/c mouse inoculated via the rectum with 2×10^5 *C. parvum* sporozoites isolated by anion-exchange chromatography.

sporozoite surface antigen of 20,000 Da has been defined by mouse monoclonal antibodies and is recognized by serum from previously infected humans, calves, and horses.[58,538]

The role that antibodies play in prevention and recovery from cryptosporidial infection can be addressed now that it is possible to isolate sporozoites and titrate their infectivity in mice.[689] Immune reagents may be incubated *in vitro* with known numbers of sporozoites prior to injection of the sporozoites into the intestinal tracts of suckling mice. Protection in such an assay defines immune reagents with the ability to neutralize sporozoite infectivity. This approach has been used to demonstrate neutralizing activity of immune bovine colostrum and immune bovine serum.[238,689] The data support the hypothesis that antibody can neutralize sporozoite infectivity. Proof for the hypothesis was obtained when mouse monoclonal antibodies, reactive with sporozoite surface determinants, protected mice against sporozoite challenge.[690] One of the antibodies reacts with a lipid or glycolipid conjugate on the sporozoite surface. The other reacts with ^{35}S-methionine-labeled sporozoite antigens of 98,000, 55,000, and 28,000 Da. Though murine monoclonal antibodies can neutralize infectivity of sporozoites in *ex vivo* assays, it was not shown that immunized and/or naturally exposed dams protect their suckling pups against oral challenge with *C. parvum* oocysts.[58,561] However, immune bovine colostrum partially protects mice and calves from oocyst challenge.[238,239] Protection of neonates by lacteal immunity is probably influenced by the titer and affinity of neutralizing antibodies, as well as other substances present in colostrum and milk.

Effect of temperature, antimicrobial agents, and disinfectants on C. parvum infectivity in mice — The pathogenicity of *C. parvum* for lambs, calves, and humans (particularly immunodeficient patients) has prompted assessment of compounds and protocols for prevention and control of cryptosporidiosis. Many of these trials have been performed in mice. Temperature extremes are deleterious to viability and infectivity of *C. parvum* oocysts. Freezing of oocysts, even in the presence of cryoprotectants, destroys infectivity for mice.[734] Oocysts lose infectivity when warmed from 9 to 55°C for 15 to 20 min, or held at 45°C for 20 min.[21] Drying of *Cryptosporidium*-containing bovine fecal specimens for 1 to 4 d progressively inhibits infectivity for infant mice.[23]

None of 15 anticoccidial prophylactic regimens or 16 antimicrobial agents has proven useful in preventing or altering cryptosporidial infections in mice challenged orally with oocysts.[35,820] The application of effective chemotherapeutic agents for prevention and control of infection is apparent and will prompt the continued evaluation of compounds. Mice provide a useful *in vivo* model for assessment of compounds that may prevent infection. However, because cryptosporidiosis is subclinical and self-limiting in suckling mice, they are of limited value for evaluating therapeutic protocols designed to resolve established and persistent clinical infections.

Disinfectants have been tested for ability to inactivate oocyst infectivity. A pentaiodide resin compound contained within a column was able to bind and retain *C. parvum* oocysts passed through the column. The retention of oocysts by pentaiodide was thought to be more significant than the ability of the compound to inactivate *C. parvum*.[846] Of the more traditional disinfectants tested, 10% formol saline and 5% ammonia were shown to render oocysts noninfectious for mice.[32,133,621]

Value of mice for future experiments — Although exposure to *C. parvum* results in subclinical infection, mice are particularly valuable for additional studies. They are susceptible to infection by *C. parvum* obtained from other mammalian hosts, and the quantitative aspects of infectivity for both oocysts and sporozoites have been determined.[233,236,689,690] Efficacy of disinfectants, as well as the value of antimicrobial compounds for preventing infection, can be assessed initially in mice. They can also be used to test the neutralizing capacity of immune reagents including serum, colostrum, and monoclonal antibodies.[58,238,635a,689,690] As investigators develop protocols for the isolation and characterization of merozoites and other stages in the life cycle of *C. parvum*, their infectivity and susceptibility to neutralizing immune reagents can be evaluted in mice.

B. *CRYPTOSPORIDIUM MURIS*

C. muris develops in the gastric glands of mice.[806,807] Infected mice grew less rapidly. Developmental stages of the organism were limited to the stomach and were accompanied by dilatation of the gastric glands and a slight increase in the number of lymphoid cells. Its stages are larger than that of *C. parvum* at almost every developmental stage of the life cycle.[808] Oocysts are 7×5.5 μm compared to 5×4.5 μm for *C. parvum* oocysts.

C. muris lacks host specificity and can be experimentally transmitted to mice, rats, guinea pigs, rabbits, dogs, and cats. *C. muris*-like oocysts have been recovered from bovine feces, and stages have been observed in glands within the abomasum (stomach) of cattle.[25,844] The substantial difference in oocyst size, as well as strict location in stomach (*C. muris*) vs. intestinal tract (*C. parvum*), allow the two species of *Cryptosporidium* infectious for mammalian hosts to be distinguished.

BIBLIOGRAPHY

Information pertinent to the subject matter of this section may be found in References 21, 23, 25, 32, 35, 57, 58, 133, 134, 189, 190, 195, 236, 238, 280, 312, 315, 328, 372, 418, 466, 561, 597, 621, 635a, 674, 689, 690, 734, 806 to 808, 811, 820, 842a, 843, 844, and 846.

III. CRYPTOSPORIDIOSIS IN RATS

A. *CRYPTOSPORIDIUM PARVUM*

Natural infection of house rats (*Rattus norvegicus*) with *C. parvum* has been documented.[371] As in mice, the life cycle is completed in the small intestine without involvement of the stomach. Infection is subclinical in both naturally infected rats and SPF rats fed *C. parvum* oocysts isolated from naturally infected house rats.[371]

Investigators have defined prepatent and patent periods for oocyst shedding in rats following oral administration of oocysts contained in bovine and rat feces. Oocyst shedding commenced 5 d and persisted for 16 d after SPF rats were fed bovine ileal homogenates.[811] A prepatent period of 7 d, followed by 13 d of oocyst shedding was observed in SPF rats fed rat fecal concentrates containing *C. parvum* oocysts.[371]

Because cryptosporidiosis is a debilitating condition in immunocompromised patients,[186,236,750,824] investigators have sought a convenient laboratory animal model of an immunosuppressed host in which cryptosporidial infection could be studied. Two protocols for immunosuppression of rats used in such studies have been described. Female Sprague-Dawley rats were treated orally each day for at least 14 d with 50 mg/kg of cyclophosphamide.[676] When subsequently dosed with 10^4 or more infectious *C. parvum* oocysts of bovine origin, at least 80% of the rats became infected. Infection was established in the ileum, without involvement of stomach, proximal small intestine, cecum, or colon. Rats continued to shed oocysts as long as they received cyclophosphamide, but intestinal infection was cleared within 7 to 10 d following termination of cyclophosphamide treatment.[676]

A second rat immunosuppression model for studying cryptosporidiosis was reported.[123] Male Sprague-Dawley rats were treated twice weekly for 8 weeks by subcutaneous injection with 25 mg hydrocortisone acetate and maintained on a low-protein diet. Oral challenge with 10^5 *C. parvum* oocysts (bovine origin) 5 weeks following initiation of treatment resulted in persistent infection of more than 38 d of duration. Organisms were present in epithelial cells of the jejunum and ileum.

B. *CRYPTOSPORIDIUM MURIS*

Natural infection of house rats (*Rattus norvegicus*) with *C. muris* has been documented.[371] As in mice, the life cycle is completed in the stomach without involvement of the intestine. Infection is subclinical in both naturally infected rats and SPF rats fed *C. muris* oocysts isolated from naturally infected house rats.[371,372]

BIBLIOGRAPHY

Information pertinent to the subject matter of this section may be found in References 123, 186, 236, 371, 372, 649, 676, 678, 811, and 824.

IV. CRYPTOSPORIDIOSIS IN GUINEA PIGS

Spontaneous infections of guinea pigs (*Cavia porcellus*) with *Cryptosporidium* sp. have been observed and the ultrastructural features of infection reported.[36,275,384,498,854,855] Because of marked variations in clinical presentation and patency of infection, speciation of *Cryptosporidium* in guinea pigs is more controversial than in other mammals, prompting some authors to claim *C. wrairi* as a valid species.[236,275,457,458,811] Transmission experiments reveal that *C. muris* of rat origin infects guinea pigs,[372] as does bovine origin *C. parvum* in some but not all studies.[36,811] An outbreak of cryptosporidiosis in a colony of guinea pigs was characterized by clinical signs including diarrhea, wasting, abdominal distention, and 2 to 5% mortality.[159a] Oocysts recovered from this outbreak were infectious for 10 weanling (2 to 3 weeks old) and 10 adult (8 to 10 weeks old) guinea pigs. Animals challenged by reinoculation with oocysts were refractory to infection.[159a] Information pertinent to the subject matter of this section may be found in References 36, 159a, 236, 275, 372, 384, 457, 458, 498, 811, 854, and 855.

V. CRYPTOSPORIDIOSIS IN HAMSTERS

Cryptosporidiosis was noted as a natural infection in a weanling Golden hamster (*Mesocricetus auratus*) with concurrent lesions of proliferative ileitis.[207] The features of experimental cryptosporidiosis were described for neonatal and adult Syrian golden hamsters inoculated with *Cryptosporidium* sp. oocysts prepared from calf feces.[411,700c] Epithelial cells of the ileum became infected, and oocysts were shed in the feces for approximately 12 d, beginning 4 d post-inoculation. Infection was subclinical, as none of the hamsters developed diarrhea. The hamster model was used to evaluate the chemotherapeutic effect of arprinocid, which reduced the number of oocysts shed in feces of treated hamsters infected with *Cryptosporidium*.[410]

Information pertinent to the subject matter of this section may be found in References 207, 410, 411, and 700c.

VI. CRYPTOSPORIDIOSIS IN OTHER RODENTS

Cryptosporidium organisms were noted in the duodenum, jejunum, ileum, and cecum of a young gray squirrel.[781] The squirrel was emaciated and had multiple cutaneous neoplasms. The immune status of the squirrel was unknown.

Cryptosporidium organisms, 2 to 5 μm in diamter were observed in the stomach, small intestine, and colon of an 8-month-old chinchilla that died following severe diarrhea of 8 d of duration.[907] The small intestine contained irregular areas of epithelial cell necrosis, infiltration of inflammatory cells, and moderate villus atrophy.

Information pertinent to the subject matter of this section may be found in References 781 and 907.

Chapter 9

CRYPTOSPORIDIOSIS IN BIRDS

David S. Lindsay and Byron L. Blagburn

TABLE OF CONTENTS

I. INTRODUCTION

Tyzzer was the first to report cryptosporidia in an avian species.[809] In 1920, he reported that a species, structurally similar to *Cryptosporidium parvum* found in mice, was present in the ceca of chickens. He did not name the organism, describe the oocysts, or developmental stages of the parasite; 35 years later, in 1955, Slavin found a structurally similar parasite in the ileum of turkey poults.[740] He named the parasite *C. meleagridis.* and partially described the endogenous life cycle. In 1986, Current and co-workers gave the name *C. baileyi* to the species of *Cryptosporidium* isolated from broiler chickens and described the complete life cycle.[194] They considered *C. baileyi* a distinct species because it differed in oocyst structure, was only moderately infectious for turkeys, and differed in site of development from *C. meleagridis.* Support for *C. baileyi* as a seperate species from *C. meleagridis* has since been provided by many experimental studies on site specificity[469,470,472,475] and oocyst morphometrics (Table 1, Figure 1).[480,901]

Two other species of *Cryptosporidium* have been named from birds: *C. anserinum* and *C. tyzzeri.* Neither of these reports gave adequate descriptions of the oocysts or provided other useful information that would support their being named new species. Therefore, they must be considered *nomen nuda* and we do not consider them valid species.

The status of a *Cryptosporidium* species in bobwhite quail is currently being investigated by numerous researchers in the U.S.[305,350,692] The oocysts are structurally distinct from *C. baileyi* and resemble those of *C. meleagridis.* It seems to differ from *C. meleagridis* because it produces a generalized small intestinal infection and is associated with extreme morbidity and mortality. Further studies are needed before the status of this parasite is fully known.

Cryptosporidiosis in birds manifests itself as respiratory disease, enteritis, or renal disease. Usually only one condition is present in an outbreak, but combinations of the three forms have been observed.[582]

II. BIOLOGY

A. LIFE CYCLES

Little is known of the development of *C. meleagridis.* Only one asexual generation was described (Table 2). In the original report, endogenously sporulated oocysts were not observed in mucosal smears or in the feces and it was believed that the oocysts were passed unsporulated in the feces.[740] Because only naturally infected birds were examined, it is possible that additional unknown aspects of the developmental cycle of *C. meleagridis* are present. Recently, it has been shown that *C. meleagridis* oocysts are passed fully sporulated in the feces of naturally or experimentally infected turkeys (Figure 1A).[480,901] The life cycle of *C. meleagridis* should be reexamined in experimentally infected turkeys.

The endogenous life cycle has been described for *C. baileyi* (Table 2).[194] Three structural types of meronts (generations) and two structural types of oocysts (thin- and thick-walled) are present (Figures 2 and 3). Because some asexual stages could apparently recycle, they were called types and not generations. Both types of oocysts sporulated endogenously (Figure 4). Thin-walled oocysts often excysted *in situ* and reinitiated infections. Thick-walled oocysts were excreted in the feces.[194]

B. SITE SPECIFICITY

Cryptosporidia have been reported to develop in many anatomic sites in various naturally infected avian hosts (Table 3). The conjunctiva, nasopharynx, larynx, trachea, bronchi, air sacs, proventriculus, small intestine, large intestine, ceca, cloaca, bursa of Fabricius (BF), kidneys, and urinary tract have supported natural infections.

Extensive studies have been conducted with *C. baileyi* to determine what factors influence infectivity and sites of development in experimentally infected chickens.[469,470,472,475] Following

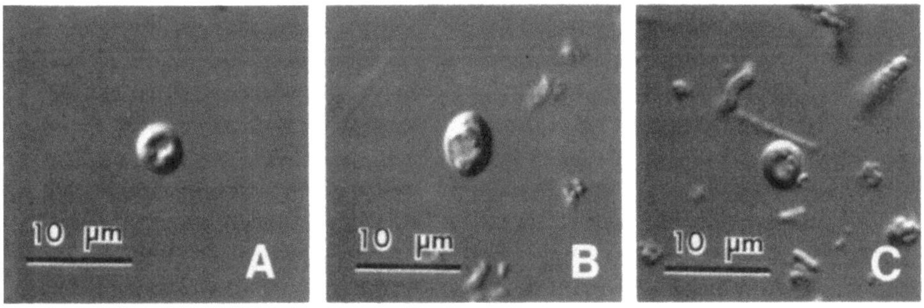

FIGURE 1. Oocysts of *Cryptosporidium* species infecting birds. (A) *C. meleagridis*; (B) *C. baileyi*; (C) *Cryptosporidium* species from bobwhite quail.

TABLE 1
Morphometric Comparison of the Oocysts of *Cryptosporidium meleagridis*, *C. baileyi*, and *Cryptosporidium* sp. from Bobwhite Quail

Species	*C. meleagridis*[480]	*C. baileyi*[480]	*Cryptosporidium*[a]
Oocyst length[b]	4.5—6.0	6.0—7.5	4.5—6.0
(5.2 + 0.08)[c]	(6.6 + 0.10)	(5.0 + 0.06)	
Oocyst width	4.2—5.3	4.8—5.7	3.6—5.6
(4.6 + 0.04)	(5.0 + 0.05)	(4.3 + 0.06)	
Length/width	1.00—1.33	1.05—1.79	1.00—1.63
(1.13 + 0.02)	(1.33 + 0.02)	(1.19 + 0.02)	
Sporozoite length	4.5—6.0	4..5—7.5	ND[d]
(5.2 + 0.10)	(6.1 + 0.10)		
Sporozoite width	1.2—1.8	1.2—1.8	ND
(1.4 + 0.03)	(1.5 + 0.03)		

[a] B. L. Blagburn, unpublished data.
[b] Measurements are in micometers.
[c] Mean ± standard error of the mean.
[d] ND = not determined.

oral inoculation of oocysts, the primary sites of development is the BF and cloaca.[469] Only in a few birds (<10%) can parasites be found in the respiratory tissues (primarily trachea). Because the oocysts are passed fully sporulated, some of these respiratory infections found after oral inoculation may be naturally acquired from fecal oocysts. Intratracheal (IT) inoculation of oocysts into chickens results in extensive parasitization of the respiratory tract.[104,469,471] Parasites are found in the nasopharynx, larynx, trachea, primary and secondary bronchi, and air sacs. Parasites are also found in the BF and cloaca following IT inoculation and represent infections that are produced due to oocysts being coughed up and swallowed following inoculation or following production in respiratory tissues. Conjunctival infections occur in some birds when oocysts are placed directly on the conjunctival sac.[472] However, more birds inoculated in the conjunctival sac have parasites in the BF and cloaca than on the conjunctiva. This is probably due to oocysts that pass out through the nasolacrimal duct and are swallowed, thus instituting oral infections. Most oocysts probably do not stay in contact with the conjunctival epithelium long enough for significant numbers to excyst and produce infections. When chickens are inoculated intracloacally with oocysts, parasites are found in the BF and cloaca.[470] Only a few birds that had been infected long enough to have naturally acquired respiratory infections had parasites in the trachea. Intravenous inoculation of oocysts does not produce infection.[475]

TABLE 2
Comparison of the Developmental Stages of *C. meleagridis*, and *C. baileyi*

Developmental stage	*C. meleagridis*[740]	*C. baileyi*[194]
1st generation meront	5.0 × 4.0[a]	5.0 × 4.9
1st generation merozoites	5.0 × 1.1	6.9 × 1.1
Number of merozoites	8	8
2nd generation meront	Not seen	5.1 × 5.1
2nd generation merozoites	Not seen	5.0 × 1.3
Number of merozoites	Not seen	4
3rd generation meront	Not seen	5.2 × 5.1
3rd generation merozoites	Not seen	3.6 × 1.1
Number of merozoites	Not seen	8
Microgamont	4.0 × 4.0	4.0 × 4.0
Microgametes	1.0 × 0.3	1.7 × 0.5
Number of microgametes	16	16
Macrogamont	5.0 × 4.0	5.8 × 5.8
Unsporulated oocyst	4.5 × 4.0	6.3 × 5.2
Sporulated oocyst	Not seen	6.2 × 4.6

[a] Measurements are in micrometers.

Intraabdominal inoculation of oocysts produces infections identical to IT induced infections.[475] Inoculated oocysts excyst in the air sacs, then the infection progresses upward. Direct inoculation of *C. baileyi* oocysts into the gall bladder of chicks produces infection of the gall bladder and bile ducts in some birds.[319a] Lesions in these birds consist of hyperplasia of the gall bladder and bile duct epithelium and infiltration of the connective tissue with lymphocytes and mononuclear leukocytes. The forgoing experiments indicate that *C. baileyi* infections will develop in many anatomic sites, but only if infectious oocysts/developmental stages gain access to the site. These studies have ruled out some type of transport mechanism which would spread the infection from intestinal tissues to respiratory or renal tissues.

Little is known about the factors that influence the establishment or site of development of *C. meleagridis* infections. Oral inoculation of poults with *C. meleagridis* oocysts produces infections in the ileum, ceca, colon, cloaca, and BF.[94] Other routes of inoculation have not been evaluated.

C. HOST SPECIFICITY

Reports of naturally occurring cryptosporidiosis generally do not provide enough information to determine which species of *Cryptosporidium* is involved. Such information is needed to understand the epidemiology of the disease.

Experimental studies indicate that most attempts to transmit *C. baileyi* isolated from chickens to other birds are successful, with the exception of bobwhite quail. *C. meleagridis* isolated from turkeys is infectious for chickens[480] and domestic ducks.[480a] *Cryptosporidium* sp. isolated from bobwhite quail is not infectious for chickens. Results of bird-to-bird transmission studies with cryptosporidia are summarized in Table 4.

Because cryptosporidia can cause serious disease in humans the results of transmission studies with oocysts isolated from birds to mammals is important to determine if a zoonotic potential is present. Results of these studies are summarized in Table 5. They indicate that no zoonotic potential is present.

D. PREVALENCE

Little is known about the prevalence of cryptosporidial infections in comercially raised

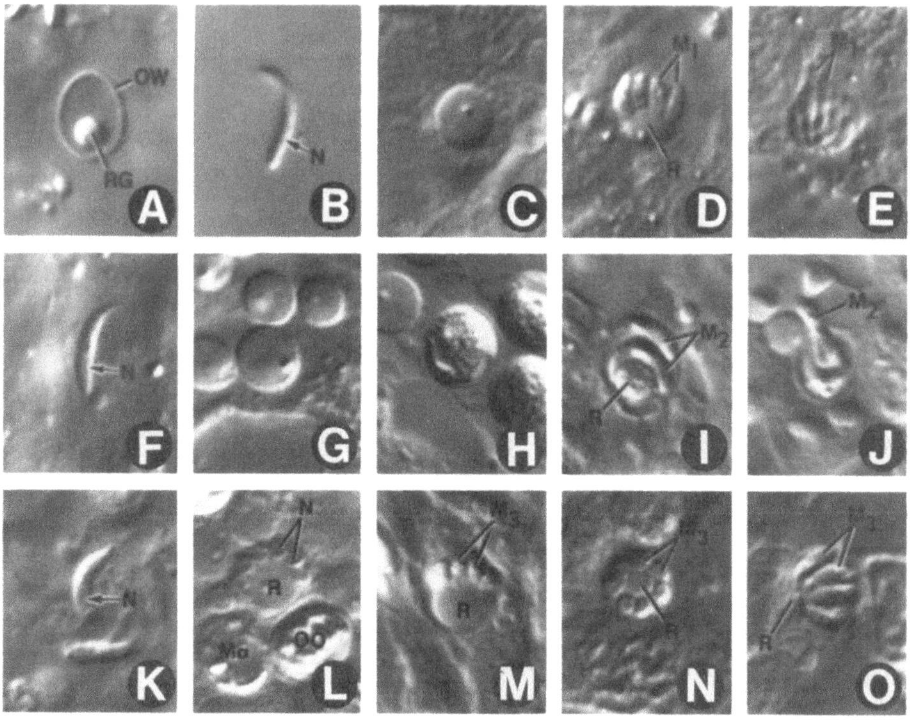

FIGURE 2. Endogenous asexual stages of *Cryptosporidium baileyi* in chickens. (A) Empty oocyst after release of sporozoites. The refractile globule (RG) is all that remains within the oocyst wall (OW). (B) Free sporozoite with nucleus (N) located in posterior third; (C) uninucleate Type 1 meront; (D) immature Type 1 meront with merozoites (M1) budding from a large residuum (R); (E) mature Type 1 meront with 8 merozoites (M1); (F) Type 1 merozoite showing posterior position of the nucleus (N); (G) uninucleate Type 2 meronts; (H) immature Type 2 meront. Note granular cytoplasm. (I) Mature Type 2 meront. Note that two of the four merozoites (M2) are in focus and the residuum (R). (J) Type 2 merozoite (M2) leaving the infected cell; (K) Type 2 merozoite with nucleus (N) located near the posterior end; (L) immature Type 3 meront with nuclei (N) arranged around what will eventually be the residuum (R). Note the macrogamont (MG) and thick-walled oocyst (OO). (M) Type 3 merozoites (M3) in the process of budding from the residuum (R); (N) Type 3 meront showing seven of the eight merozoites (M3) and the residuum (R); (O) nearly mature Type 3 meront showing four of the eight merozoites (M3) still attached to the residuum (R). (All NIC microscopy; magnification × 2100.) (From Current, W. L., Upton, S. J., and Haynes, T. B., J. Protozool., 33, 289, 1986. With permission.)

chickens in the U.S. One study examined fecal samples collected from chickens at necropsy in a diagnostic laboratory in North Carolina and demonstrated infection in 27.3% (9/33) of broilers, 10% (3/30) of broiler breeders, and 5.9% (1/75) of layers.[465] In another study, 6.4% of 1065 histological cases examined from chickens at a diagnostic laboratory in Georgia were positive for cryptosporidia.[287] Prevalence data indicated a higher incidence of cryptosporidiosis in chickens in Georgia in recent years.[290f] Serologic studies indicate that 22 to 50% of 18 flocks from the Delmarva Peninsula (encompassing portions of Delaware, Maryland, and Virginia) examined 49 to 63 d posthatch were positive for *Cryptosporidium*-specific antibodies.[748]

Information on the prevalence of infection in chickens in other countries is also scarce. Cryptosporidia were seen in 26 of 139 (18.7%) histological sections of BF from broilers in Scotland.[667] Tissues from 17 of 70 (24%) broilers from Greece were positive.[608]

Cryptosporidium meleagridis oocysts were found in the feces of 0 of 20 (0%) 5-d-old, 16 of 20 (80%) 17-d-old, 38 of 100 (38%) 24-d-old, 0 of 10 (0%) 60-d-old, and 0 of 10 (0%) 9-week-old turkeys in a survey of a poultry farm in Iowa.[901]

Nothing is known concerning the prevalence of cryptosporidia in wild bird populations.

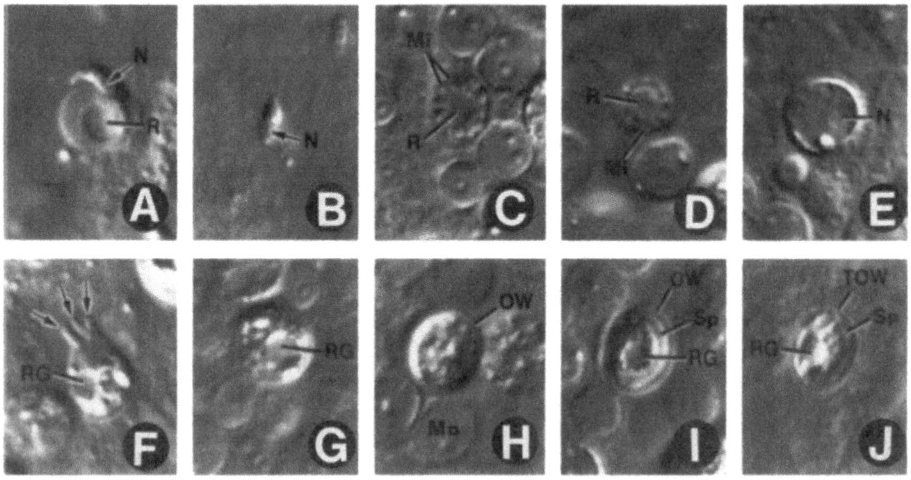

FIGURE 3. Asexual and sexual stages and oocysts of *Cryptosporidium baileyi* in chickens. (A) Type 3 merozoite leaving a host cell. Note the posteriorly located merozoite nucleus (M3) and the residuum (R). (B) Type 3 merozoite showing posteriorly located nucleus (N); (C) microgamont with microgametes (Mi) budding from a residuum (R); (D) mature microgamont with microgametes (Mi) surrounding the residuum (R); (E) immature macrogamont with nucleus (N) near the periphery; (F) macrogamont with three attached microgametes (arrows). Note the prominent refractile globule (RG). (G) Macrogamont with granular cytoplasm and large refractile globule (RG); (H) unsporulated oocyst with a thick oocyst wall (OW). Note the adjacent immature macrogamont (Ma). (I) Thick-walled, sporulated oocyst showing the refractile globule (RG) and one of four sporozoites (Sp) surrounded by a thick oocyst wall (OW); (J) Thin-walled autoinfective oocyst showing the refractile globule (RG) and one of four sporozoites (Sp) surrounded by a thin oocyst wall (TOW). (All Nomarski interference-contrast [NIC] microscopy; magnification ×2100.) (From Current, W. L., Upton, S. J., and Haynes, T. B., J. Protozool., 33, 289, 1986. With permission.)

E. IMMUNITY

It is important to remember that the life cycle of *C. baileyi* includes asexual stages that recycle, autoinfective oocysts, and fully sporulated fecal oocysts. The presence of these developmental stages and the fact that oocysts are passed fully sporulated in the feces make studies of immunity to *C. baileyi* challenging.

Age-related resistance to clinical disease and ability to clear *C. baileyi* infections has been examined in broiler chickens.[478] Chickens inoculated orally or IT at 2, 14, 28, or 42 d of age became infected and passed oocysts in their feces. Clinical respiratory disease occurred only in the 2- and 14-d- old birds. Chickens inoculated at 2 or 14 d of age excreted oocysts sooner, excreted oocysts for a longer time, and had greater numbers of developmental stages in their tissues than did chickens inoculated at 28 or 42 d of age. Similar results were obtained in 26-d-old birds.[290g] The older birds probably have better functioning immune systems and are therefore better able to mount an effective immune response and clear the parasites.

Chickens that have been orally inoculated with *C. baileyi* oocysts and cleared the infection are completely resistant to secondary challenge oral inoculations.[196] These birds pass no oocysts and have no parasites in their tissues. Serum antibodies to oocyst/sporozoite antigens are found as early as 7 days postinoculation (DPI) using an indirect immunoflourescent antibody assay or an ELISA. These serum antibodies recognize *C. baileyi* sporozoite/oocyst antigens of 198, 190, 128, 97, 78, 72, 54, 47, 42, 34, and 25 kDa.[196] The significance of the serum antibody response to *C. baileyi* in clearing or limiting the infection has not been established. These specific antibodies are important in serological surveys and flock monitoring programs to help determine the prevalence of chicken cryptosporidiosis.[748]

The effect of an existing 7-d-old *C. baileyi* infection on the ability of chickens to mount an

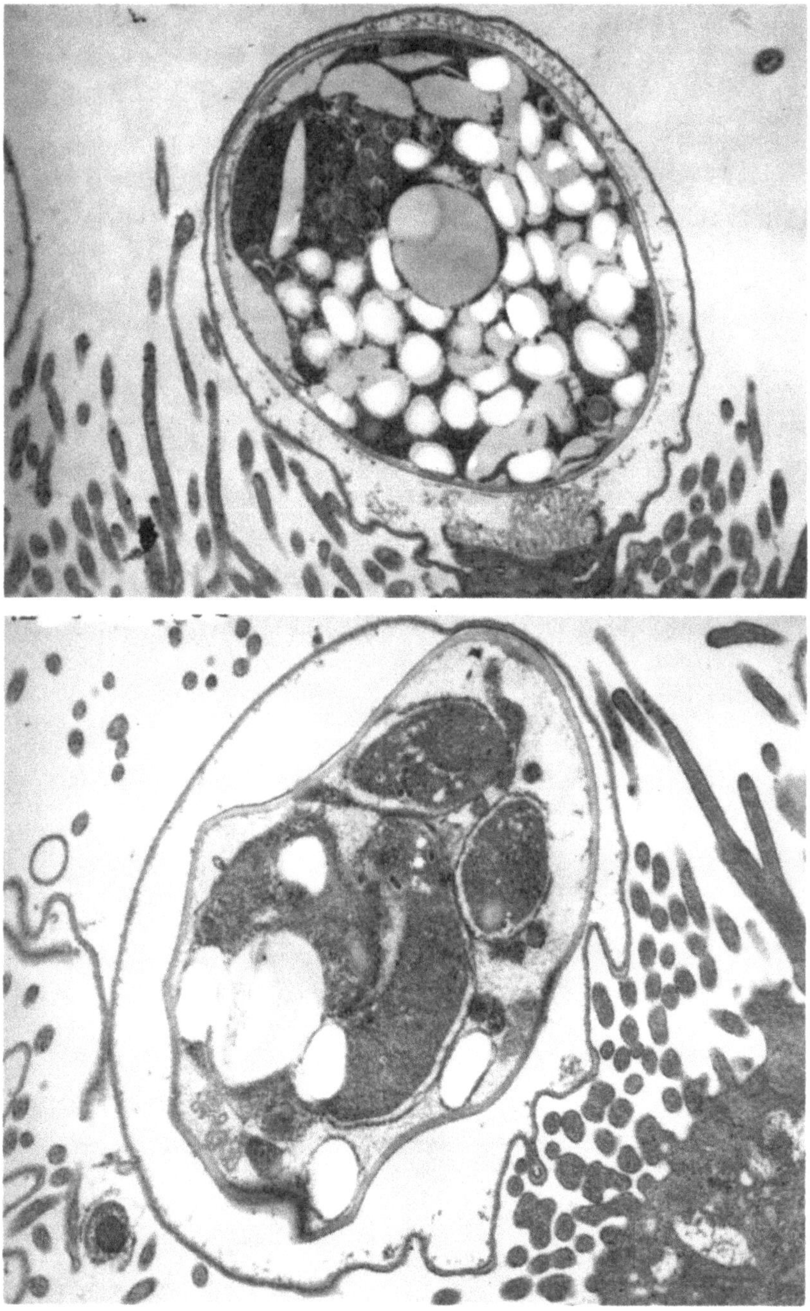

FIGURE 4. Transmission electron micrographs of the thick-walled oocysts of *Cryptosporidium baileyi* in chickens. (Upper) Unsporulated oocyst. (Magnification × 20,000.) (Lower) Fully sporulated oocyst. (Magnification × 21,000.) ([Lower] From Lindsay, D. S., Blagburn, B. L., Sundermann, C. A., Hoerr, J. F., and Ernest, J. A., *Am. J. Vet. Res.*, 47, 876, 1986. With permission.)

antibody response to vaccination with Newcastle disease virus (NDV) and infectious bursal disease virus (IBDV) and their ability to mount a delayed type hypersensitivity reaction (DH) to avian oil tuberculin has been examined.[104] No significant differences were seen in IBDV

TABLE 3
Anatomical Distribution of Cryptosporidia in Naturally Infected Birds

Host	Location[a]	Disease	Ref.
Chicken	Ceca	?	809
Chicken	Ceca, BF	No	251
Chicken	T	Yes	217
Chicken	BF, T		667
Chicken	BF, cloaca	?	803
Chicken	BF, NF, larynx, T, ceca	Yes	374
Chicken	BF	?	467
Chicken	SI	?	273
Chicken	BF, T	?	592
Chicken	BF	Yes	291
Chicken	T, bronchi, air sacs, K, ureters, BF	Yes	582
Chicken	SI	?	289
Chicken	Conjunctiva, NF, T, lungs, SI, BF	Yes	286
Chicken	T	Yes	441
Chicken	T, BF, SI	Yes	608
Chicken	BF	Yes	290
Turkey	Ileum	Yes	740
Turkey	T, bronchi, NF, ceca	Yes	349
Turkey	NF, T, BF	Yes	279
Turkey	T	Yes	789
Turkey	T, NF	Yes	666
Turkey	Jejunum, ileum	Yes	288
Turkey	Ileum	Yes	862
C. quail[b]	NF, T, bronchi, BF	Yes	795
C. quail	NF, T, BF	Yes	596
B. quail[c]	SI	Yes	350
B. quail	SI	Yes	692
Goose	Large intestine	No	661
Duck	BF	?	803
Duck	Conjunctiva	No	521
Duck	NF, T, BF	Yes	596
Pheasant	Larynx, T	Yes	882
Pheasant	Conjunctiva	?	669
Pheasant	NF, T, BF	Yes	596
Peafowl	Conjunctiva, NF, T	Yes	519
Parrot	Colon, cloaca	No	222
Lovebird	Jejunum, ileum	Yes	84
Finch	Kidney	?	266
Canary	Proventriculus	?	803
Budgerigar	T	Yes	596
Junglefowl	K, NF		668

[a] BF = bursa of Fabricius, T = trachea, NF = nasopharynges, K = kidney, SI = small intestine, ? = undetermined.
[b] Common quail.
[c] Bobwhite quail.

antibody titers compared to controls. However, significantly increased NDV antibody titers and decreased DH reaction were seen in chickens that were *C. baileyi* infected. Additional work needs to be done to determine the influence of *C. baileyi* on the ability of chickens to respond to other pathogens.

Nothing is known about resistance or immunity to *C. meleagridis* infection.

TABLE 4
Transmission Experiments with Avian Cryptosporidia to Other Birds

Host	*Cryptosporidium* sp.[a]	Recipient	Result[b]	Ref.
Chicken	CB	Turkey	POS	194, 468
Chicken	CB	Goose	POS	194
Chicken	CB	Domestic duck	POS	194, 479
Chicken	CB	Muscovy duck	POS	468
Chicken	CB	Bobwhite quail	NEG	194, 468
Chicken	CB	Pheasant	POS	480a
Chicken	CB	Guinea fowl	POS	480a
Chicken	CB	Chuckar partridge	POS	480a
Turkey	CM	Chicken	POS	480, 901
Turkey	CM	Domestic duck	POS	480a
Bobwhite quail	UN	Chicken	NEG	350, 480a
Common quail	UN	Chicken	POS	597
Pheasant	UN	Chicken	POS	597

[a] CB = *Cryptosporidium baileyi*, CM = *Cryptosporidium meleagridis*, UN = unknown species of *Cryptosporidium*.
[b] POS = positive, NEG = negative.

F. CHEMOPROPHYLAXIS AND CONTROL

Most reports of naturally occurring cryptosporidiosis in birds fail to mention if anticoccidial drugs were present in the feed during the outbreak or if treatment was attempted. Oxytetracycline was not effective in treating peafowl chicks which had respiratory cryptosporidiosis.[519] Amprolium and chlortetracycline had no effect on respiratory (sinusitis) cryptosporidiosis in turkeys.[279] Likewise, oxytetracycline, neomycin, and furazolidone had no effect on intestinal cryptosporidiosis in bobwhite quail.[350]

Experimental studies indicate that most commonly used anticoccidials do not prevent or reduce respiratory disease in chickens inoculated with *C. baileyi* oocysts. Halofuginone, salinomycin, lasalocid, and monensin have been evaluated and do not protect chickens.[471] Clinical respiratory disease, gross lesions, and tissue parasite densities are similar in drug-treated and in untreated inoculated controls.

Because treatment is of little value, control methods may be useful in limiting or preventing avian cryptosporidiosis. Several commonly used disinfectants were evaluated for the ability to kill *C. baileyi* oocysts in an excystation assay.[783] None of the disinfectants were effective at concentrations recomended by the producers. Commercially available ammonia compounds when used at 50% (v/v) were effective, with <5% of oocysts remaining viable. A similar concentration of commercial bleach (5.25 % sodium hypochlorite) was somewhat effective, with <15% of oocysts remaining viable.

One study reported that outbreaks of diarrhea and mortality in young bobwhite quail stopped after a control program was instituted.[350] The program consisted of bleach cleaning metal brooders, feeders, and waterers and then exposing these items to direct sunlight for 3 d. Concrete floor pens were cleaned and new shavings spread.

Adaptations of these types of control programs using ammonia and bleach compounds with improved sanitation may be useful on small farms or in pet bird production facilities.[783] Because of costs and size problems, these types of control programs may not be practical on large farms.

G. DEVELOPMENT *IN VITRO*

Excystation of sporozoites from *C. baileyi* oocysts has been described (Figure 5).[782] Prior to excystation, the sporozoites begin to move within the oocyst and rearrange themselves. An opening forms in the oocyst wall, and the sporozoites are released from the oocyst. The residuum may remain inside or be expelled with the sporozoites. The optimum temperature for excystation

TABLE 5
Transmission Experiments with Avian Cryptosporidia to Mammals

Host	*Cryptosporidium* sp.[a]	Recipient[b]	Results	Ref.
Chicken	CB	Mouse	NEG	194, 468
Chicken	CB	White rat	NEG	468
Chicken	CB	Cotton rat	NEG	468
Chicken	CB	Gerbil	NEG	468
Chicken	CB	Guinea pig	NEG	468
Chicken	CB	Hamster	NEG	468
Chicken	CB	Pig	NEG	468
Chicken	CB	Goat	NEG	194
Chicken	CB	Calf	NEG	105a
Common quail	UN	Mouse	NEG	597
Pheasant	UN	Mouse	NEG	597

[a] CB = *Cryptosporidium baileyi*, UN = unknown species of *Cryptosporidium*.
[b] NEG = negative.

was 40°C. Excystation did not occur in Hank's balanced salt solution (HBSS) or 0.25% trypsin in HBSS. Excystation solutions containing 0.75% sodium taurocholate (ST) in HBSS or 0.75% ST plus 0.25% trypsin in HBSS caused release of sporozoites. Bile collected from the gallbladders of chickens, turkeys, pigs, or goats could be substituted for ST and cause release of sporozoites.[782] This indicated that lack of mammalian infectivity of *C. baileyi* was not due to inability of host bile to initiate excystation. *C. meleagridis* oocysts release sporozoites if treated similarly as *C. baileyi* oocysts.[480]

C. parvum from mammals has been grown in cell cultures.[189,576] However, attempts to obtain development of *C. baileyi* in cell cultures have been unrewarding.[477] When sporozoites were inoculated onto six types of primary avian cell cultures, six types of primary mammalian cell cultures or four types of mammalian cell lines, no development was observed. Additional studies using different cell types and culture conditions have proven unsuccessful.[783a] Coccidial parasites differ in their ability to develop in cell cultures, and it appears that *C. baileyi* is more fastidious than *C. parvum*. No studies have been conducted to determine if *C. meleagridis* will develop in cell culture.

C. baileyi will complete development in the chorioallantoic membrane (CAM) of chicken embryos when sporozoites or oocysts are inoculated into the allantoic cavity.[193,477] No deaths or gross lesions due to infection were seen. Serial passage of *C. baileyi* has been performed for up to 20 times in chicken embryos.[477] Oocysts collected from allantoic fluid are infectious for chickens and turkeys. Embryo-passaged oocysts retain their ability to produce disease in chickens and turkeys following IT inoculation.[477] No embryo host specificity exists for *C. baileyi*, and it has been shown to complete development in the CAM of turkey, domestic duck, muscovy duck, chukar partridge, guinea fowl, ring-necked pheasant, bobwhite quail, and common (Japanese) quail embryos. Infections are nonpathogenic in these embryos.[477]

III. LESIONS

A. NATURALLY OCCURRING DISEASE

Naturally occurring cryptosporidiosis in birds manifests itself in three clinical forms: respiratory disease, enteritis, and renal disease. Usually only one form of the disease is present in an outbreak. Secondary bacteria and viruses may also be present in natural cases and add to the severity of disease.

Respiratory disease has been reported in chickens,[217,242a,286,374,441,582] turkeys,[279,349,666,789] common

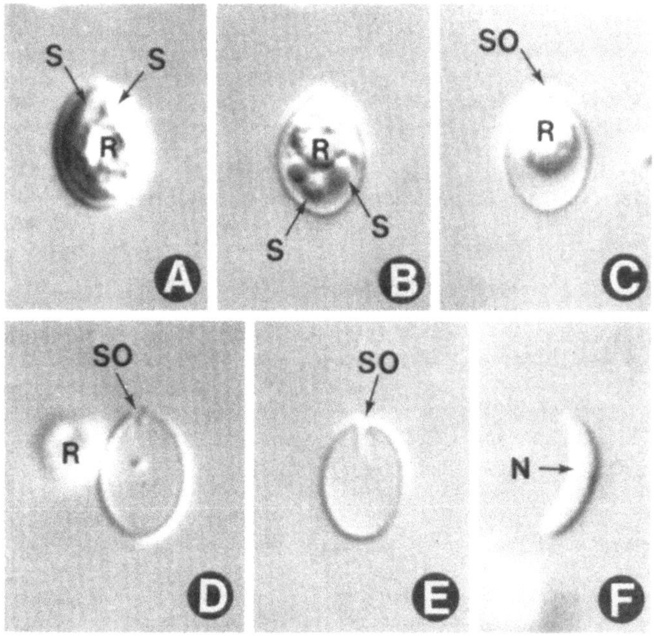

FIGURE 5. NIC photomicrographs of various stages of excystation of *Cryptosporidium baileyi*. (A) Sporulated oocyst with sporozoites (S) and oocyst residuum (R); (B) Same oocyst pictured in A, but photographed several minutes later. Note that the four sporozoites have become repositioned. (C) Oocyst after excystation; the sutural opening (SO) in the oocyst is apparent, and the residuum (R) is inside. (D) Oocyst in which the residuum (R) has just been expelled through the sutural opening (SO); (E) oocyst devoid of sporozoites and residuum. Note sutural opening (SO). (F) Excysted sporozoite; note the pointed anterior end, nucleus (N), and absence of refractile bodies. (All magnifications × 5800.) (From Sundermann, C. A., Lindsay, D. S., and Blagburn, B. L., *J. Protozool.*, 34, 28, 1987. With permission.)

quail,[597,795] mallard ducks,[597] ring-necked pheasants,[597,882] and budgerigars.[597] Clinical signs are generally similar in all avian species and consist of rales, coughing, sneezing, and dyspnea. Grossly, there may be excess mucous in the trachea and nasal cavities. Airsacculitis may be present. Microscopic lesions generally consist of hypertrophy and hyperplasia of infected epithelial surfaces. Infiltrates of macrophages, heterophils, lymphocytes, and plasma cells usually are present. Cilia are generally reduced or absent on ciliated epithelial surfaces.

Enteritis associated with cryptosporidia has been reported in turkeys,[288,740,862] bobwhite quail,[305,350] budgerigars,[290c] cockatiels,[290c,480a] and lovebirds.[84] High morbidity and mortality are associated with disease in bobwhite quail and lovebirds, whereas only mild disease occurs in turkeys. Gross lesions are confined to the intestinal tract and consist of small intestinal distention due to mucoid intestinal contents and gas. Similar gross lesions may be seen in the ceca. Microscopic lesions consist of villous atrophy, villous fusion, and crypt hyperplasia. Infiltrates of lymphocytes, heterophils, macrophages, and plasma cells may be present. Cryptosporidia generally are on the distal two thirds of the villi and are usually not seen in the ceca.

Renal cryptosporidiosis has been reported in black-throated finches,[266] jungle fowl,[668] and chickens.[582] Grossly, the kidneys are pale and enlarged. Occasionally urate crystals can be seen in surface tubules. The epithelial cells of the collecting ducts, collecting tubules, distal convoluted tubules, and ureters are hypertrophic and hyperplastic. Interstital tissues may be infiltrated by lymphocytes, macrophages, heterophils, and plasma cells. Fibrotic areas may be present.

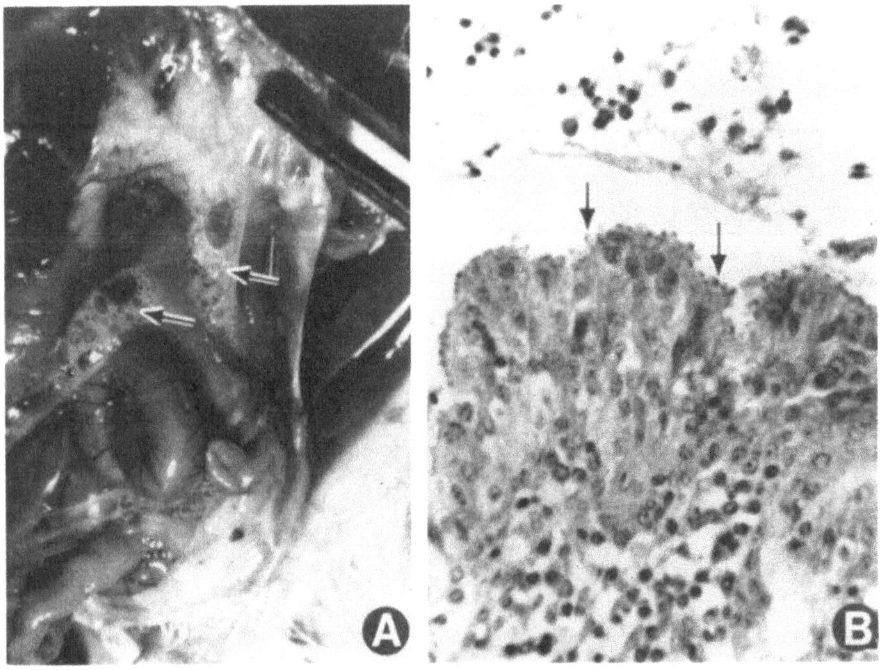

FIGURE 6. Gross and microscopic lesions of experimentally produced respiratory cryptosporidiosis in broiler chickens intratracheally inoculated with *Cryptosporidium baileyi* oocysts. (A) Abdominal airsacculitis 14 DPI. The air sacs are covered with a foamy, gray, fluid exudate (arrows). (B) Secondary bronchus demonstrating exudate in the lumen, developmental stages of *C. baileyi* on the epithelial surface (arrows), and mucosal thickening due to epithelial hyperplasia and infiltration of inflamatory cells. (H. & E.; magnification × 290.) (From Blagburn, B. L., Lindsay, D. S., Giambrone, J. J., Sundermann, C. A., and Hoerr, F. J., *Poult. Sci.*, 66, 442, 1987. With permission.)

B. EXPERIMENTALLY INDUCED DISEASE

Most studies on experimentally induced crytosporidiosis in birds have been done with *C. baileyi* isolated from chickens. Oral inoculation of chickens[194,306,463,468,469,478] or turkeys[473] with C. baileyi oocysts generally does not produce clinical signs. Some birds may appear depressed and huddle, but otherwise are normal. Weight gains of infected birds may be reduced 1 to 2 weeks postinoculation.[104,306,463] Parasite-induced lesions are generally confined to the BF and cloaca. These lesions consist of epithelial hyperplasia and hypertrophy with underlying inflammatory response. Bursal folical atrophy may also be observed. Coinfection of chickens with reovirus has little effect on oral *C. baileyi* infections, but shedding of both virus and oocysts appear to be enhanced by some unknown mechanism.[306] Coinfection of chickens with infectious bursal disease virus and *C. baileyi* produces more severe bursal lesions, more infected birds, and a greater parasite density in infected birds.[463]

Several studies have demonstrated that *C. baileyi* can cause respiratory disease in chickens[469,471,477,478,606a] (Figures 6 and 7) and turkeys[473] following IT inoculation of oocysts. Clinical signs are first seen 5 to 9 DPI and consist of moist rales, reluctance to move, dyspnea, sneezing, and nasal discharge. Gross lesions consist of excess mucus in the trachea, tracheal mucosal thickening, and airsacculitis. Air sacs are thickened, white, and contain foamy exudate. The exudate is composed of developmental stages, oocysts, sloughed cells, and inflammatory cells. The lungs are wet, firm, and discolored gray. Lesions in the respiratory tissues are characterized by epithelial hyperplasia, mucocellular exudate, deciliation, and mucosal thick-

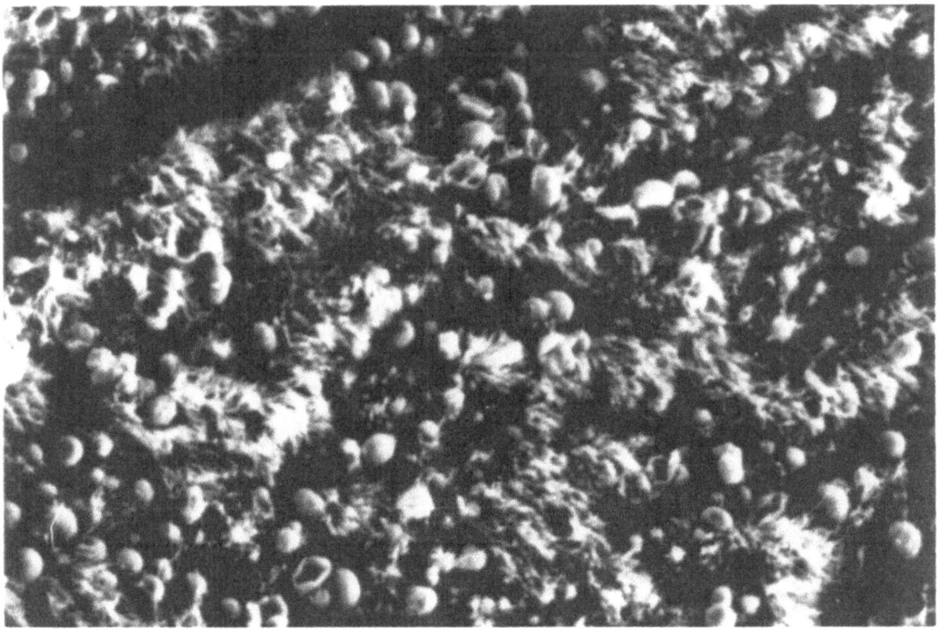

FIGURE 7. Scanning electron micrograph of the trachea from a chicken that was inoculated intratracheally with *Cryptosporidium baileyi* oocysts. Note the cilia and developmental stages. (Magnification × 1000.) (Courtesy of Dr. C. A. Sundermann, Department of Zoology and Wildlife Sciences, Auburn University, Alabama.)

ening due to lymphocytic infiltrates. Bronchopeumonia is a sequela to the damaged respiratory tissues and the subsequent loss of mucocilliary function. Tertiary bronchi, atria, and air capillaries become filled with parasites, heterophils, mucus, and fibrin. Air capillary interstices become expanded by macrophages, mononuclear cells, and heterophils. Respiratory portions of the air sacs demonstrate lesions similar to those seen in other respiratory tissues.

Attempts to produce attenuated strains of *C. baileyi* by serial passage in chicken embryos have not succeeded.[477] Oocysts that have been passed 10 or 20 times still produce disease in chicks following IT inoculation. Clinical signs and gross lesions are identical to those seen in inoculated controls.

Several studies have demonstrated that turkeys and chickens are susceptible to infection following oral inoculation with *C. meleagridis* oocysts.[94,480,901] Only one of these studies was designed to examine the pathogenicity of the experimentally produced infection.[94] Inoculation of 5-d-old poults with 5×10^5 oocysts did not produce clinical signs of disease or mortality. No effect was seen on body weight. Parasites were seen in the ileum, ceca, and BF. No microscopic lesions were present in these tissues.

Severe disease results from inoculation of 5-d-old bobwhite quail with 7.6×10^4 oocysts of *Cryptosporidium* sp. isolated from bobwhite quail.[305] Clinical signs are first seen 3 DPI and are associated with diarrhea, increased mortality, and growth depression. Grossly, the small intestines and ceca were thin-walled, distended with gas, and contained gray, watery fluid. Crpyptosporidia were found throughout the small intestine and rarely in the ceca, but not other intestinal tissues, BF, or respiratory tissues. Microscopic lesions were villous atrophy, increased villus width, crypt hyperplasia, and infiltration of the lamina propria with large mononuclear cells and occasional heterophils. Coinfection of bobwhite quail with a reovirus produced a synergistic effect which consisted of greater oocyst shedding, greater parasite density in the intestine, and systemic spread of the reovirus infection.

IV. CRYPTOSPORIDIOSIS IN CHICKENS

Cryptosporidiosis in chickens most often manifests itself as respiratory disease.[217,242a,286,290a,290b,374,441,582] Only occasionally is renal[582] or small intestinal involvement present.[273,289] Respiratory distress is the major presenting clinical sign. Gross lesions consist of excess mucus in trachea, petechiae of the nasal sinuses, and seldomly airsacculitis. Microscopic lesions vary from mild epithelial hyperplasia in mild cases to more extensive lesions including marked epithelial hyperplasia, deciliation, cellular exudation, and mucosal thickening due to underlying inflammatory responses. Bronchiopneumonia may be present. Cellular infiltrates are composed of macrophages, lymphocytes, plasma cells, and heterophils. Cryptosporidia stages have been reported to be phagocytosed by macrophages. Data on the prevalence of respiratory cryptosporidiosis in chickens is lacking. Losses on individual flocks may be heavy or only incidental.

Renal cryptosporidiosis is infrequently reported in chickens.[582] Clinical signs are not known because they were masked by other disease signs in the birds reported. Grossly, white foci were present in the parenchyma of infected kidneys. Microscopically, the epithelial cells of the collecting ducts, collecting tubules, and rarely the distal convoluted tubules were hyperplastic and hypertrophic and contained cryptosporidia. Infiltrates of lymphocytes and macrophages were present in the interstitium around the collecting ducts in the renal medulla. The epithelial lining of infected ureters was hyperplastic and infiltrated with lymphocytes and macrophages.

Small intestinal cryptosporidiosis has been reported in a 12-week-old pet Bantam-type chicken.[289] The bird had green watery diarrhea, was weak, and failed to gain weight. At necropsy, the small and large intestines contained thick green contents. Microscopically, the villi in all small intestinal segments were atrophic and misshapen and the crypts were hypertrophic. The lamina propria was infiltrated by heterophils, lymphocytes, plasma cells, and macrophages. Other disease-causing agents were present; therefore, the role of cryptosporidia as a cause of enteritis in chickens is still unclear. Another case of small intestinal cryptosporidiosis was seen in a 6-month-old rooster with severe chronic diarrhea.[273] At necropsy, there was marked thickening of the intestinal wall and intestinal contents were mucoid. Lesions consisting of epithelial and gladular hyperplasia were present in the intestine. The lamina propria was infiltrated with plasma cells, lymphocytes, and macrophages. No other infectious agents were isolated from this bird.

V. CRYPTOSPORIDIOSIS IN TURKEYS

Respiratory cryptosporidiosis is more commonly reported in turkeys than intestinal disease. Renal involvement has not been reported.

Clinical signs seen in turkeys with respiratory cryptosporidiosis are conjunctival exudates, rales, coughing, sneezing, and dyspnea.[279,349,666,789] Gross lesions generally include excess mucus in the nasal cavities and trachea. Airsacculitis has been seen in cases with secondary *Escherichia coli* involvement. Mortalities range from 1 to 20%. Microscopic lesions consist of hyperplasia, hypertrophy, and deciliation of respiratory epithelial surfaces. Submucosal tissues are infiltrated with lymphocytes, macrophages, plasma cells, and heterophils. These infiltrates may lead to considerable thickening of the submucosal tissues. Bronchopneumonia may be present in severe cases.

Diarrhea is the major clinical sign seen in cryptosporidial enteritis in turkeys.[288,740,862] It is of mild to moderate severity. Blood is not present in the feces. Birds may be depressed and huddle in groups. Gross lesions are seen in the small intestine and occasionally the ceca. These tissues are distended with cloudy mucoid material and gas. Microscopic lesions consist of villous atrophy, crypt hyperplasia, and infiltration of the lamina propria by plasma cells, lymphocytes, macrophages, and heterophils. Parasites are seen on the distal two thirds of the villi in the ileum and occasionally in the ceca and BF. A natural case of intestinal cryptosporidiosis and reoviral induced hepatitis in turkeys has been reported.[862] The immunosupressive effects of the reovirus was thought to have enhanced the intestinal cryptosporidiosis observed in the birds.

VI. CRYPTOSPORIDIOSIS IN BOBWHITE AND COMMON QUAIL

Fatal intestinal infections occur in young bobwhite quail.[350,692] Deaths occur after the birds develop diarrhea, and mortality may reach 90 to 100%. At necropsy there is marked carcass dehydration and the small intestines and ceca are distended with watery fluid. Cryptosporidia are located throughout the small intestine and are rarely found in the colon. Microscopic lesions consist of villous atrophy, villous fusion, and sloughing of apical enterocytes. Experimental studies have demonstrated that cryptosporidia are a primary pathogen in bobwhite quail. When coinfections with cryptosporidia and reovirus are present, a synergistic association exists.

Respiratory disease is reported in naturally infected common quail.[597,795] Up to 50% of a flock may show clinical signs, and mortalities may reach 10%. Gross lesions consist of excess mucus in the trachea, nasal mucosal congestion, and atrophic BF. Cryptosporidia are found in the nasopharynges, trachea, bronchi, and BF. They are not seen in the small intestine. Microscopic changes consist of epithelial deciliation and hyperplasia with an underlying inflammatory response. Bronchopneumonia may be present in severely infected birds.

Transmission studies indicate that cryptosporidia isolated from bobwhite quail are not infectious for chickens,[350,105a] whereas cryptosporidia isolated from common quail are infectious for chickens.[597] This host specificity data and the very marked differences in clinical manifestations indicate that the species causing disease in bobwhite quail is different from the species causing disease in common quail.

VII. CRYPTOSPORIDIOSIS IN OTHER AVIAN SPECIES

A. DUCKS

Respiratory disease has been reported in naturally infected mallard ducks.[597] Gross and microscopic lesions were similar to those reported in other avian species. Cryptosporidia were present in the nasal cavities, trachea, and BF of infected mallards. Cryptosporidia have also been seen on the conjunctiva of a duck with conjunctivitis[521] and in the BF of ducks with diarrhea.[803] Neither of the reports clearly associated disease with the presence of cryptosporidia in these ducks. Ducks are susceptible to oral and IT inoculation with *C. baileyi* oocysts.[479] The site of development is identical to that seen in chickens and turkeys. Ducks appear to be more resistant to experimentally induced clinical respiratory disease because only mild clinical signs and no deaths occurred following IT inoculation of *C. baileyi* oocysts.

B. GEESE

Cryptosporidia have not been associated with disease in geese. A naturally infected 25-d-old gosling had cryptosporidia in the large intestine.[661] A new species (*C. anserinum*) was named for the parasites found in geese, but oocysts were not described and the validity is questionable.

Goslings which were 2-d-old developed patent infections following oral inoculation of *C. baileyi* oocysts, but did not develop clinical signs.[194] Parasites were found only in the BF.

C. PHEASANTS

Severe respiratory disease and associated mortalities have been reported in ring-necked pheasants. Occulonasal involvement may be present[597] or absent.[882] Gross and microscopic lesions are similar to those in other avian species. Conjunctivitis associated with cryptosporidia in pheasants has been reported, although a causative relationship between parasite and lesions was not conclusively demonstrated.[669]

Ring-necked pheasants are susceptible to oral *C. baileyi* inoculation.[480a] Parasites are found in the same anatomic locations as in chickens. Disease does not occur following oral inoculation of ring-necked pheasants.

D. MISCELLANEOUS BIRDS

Fatal respiratory cryptosporidiosis has been reported in peafowl.[519] A hatch of chicks developed respiratory disease at 2 weeks of age and all died. Clinical signs and gross and microscopic lesions were similar to other avian species. Cryptosporidia were present on the conjunctiva, nasal cavity, and trachea of the chick examined.

Renal and nasal cryptosporidiosis was reported in an 8-week-old jungle fowl (*Gallus sonneratii*).[668] The bird died suddenly without showing clinical signs. Respiratory lesions were typical of cryptosporidiosis. The epithelium lining the ureteral branches and collecting ducts was hyperplastic and dilated. Parasites were confined to the collecting ducts.

Renal cryptosporidiosis has been reported in a 4-month-old black-throated finch (*Poephila cincta*), found dead of unknown causes.[266] Grossly, the kidneys were enlarged, pale, and firm. Tubular epithelium was hyperplastic and contained cryptosporidia. The interstitial tissues were infiltrated with plasma cells, lymphocytes, macrophages, and heterophils.

An outbreak of enteric cryptosporidiosis was seen in 7- to 14-d-old lovebirds (*Agapornis* sp.).[84] Almost all chicks in the aviary died within a month of hatching. At necropsy, the chicks were dehydrated and had dilated, yellow fluid filled intestinal tracts. Cryptosporidia were seen in the jejunum and ileum and associated with villus atrophy and villous fusion. No parasites were seen in the respiratory tract. The high mortality and clinical manifestations of the disease outbreak seen in lovebirds is similar to that seen only previously in bobwhite quail.

Respiratory cryptosporidiosis has been observed in a budgerigar (*Melopsittacus undulatus*).[596] The bird was found dead, and no clinical signs were observed. Grossly, the lungs were congested and there was excess mucus in the trachea. Microscopic lesions were subacute tracheitis, fibrinous airsacculitis, and pulmonary congestion. Cryptosporidia were seen only in the trachea.

Cryptosporidia have been seen in the small intestines of a budgerigar that had diarrhea.[290c] The lamina porpria of the small intestine was infiltrated with sheets of plasma cells and lymphocytes, and cryptosporidia were present on the distal protions of the villi.

Small intestinal cryptosporidiosis has been seen in cockatiels (*Nymphocus hollandicus*).[290c,480a] Diarrhea was reported in both cases. In one study,[290c] the small intestine was infiltrated with lymphocytes and plasma cells. In the other study,[480a] no inflammatory response was present.

Cryptosporidia have been seen in the proventriculus of 4 week old canaries.[803] These birds also had septicemia due to *Salmonella* species, and the significance of the cryptosporidial infection is unknown.

Cryptosporidia have been observed as incidental findings in the cloacal corprodeum of a red-lored parrot (*Amazona autumnalis*)[222] and in the feces of a 7-year-old macaw (*Ara* sp.)[465] and an adult Tundra swan (*Cygnus* sp.).[465]

BIBLIOGRAPHY

Information pertinent to the subject matter of this chapter may be found in References 84, 94, 104, 105a, 189, 193, 194, 196, 217, 222, 242a, 251, 266, 273, 279, 286 to 291, 305, 306, 309a, 349, 350, 374, 441, 463, 465, 467 to 473, 475, 477 to 480a, 519, 521, 575, 582, 592, 597, 606a, 608, 666 to 669, 692, 740, 748, 782, 783, 789, 795, 803, 809, 862, 882, and 901.

Chapter 10

CRYPTOSPORIDIUM SPP. IN LOWER VERTEBRATES

S. J. Upton

TABLE OF CONTENTS

I. HISTORY

Early reports of the genus *Cryptosporidium* in snakes and lizards, where *C. ameivae*,[51] *C. crotali*,[802] *C. ctenosauris*,[228] and *C. lampropeltis*[28] were named as new species, are either nomen nuda or species of *Sarcocystis*.[456,461,529,844,848] The first valid report of the genus in lower vertebrates appears to be in 1977, when the parasite was reported to be the causative agent of severe, chronic, hypertrophic gastritis in 14 captive snakes.[128] The name *C. serpentis* was later assigned to the coccidium,[457] which is currently the only valid species from reptiles.

A *Cryptosporidium* sp. was reported from an 8-cm-long naso tang, *Naso lituratus*, purchased from a pet store in Ontario, Canada, as well as from another unnamed marine tropical fish.[360] The name *C. nasorum* was given to the parasite, which is the only named species of the genus reported from fish. Several other reports have since described the parasite from additional piscine hosts (Table 1).

II. BIOLOGY

A. REPTILIAN CRYPTOSPORIDIOSIS

Most cases of cryptosporidiosis in reptiles appear to be a problem in captive animals. The parasite was found in various wild natural populations, but no apparent clinical signs of disease were noted.[848] Dr. Elliott Jacobson[377a] has reported that he has observed *Cryptosporidium* sp., with associated gut lesions, in both a wild-caught juvenile corn snake and a juvenile Texas rat snake, suggesting that some clinical cryptosporidiosis may exist in natural populations as well.

The prevalence of the parasite in reptiles appears to be relatively low. Of 528 reptiles collected from three continents, the prevalence of *Cryptosporidium* spp. based on identification of oocysts was about 3%.[848] These data are somewhat misleading because many infected specimens in the study represented captive animals at zoos. When such data are omitted, the overall prevalence in natural populations appears to be <2%.

In snakes, *Cryptosporidium* sp. has been reported to infect boids, colubrids, elapids, and viperids (Table 1), localizing in the gastric mucosa (Figure 1). Although oocysts may be passed in the feces for many months or even years,[848] it is most often diagnosed because of a clinical problem. Clinical signs might include anorexia, persistent or intermittent postprandial regurgitation, lethargy, firm midbody swelling, and progressive weight loss. Positive-contrast gastrography and fluoroscopy have revealed narrowing of the gastric lumen, with circumferential swelling suggestive of a thickened stomach wall.[282] Microscopic lesions consist of hyperplasia and hypertrophy of gastric glands, concomitant atrophy of the granular cells, edema of the submucosa and lamina propria, reduction in lumenal diameter, and inflammation of gastric mucosa.[118,128,282,534,786] Mucosal petechiae, aggregations of lymphocytes and heterophils, exaggeration of longitudinal rugae, excess mucus production, and multifocal necrosis can result from infection.[128,786] Secondary invasion of the mucosa with Gram-negative bacteria has been reported also.[534] Adenovirus-like intranuclear inclusions occur in the glandular epithelium of some infected animals[377a] and suggests the possibility that a virus might predispose snakes to a more severe infection.

Cryptosporidiosis also has been reported in lizards.[220,255,298] In the most severe case reported, a subadult chameleon, *Chameleo senegalensis* (=*C. chameleo senegalensis*), developed clinical signs including anorexia, weight loss, and lethargy at a zoo in Birmingham, AL.[220] The chameleon eventually died, and necropsy revealed the stomach to be contracted and turgid, with prominent rugae covered by a thick layer of mucus. Although it was unknown if *Cryptosporidium* sp. was the sole contributor to death, histological sections revealed numerous organisms lining the gastric mucosa (Figure 1). The mucosa was necrotic, contained inflammatory cells, and the epithelium was sloughing. The submucosa was edematous and fibrotic and contained inflammatory cells and had congested blood vessels.

TABLE 1
Species of Lower Vertebrates Reported to be Infected with *Cryptosporidium* spp.[a]

Host species	Locality	Ref.
Fish		
Cyprinus carpio	Bohemia	618
Lates calcarifer	Australia	278a
Naso lituratus	Canada	360
Oreochromis hybrids	Israel	437
Oreochromis aureus	Israel	610
Salmo trutta	U.K.	704
Unnamed tropical fish	Canada	360
Amphibia		
Ceratophrys ornata	Canada	180a
Sauria		
Agama aculeata	Namibia	848
Agama planiceps	Namibia	848
Chamaeleo senegalensis	Alabama	220
Chondrodactylus angulifer	Namibia	848
Hemidactylus turcicus	Texas	848
Lacerta sp.	Tennessee	255
Naultinus grevi	Tennessee	255
Phelsuma madagascariensis	Madagascar	845, 848
Varanus exanthematicus	Texas	848a
Unnamed species of chameleon	California	298
Testudines		
Geochelone carbonaria	Tennessee	255
Geochelone elegans	California	340
Serpentes		
Bitis gabonica	Australia	118
Bothrops nummifer	Canada	180a
Corallus caninus	Tennessee	255
Crotalus cerastes	Canada	180a
Crotalus durissus	Texas	848
Crotalus horridus	Maryland	128
	California	340
Crotalus spp.	California	276
Elaphe guttata	Australia	118
	Maryland	128
	California	276
Elaphe longissima	Switzerland	331
Elaphe obsoleta	Australia	118
	Maryland	128
	Tennessee	255
	Illinois	786
	Texas	848
Elaphe subocularis	Maryland	128
	California	276
Elaphe vulpina	Georgia	848
Gonyosoma oxycephala	Tennessee	255
Lampropeltis triangulum	Texas	848
Nerodia harteri	Texas	848
Nerodia rhombifera	Texas	848
Oxyuranus scutellatus	Australia	118
Pituophis melanoleucus	Tennessee	255
	Illinois	282
Pseudechis porphyriacus	Australia	534
Sanzina madagascariensis	Maryland	128
Sistrurus catenatus	Canada	180a

TABLE 1 (continued)
Species of Lower Vertebrates Reported to be Infected with *Cryptosporidium* spp.[a]

Host species	Locality	Ref.
Thamnophis sirtalis	Canada	180a
Unnamed species of Boidae	Illinois	786
Unnamed species of Crotalids	California	298

[a] Dr. Elliott Jacobson[377a] also reports finding *Cryptosporidium* sp. in *Boa constrictor occidentalis*, *Corallus canina*, *Crotalus horridus*, *C. tigris*, *Elaphe gutatta gutatta*, *E. obsoleta lindheimeri*, *Lampropeltis zonata pulchra*, *Pituophis melanoleucus affinis*, *P. m. lodingi*, and *Tupinambis teguixin*.

Although no clinical signs have been noted in most lizards, infections in *Phelsuma madagascariensis grandis* are noteworthy. These animals were part of a shipment of geckos imported illegally into North America from Madagascar. Numerous oocysts and developmental stages of a *Cryptosporidium* sp. were found in the cloaca of several of them.[845,848] Although many died or were killed while moribund shortly after import, the precise role of cryptosporidiosis in mortality could not be determined because of concurrent infection with two other species of coccidia. In addition, no attempts were made to identify pathogenic bacteria or viruses. However, this report of the parasite infecting the cloaca is the only mention of the parasite in reptiles colonizing a sight other than gastric mucosa, suggesting that this might be a species of *Cryptosporidium* not reported in reptiles previously.

Two reports have described *Cryptosporidium* in Chelonia. A star tortoise, *Geochelone elegans*, had clinical signs of gastritis, including regurgitation.[340] A red-footed tortoise (*G. carbonaria*) was eventually euthanized following steady decline due to the disease.[255]

B. PISCINE CRYPTOSPORIDIOSIS

The first report of *Cryptosporidium* sp. in fish was in 1981.[360] After 2 months of progressive illness characterized by intermittent anorexia, emaciation, regurgitation, and passage of feces containing undigested food, an 8-cm-long tropical marine fish (*Naso lituratus*) was killed and tissues were examined histologically.[360] Although skeletal muscle fiber atrophy was noted, morphologic changes in the intestine were not found. However, developmental stages of *Cryptosporidium* sp. were found in the microvillus surface of the intestinal tract, and the authors termed the parasite *Cryptosporidium nasoris* (Figure 2). In an addendum to the paper, the authors also noted a similar infection in another, unnamed marine tropical fish.

Cryptosporidium sp. was reported in 5 of 35 carp (*Cyprinus carpio*) in Czechoslovakia.[618] Developmental stages were found along the villi in the middle portion of the intestinal tract. *Cryptosporidium* sp. was also described from intestinal contents of five brown trout (*Salmo trutta*) from a reservoir near Sheffield, U.K.[704] Neither the carp nor trout were described as having the clinical signs noted by Hoover et al.[360]

Cryptosporidium sp. was found in the stomachs of hatchery-reared cichlids (crosses between *Oreochromis aureus* and *O. niloticus*).[437] Infections in fry and fingerlings 11 to 55 mm in length were prevalent, whereas older fish 56 to 152 mm in length were not found to be infected. Ultrastructurally, meronts and gamonts were found localized at the microvillous zone, where they caused few lesions, whereas oocysts were found intracellularly causing necrosis of host cells. Oocysts were usually enclosed within degenerating or necrotic cells located near the surface. In some cases, sloughing of cells enabled oocysts to be released into the stomach lumen. Developmental stages in the epithelium were covered by a unique microvillus membrane, not reported previously for any other species of *Cryptosporidium* (Figure 2).

Cryptosporidium sp. of fish might provide clues to the evolutionary link between *Cryptospo-*

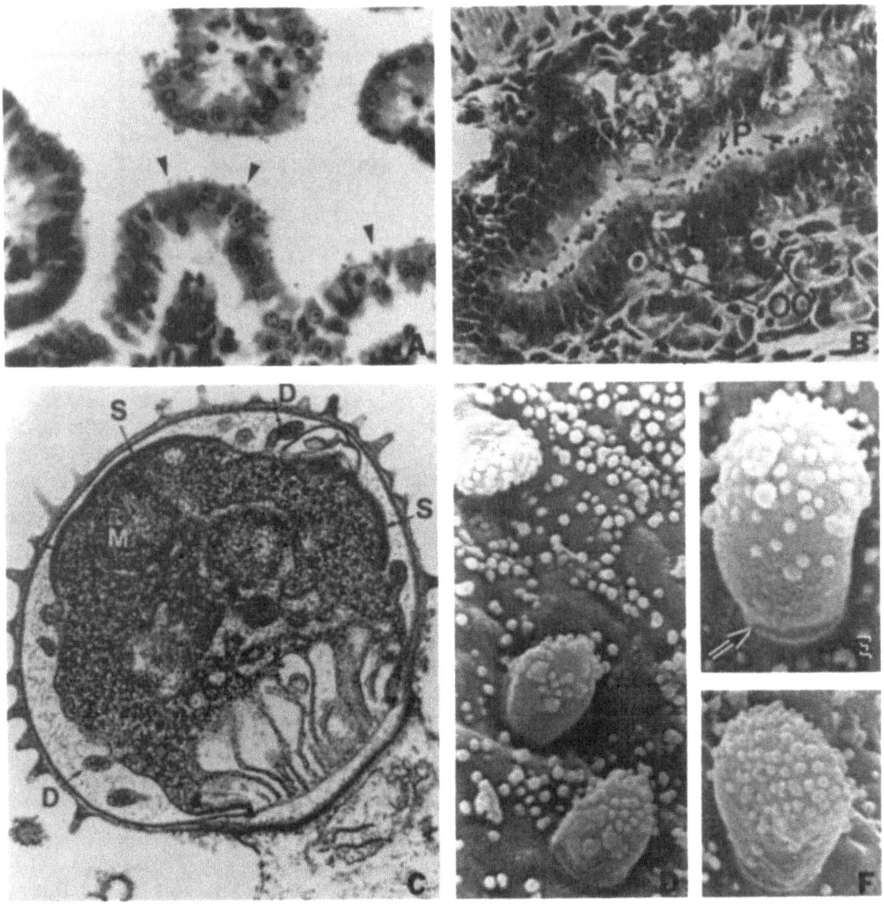

FIGURE 1. *Cryptosporidium* spp. from the intestine of reptiles. (A) *Cryptosporidium* sp. (arrowheads) developing on the surface of the gastric mucosa of a chameleon. (H. & E.; magnification × 1000.) (From Dillehay, D. L., Boosinger, T. R., and MacKenzie, S., *J. Am. Vet. Med. Assoc.*, 189, 1139, 1986. With permission.) (B) Developmental stages along border of hypertrophied gastric epithelium of a snake. Attachment zone (a); macrogametes (M); meronts (S); oocysts (O); trophozoites (T). (Magnification × 1300.) (C) Transmission electron micrographs of parasite along hyperplastic gastric gland of a snake. Attachment zone (a); host cell membrane (h); parasitophorous vacuole (p); membrane-bound vesicle (V). Magnification × 9300.) ([B and C] From Brownstein, D. G., Strandberg, J. D., Montali, R. J., Bush, M., and Fortner, J., *Vet. Pathol.*, 14, 606, 1977. With permission.) (D) Swollen stomach of a black rat snake infected with *Cryptosporidium* sp. exteriorized through surgical incision. (Magnification × 0.6.) (From Szabo, J. R. and Moore, K. *Vet. Med.*, 1984, 98, 1984. With permission.)

ridium and the Eimeriidae. As described previously,[437,610] most developmental stages of *Cryptosporidium* sp. in fish are typical of the genus, but zygotes and oocysts develop within a parasitophorous vacuole deep within the epithelial cell. In some piscine *Eimeria* (syn. *Epieimeria*), parasites develop in a parasitophorous vacuole that is about half embedded in the epithelial cell, while half projects into the intestinal lumen.[551]

C. Experimental Transmission Studies

Cross-transmission data and studies of endogenous development are needed to determine if species of *Cryptosporidium* from lower vertebrates are capable of infecting higher vertebrates. It was recently hypothesized that snakes could acquire cryptosporidiosis from infected mice or other prey.[255] Although possible, experimental data supporting this hypothesis are lacking.

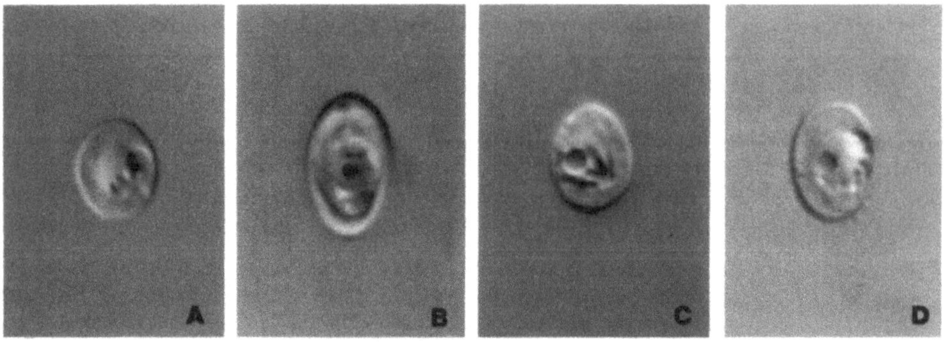

FIGURE 2. *Cryptosporidium* spp. from the intestine of fish. (A) *Cryptosporidium nasorum* (arrowheads) along the intestine of *Naso lituratus*. (H. & E.; magnification × 30.) (From Hoover, D. M., Hoerr, F. J., Carlton, W. W., Hinsman, E. J., and Ferguson, H. W., *J. Fish Dis.*, 4, 425, 1981. With permission.) (B and C) *Cryptosporidium* sp. infecting the intestinal epithelium of a hybrid cichlid, *Oreochromis aureus* × *O. niloticus*. ([B] H. & E.; magnification × 535.) Intracellular oocysts (OO); epithelial parasites (P). (C) Transmission electron micrograph of developing meront. (Magnification × 17,700.) Plasmalemma of parasite (CW); pseudopodium-like process (D); rough endoplasmic reticulum (ER); feeder organelle (F); host cell (H); developing merozoite (M); nucelus (N); parasitophorous vacuole (PE); subpellicular layer (S). ([B and C] from Landsberg, J. H. and Paperna, I., *Dis. Aquat. Org.*, 2, 13, 1986. With permission.) (D to F) Scanning electron micrographs of *Cryptosporidium* sp. in the intestinal epithelium of *Oreochromis aurea*. Arrow in E points to collar near end of attachment. ([D] Approximate magnification × 3000. [E and F] approximate magnification × 4000.) (From Paperna, I., *Dis. Aquat. Org.*, 3, 231, 1987. With permission.)

Indeed, previous attempts to experimentally transmit *Cryptosporidium* spp. between mammalian and avian hosts have been unsuccessful. In addition, measurements of oocysts collected from various reptilian hosts suggest that reptilian *Cryptosporidium* spp. are distinct from all known mammalian and avian isolates.[848]

Two attempts to cross-transmit *Cryptosporidium* between higher and lower vertebrates have gone unpublished. Dr. B. C. Anderson[276] allowed "guppies" (ca. 1.5 cm in length) to ingest calf feces containing oocysts of *C. parvum* at an approximate concentration of 1.0×10^6 oocysts per milliliter. Fish were killed at daily intervals for 9 d postinoculation, and the entire fish was immersed in 10% neutral buffered formalin. Examination of H. & E.-stained sections failed to reveal the presence of parasite stages within the intestinal tract of the fish, whereas suckling mice inoculated orally with the same isolate developed infections.

Oocysts from the feces of *Varanus exanthematicus* were placed in 2.5% (w/v) aqueous $K_2Cr_2O_7$ solution and stored at 4°C for 2 months, then purified by CsCl gradients and stored at 4°C for an additional month in phosphate-buffered saline.[848a] A total of 12 5-d-old ICR outbred mice were each inoculated orally with 1×10^4 oocysts of this 3-month old isolate, but none became infected. Although these results suggest that the reptilian isolate is noninfectious to suckling mice, results must be interpreted with caution. Although the oocysts from *Varanus* appeared normal, uninfected lizards were unavailable for inoculations and, thus, it cannot be stated for certain that the reptilian-derived oocysts were still viable after 3 months of storage.

D. EVIDENCE FOR MULTIPLE SPECIES IN LOWER VERTEBRATES

Cryptosporidium spp. in lower vertebrates appears to consist of several species. The ultrastructural studies of the forms infecting *Oreochromis* spp. reveal morphologic and developmental characteristics that distinguish the organism from species infecting mammalian and avian hosts.[437,610] Whether this is the same species that was described by Hoover et al.[360] is not known. However, the use of the name *C. nasorum* for isolates from fish is recommended until cross-transmission studies can be performed.

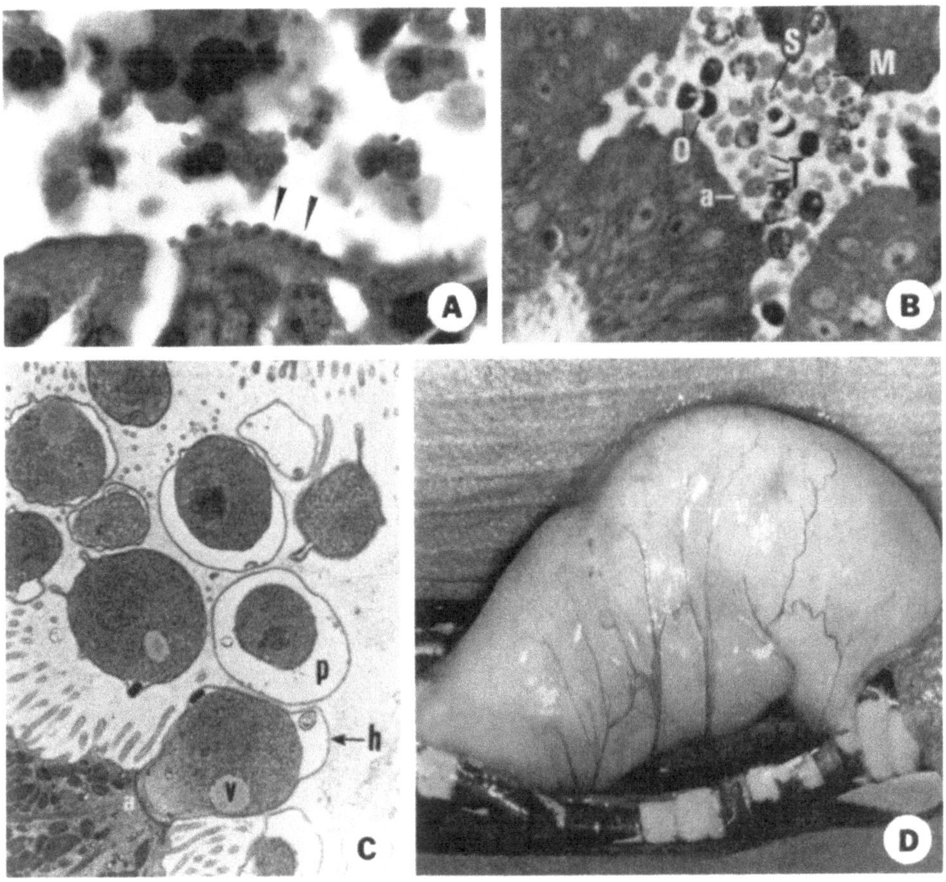

FIGURE 3. Oocysts of *Cryptosporidium* spp. from the feces of various reptiles. (A) From *Elaphe obsoleta*; (B) from *Elaphe vulpina*; (C) from *Hemidactylus turcicus*; (D) from *Nerodia harteri*. (All magnifications × 2500.) (From Upton, S. J., McAllister, C. T., Freed, P. S., and Barnard, S. M., *J. Wildlf. Dis.*, 25, 20, 1989. With permission.)

The present author is aware of only one published account of the genus *Cryptosporidium* from amphibia, where the organism was reported from a frog, *Ceratophrys ornata*, from the Metropolitan Toronto Zoo.[180a] However, all evidence suggests that the prevalence of *Cryptosporidium* sp. in Amphibia is quite low. A recent study specifically surveyed Anura for coccidia, including *Cryptosporidium* spp., but found no specimens infected with *Cryptosporidium* spp.[847] Indeed, we have surveyed over 500 specimens of Amphibia representing 29 species in 15 genera in North and South America to date with no evidence of *Cryptosporidium* oocysts.

In reptiles, the species originally reported to cause hypertrophic gastritis in snakes has been named *C. serpentis*.[457] This organism appears responsible for a number of cases of cryptosporidiosis in captive reptiles and is a true pathogen. A recent morphologic study attempted to determine the number of species of *Cryptosporidium* in reptiles and concluded that there were at least five morphologic types based on oocyst size and shape representing multiple species (Figure 3).[848]

Several isolates infecting terrestrial snakes at the Houston Zoo (*Crotalus durissus*, *Elaphe obsoleta*, and *Lampropeltis triangulum*) all had oocysts with mean measurements of 6.1 to 6.2 × 5.3 μm. A more recent isolate obtained from the same locality was found in the feces of a captive *Varanus exanthematicus* (see Section II.C) and is similar in size, with a mean of 6.0 × 5.4 μm (n = 30). In all probability, these organisms represent *C. serpentis*, whereas those found in other hosts (see below) represent additional, unnamed species.

An isolate obtained from *Elaphe vulpina vulpina* in the Atlanta Zoo had very large oocysts, averaging 8.0 × 5.6 μm.[848] Oocysts from a wild-caught *Agama planiceps* from Namibia had oocysts measuring 5.8 × 5.0 μm, whereas those collected from *Hemidactylus turcicus* in the Houston area and from *Phelsuma madagascariensis grandis* from Madagascar were 6.3 × 5.5 to 5.6 μm. These gekkonid isolates probably represent species separate from *C. serpentis* based on size and because developmental stages infected cloaca rather than stomach in *Phelsuma*.[845,848] Oocysts recovered from individual water snakes, *Nerodia harteri* and *N. rhombifera*, were 6.7 × 5.6 μm and 6.5 × 5.6 μm, respectively, and had a relatively thin wall that collapsed around the sporozoites following flotation in sucrose solutions. Measurement of oocysts from several additional *N. harteri* continue to show these forms to be larger than those found in reptiles housed at the Houston Zoo.

BIBLIOGRAPHY

Information pertinent to the subject matter contained within this chapter may be found in References 27b, 28, 51, 118, 128, 180a, 220, 228, 255, 276, 282, 298, 331, 340, 360, 437, 456, 457, 461, 529, 534, 551, 610, 618, 704, 786, 802, 844, 845, 847, and 848.

REFERENCES

1. **Aboul-Magd, L. A., El-Ridi, A. M. S., and Michael, S. A.**, Biological and haematological changes in experimental cryptosporidiosis, *J. Egypt. Soc. Parasitol.*, 19, 49, 1989.
2. **Addy, P. A.-K. and Atkins-Bekoe, P.**, Cryptosporidiosis in diarrhoeal children in Kumasi, Ghana, *Lancet*, 1, 735, 1986.
3. **Adjei, O., Agbemadzo, T., and Addy, P. A. K.**, The occurrance of Cryptosporidium oocysts in Ghanaian patients with diarrhoea, *East Afr. Med. J.*, 64, 108, 1987.
4. **Ahourai, P., Ezzi, A., Gholami, M. R., Vandyoösefi, J., Kargar, R., and Maalhagh, N.**, Cryptosporidium spp. in new born lambs in Iran, *Trop. Anim. Health Prod.*, 17, 6, 1985.
5. **Ahourai, P., Ezzi, A., Gholami, M. R., Vandyoösfi, J., Kargar, R., and Maalhagh, N.**, Cryptosporidium spp. in new born lambs in Iran, *Arch. Inst. Razi*, 36 and 37, 15, 1986.
6. **Albert, M. J.**, Significance of Cryptosporidium and other enteric pathogens in developing countries, *Lancet*, 1, 921, 1986.

6a. **Alcaino, H., Laval, E., Gorman, T., and Pinochet, L.**, Isosporosis y criptosporidiosis en cerdos de criaderos industriales de la region metropolitana de Chile, *Arch. Med. Vet.*, 21, 131, 1989.

7. **Alonso, J. F., Becerra, E. A., Heras, M. M. J., Rodríguez, J. L. V., Rodríguez, M. A. J., and Leal, J. A. L.**, Criptosporidiosis intestinal con diseminación pancreatobiliar y bronquial en una niña afecta de síndrome de inmunodeficiencia adquirida, *Med. Clin. (Barcelona)*, 89, 335, 1987.
8. **Alonso, M., Blandino, T., and Gómez, E.**, Protozoos intestinales en terneros: extensión de invasión en heces fecales, *Rev. Salud. Anim.*, 8, 313, 1986.
9. **Alós, J. I. and Bouza, E.**, *Cryptosporidium* y criptosporidiasis, *Rev. Clín. Esp.*, 180, 385, 1987.
10. **Alpert, G., Bell, L. M., Kirkpatrick, C. E., Budnick, L. D., Campos, J. M., Friedman, H. M., and Plotkin, S. A.**, Cryptosporidiosis in a day-care center, *N. Engl. J. Med.*, 311, 860, 1984.
11. **Alpert, G., Bell, L. M., Kirkpatrick, C. E., Budnick, L. D., Campos, J. M., Friedman, H. M., and Plotkin, S. A.**, Outbreak of cryptosporidiosis in a day-care center, *Pediatrics*, 77, 152, 1986.
12. **Anderson, B. C.**, Patterns of shedding of cryptosporidial oocysts in Idaho calves, *J. Am. Vet. Med. Assoc.*, 178, 982, 1981.
13. **Anderson, B. C. and Bulgin, M. S.**, Enteritis caused by *Cryptosporidium* in calves, *Vet. Med. Small Anim. Clin.*, 76, 865, 1981.
14. **Anderson, B. C.**, Is cryptosporidial infection responsible for diarrhea?, *Calif. Vet.*, 36, 9, 1982.
15. **Anderson, B. C.**, Cryptosporidiosis in Idaho lambs: natural and experimental infections, *J. Am. Vet. Med. Assoc.*, 181, 151, 1982.
16. **Anderson, B. C.**, Cryptosporidiosis: a review, *J. Am. Vet. Med. Assoc.*, 180, 1455, 1982.
17. **Anderson, B. C., Donndelinger, T., Wilkins, R. M., and Smith, J.**, Cryptosporidiosis in a veterinary student, *J. Am. Vet. Med. Assoc.*, 180, 408, 1982.
18. **Anderson, B. C. and Hall, R. F.**, Cryptosporidial infection in Idaho dairy calves, *J. Am. Vet. Med. Assoc.*, 181, 484, 1982.
19. **Anderson, B. C.**, Cryptosporidiosis, *Lab. Med.*, 14, 55, 1983.
20. **Anderson, B. C.**, Location of cryptosporidia: review of the literature and experimental infections in calves, *Am. J. Vet. Res.*, 45, 1474, 1984.
21. **Anderson, B. C.**, Moist heat inactivation of *Cryptosporidium* sp., *Am. J. Public Health*, 75, 1433, 1985.
22. **Anderson, B. C.**, Quick and easy diagnosis of cryptosporidiosis and Johne's disease, *Vet. Med. Food Anim. Pract.*, 6, 87, 1985.
23. **Anderson, B. C.**, Effect of drying on the infectivity of cryptosporidia-laden calf feces for 3- to 7-day-old mice, *Am. J. Vet. Res.*, 47, 2272, 1986.
25. **Anderson, B. C.**, Abomasal cryptosporidiosis in cattle, *Vet. Pathol.*, 24, 235, 1987.
26. **Anderson, B. C.**, Gastric cryptosporidiosis of feeder cattle, beef cows, and dairy cows, *Bovine Pract.*, No. 23, 99, 1988.
27. **Anderson, B. C.**, Cryptosporidium spp. in cattle, in *Proc. 1st Int. Workshop on Cryptosporidiosis*, Angus, K. W. and Blewett, D. A., Eds., September 7 to 8, 1988, Edinburgh, 1989, 55.

27a. **Anderson, W. H., Gerstoft, J. and Henriksen, S.**, Human cryptosporidiosis, *N. Engl. J. Med.*, 309, 1325, 1987.

27b. **Anderson, B. C.**, Personal communication.

28. **Anderson, D. R., Duszynski, D. W., and Marquardt, W. C.**, Three new coccidia (Protozoa: Telosporea) from kingsnakes, *Lampropeltis* spp., in Illinois, with a redescription of *Eimeria zamenis* Phisalix, 1921, *J. Parasitol.*, 54, 577, 1968.
29. **Andreani, T., Modigliani, R., le Charpentier, Y., Galian, A., Brouet, J.-C., Liance, M., Lachance, J.-R., Messing, B., and Vernisse, B.**, Acquired immunodeficiency with intestinal cryptosporidiosis: possible transmission by Haitian whole blood, *Lancet*, 1, 1187, 1983.
30. **Angus, K. W., Campbell, I., Gray, E. W., and Sherwood, D.**, Staining of faecal yeasts and cryptosporidium oocysts, *Vet. Rec.*, 108, 173, 1981.

31. **Angus, K. W., Appleyard, W. T., Menzies, J. D., Campbell, I., and Sherwood, D.,** An outbreak of diarrhoea associated with cryptosporidiosis in naturally reared lambs, *Vet. Rec.*, 110, 129, 1982.
32. **Angus, K. W., Sherwood, D., Hutchison, G., and Campbell, I.,** Evaluation of the effect of two aldehyde-based disinfectants on the infectivity of faecal cryptosporidia for mice, *Res. Vet. Sci.*, 33, 379, 1982.
33. **Angus, K. W., Tzipori, S., and Gray, E. W.,** Intestinal lesions in specific-pathogen-free lambs associated with a Cryptosporidium from calves with diarrhea, *Vet. Pathol.*, 19, 67, 1982.
34. **Angus, K. W.,** Cryptosporidiosis in man, domestic animals and birds: a review, *J. R. Soc. Med.*, 76, 62, 1983.
35. **Angus, K. W., Hutchison, G., Campbell, I., and Snodgrass, D. R.,** Prophylactic effects of anticoccidial drugs in experimental murine cryptosporidiosis, *Vet. Rec.*, 114, 166, 1984.
36. **Angus, K. W., Hutchison, G., and Munro, H. M. C.,** Infectivity of a strain of *Cryptosporidium* found in the guinea-pig *(Cavia porcellus)* for guinea-pigs, mice and lambs, *J. Comp. Pathol.*, 95, 151, 1985.
37. **Angus, K. W.,** Survival of cryptosporidium in excreta: zoonotic aspects of infections and possible implications for spread by the agricultural use of sewage sludge, in *Epidemiological Studies of Risks Associated with the Agricultural use of Sewage Sludge: Knowledge and Needs*, Block, J. C., Havelaar, A. H., and L'Hermite, P., Eds., Elsevier, London, 1986, 98.
38. **Angus, K. W.,** Cryptosporidiosis in domestic animals and humans, *In Practice*, 9, 47, 1987.
39. **Angus, K. W.,** Cryptosporidiosis in red deer, *Public Vet. Deer Soc.*, 3, 3, 1988.
40. **Angus, K. W.,** Mammalian cryptosporidiosis: a veterinary perspective, in *Proc. 1st Int. Workshop on Cryptosporidiosis*, Angus, K. W. and Blewett, D. A., Eds., September 7 to 8, 1988, Edinburgh, 1989, 43.
41. **Anon.,** Immunodeficiency and cryptosporidiosis. Demonstration at the Royal College of Physicians of London, *Br. Med. J.*, 281, 1123, 1980.
41a. **Anon.,** Human cryptosporidiosis — Alabama, *Morbid. Mortal.*, 31, 252, 1982.
42. **Anon.,** Cryptosporidiosis: assessment of chemotherapy of males with acquired immune deficiency syndrome (AIDS), *Morbid. Mortal.*, 31, 589, 1982.
43. **Anon.,** Cryptosporidiosis among children attending day-care centers — Georgia, Pennsylvania, Michigan, California, New Mexico, *Morbid. Mortal.*, 33, 599, 1984.
44. **Anon.,** Update: treatment of cryptosporidiosis in patients with acquired immunodeficiency syndrome (AIDS), *Morbid. Mortal.*, 33, 117, 1984.
44a. **Anon.,** Cryptosporidiosis, *Lancet*, 1, 492, 1984.
45. **Anon.,** Case records of the Massachusetts General Hospital. Case No. 39, *N. Engl. J. Med.*, 313, 805, 1985.
46. **Anon.,** Cryptosporidiosis — New Mexico, 1986, *Morbid. Mortal. Rep.*, 36, 561, 1986.
47. **Anon.,** Troubled waters, *Lancet*, 2, 251, 1989.
48. **Antony, M. A., Brandt, L. J., Klein, R. S., and Bernstein, L. H.,** Infectious diarrhea in patients with AIDS, *Digest. Dis. Sci.*, 33, 1141, 1988.
49. **Araya, J., González, J., Sagua, H., Olivares, W., Rimassa, C., and Videla, M.,** Cryptosporidiosis en el Norte de Chile. I. Prevalencia en animales doméსticos, sinantrópicos y silvestres, *Bol. Chil. Parasitol.*, 42, 7, 1987.
50. **Araya, J., Hartard, M., Sagua, H., González, J., Arriagada, J., and Palma, J.,** Cryptosporidiosis en el Norte de Chile. II. Prevalencia en lactantes con sindrome diarreico agudo, *Bol. Chil. Parasitol.*, 42, 12, 1987.
51. **Arcay de Peraza, L. and Bastardo de San Jose, T.,** *Cryptosporidium ameivae* sp. nov. coccidia Cryptosporidiidae del intestino delgado de *Ameiva ameiva* de Venezuela. *Acta Cient. Venez.*, 20, 125, 1969.
51a. **Argenzio, R. A., Liacos, J. A., Levy, M. L., Meuten, D. J., Lecce, J. G., and Powell, D. W.,** Villous atrophy, crypt hyperplasia, cellular infiltration and impaired glucose-Na absorption in enteric cryptosporidiosis of pigs, *Gastroenterology*, 1990, in press.
52. **Armstrong, M.,** Cryptosporidiosis, *Med. Lab. Sci.*, 44, 280, 1987.
53. **Arnaud-Battandier, F., Naciri, M., Fischer, A., Ricour, C., Griscelli, C., and Yvore, P.,** Cryptosporidiose intestinale: une cause nouvelle de diarrhée chez l'homme?, *Gastroenterol. Clin. Biol.*, 6, 1045, 1982.
54. **Arnaud-Battandier, F.,** Cryptosporidiose isosporose, lambiase au cours des déficits immuntaires, *Arch. Fr. Pediatr.*, 42, 959, 1985.
55. **Arnaud-Battandier, F., Naciri, M., and Maurage, C.,** Cryptosporidiosis in immunocompetent patients, *N. Engl. J. Med.*, 313, 1019, 1985.
56. **Arrowood, M. J. and Sterling, C. R.,** Isolation of *Cryptosporidium* oocysts and sporozoites using discontinuous sucrose and isopycnic percoll gradients, *J. Parasitol.*, 73, 314, 1987.
57. **Arrowood, M. J. and Sterling, C. R.,** Comparison of conventional staining methods and monoclonal antibody-based methods for *Cryptosporidium* oocyst detection, *J. Clin. Microbiol.*, 27, 1490, 1989.
58. **Arrowood, M. J., Mead, J. R., Mahrt, J. L., and Sterling, C. R.,** Effects of immune colostrum and orally administered antisporozoite monoclonal antibodies on the outcome of *Cryptosporidium parvum* infections in neonatal mice, *Infect. Immum.*, 57, 2283, 1989.
59. **Asahi, H., Kumada, M., Kato, K., and Koyama, T.,** A simple staining method for cryptosporidian oocysts and sporozoites. *Jpn. J. Med. Sci. Biol.*, 41, 117, 1988.
60. **Atterholm, I., Castor, B., and Norlin, K.,** Cryptosporidiosis in southern Sweden, *Scand. J. Infect. Dis.*, 19, 231, 1987.

61. **Augustin-Bichl, G., Boch, J., and Henkel, G.,** Kryptosporidien-Infektionen bei Hund und Katze, *Berl. Muench. Tieraerztl. Wochenschr.*, 97, 179, 1984.
62. **Ayeni, A. O., Olubunmi, P. A., and Abe, J. O.,** The occurrence and effect of *Cryptosporidium* species on livestock in Ile-Ife, Nigeria, *Trop. Vet.*, 3, 96, 1985.
63. **Azevedo, C.,** Ultrastructural observations of *Cryptosporidium* sp. parasite of *Ruditapes decussatus* (Mollusca, Bivalvia), *J. Invertebr Pathol.*, 54, 23, 1989.
64. **Babb, R. R., Differding, J. T., and Trollope, M. L.,** Cryptosporidia enteritis in a healthy professional athlete, *Am. J. Gastroenterol.*, 77, 833, 1982.
65. **Bannister, P. and Mountford, R. A.,** *Cryptosporidium* in the elderly: a cause of life-threatening diarrhea, *Am. J. Med.*, 86, 507, 1989.
66. **Barker, I. K. and Carbonell, P. L.,** *Cryptosporidium agni* sp. n. from lambs and *Cryptosporidium bovis* sp. n. from a calf with observations on the oocyst, *Z. Parasitenkd.*, 44, 289, 1974.
67. **Baron, E. J., Schenone, C., and Tanenbaum, B.,** Comparison of three methods for detection of *Cryptosporidium* oocysts in a low-prevalence population, *J. Clin. Microbiol.*, 27, 223, 1989.
68. **Barrat, J., Gerard, Y., Schwerg, A., Thiry, E., Dubuisson, J., and Blancou, J.,** Serological survey in free-living red deer in France, in *The Management and Health of Farmed Deer*, Reid, H. W., Ed., Kluwer Academic, Boston, 1988, 123.
69. **Baskin, G. B., Murphey-Corb, M., Watson, E. A., and Martin, L. N.,** Necropsy findings in rhesus monkeys experimentally infected with cultured simian immunodeficiency virus (SIV)/Delta, *Vet. Pathol.*, 25, 456, 1988.
70. **Baxby, D., Hart, C. A., and Taylor, C.,** Human cryptosporidiosis: a possible case of hospital cross infection, *Br. Med. J.*, 287, 1760, 1983.
71. **Baxby, D. and Blundell, N.,** Sensitive, rapid, simple methods for detecting Cryptosporidium in faeces, *Lancet*, 2, 1149, 1983.
72. **Baxby, D., Blundell, N., and Hart, C. A.,** The development and performance of a simple, sensitive method for the detection of *Cryptosporidium* oocysts in faeces, *J. Hyg.*, 92, 317, 1984.
73. **Baxby, D. and Hart, C. A.,** Cryptosporidiosis, *Br. Med. J.*, 289, 1148, 1984.
74. **Baxby, D., Getty, B., Blundell, N., and Ratcliffe, S.,** Recognition of whole *Cryptosporidium* oocysts in feces by negative staining and electron microscopy, *J. Clin. Microbiol.*, 19, 566, 1984.
75. **Baxby, D., Hart, C. A., and Blundell, N.,** Shedding of oocysts by immunocompetent individuals with cryptosporidiosis, *J. Hyg. Camb.*, 95, 703, 1985.
76. **Baxby, D. and Hart, C. A.,** The incidence of cryptosporidiosis: a two-year prospective survey in a children's hospital, *J. Hyg.*, 96, 107, 1986.
77. **Baxby, D., Blundell, N., and Hart, C. A.,** Excretion of atypical oocysts by patients with cryptosporidiosis, *Lancet*, 2, 974, 1987.
78. **Baxby, D. and Blundell, N.,** Recognition and laboratory characteristics of an atypical oocyst of *Cryptosporidium*, *J. Infect Dis.*, 158, 1038, 1988.
79. **Beauvais, B., Sarfati, C., Derouin, F., Garin, Y. J.-F., Larivière, M., and Deletoille, P.,** Evaluation comparative de deux méthodes de dépistage des oocystes de cryptosporidies dans les feces humaines, *Ann. Biol. Clin.*, 47, 45, 1989.
80. **Beier, T. V., Sidorenko, N. V., Pashkin, P. I., and Ponizovskii, A. K.,** Cryptosporidiosis in animals (Distribution, clinical symptoms, prophylaxis and treatment) [In Russian], *Veterinariya*, 3, 52, 1987.
81. **Belchev, L., Arnaudov, D., Ignatov, G., and Boyadzhiev, S.,** Cryptosporidium infection in neonate pigs with diarrhoea, [In Russian] O Nauchni Trudove, Plovdivski Universitet "Paisii Khilendarski" Biologiya, 22, 177, 1984.
83. **Belchev, L., Ignatov, D., Ignatov, G., and Boyadzhiev, S.,** A mixed coli-rotavirus and *Cryptosporidium* infection in newborn pigs, *Vet. Med. Nauki*, 24, 3, 1987.
84. **Belton, D. J. and Powell, I. B.,** Cryptosporidiosis in lovebirds (*Agapornis* sp.), *N. Z. Vet. J.*, 35, 15, 1987.
85. **Benkov, K. J., Stawski, C., Sirlin, S. M., Klapholz, M. B., Siegal, F., and LeLeiko, N. S.,** Atypical presentation of childhood acquired immune deficiency syndrome mimicking Crohn's disease: nutritional considerations and management, *Am. J. Gastroenterol.*, 80, 260, 1985.
86. **Bennett, M., Baxby, D., Blundell, N., Gaskell, C. J., Hart, C. A., and Kelly, D. F.,** Cryptosporidiosis in the domestic cat, *Vet. Rec.*, 116, 73, 1985.
87. **Berg, I. E., Peterson, A. C., and Freeman, T. P.,** Ovine cryptosporidiosis, *J. Am. Vet. Med. Assoc.*, 173, 1586, 1978.
88. **Bergeland, M. E.,** Necrotic enteritis in nursing piglets, *Proc. Am. Assoc. Vet. Lab. Diagn.*, 20, 151, 1977.
89. **Bergeland, M. E., Johnson, D. D., and Shave, H.,** Bovine cryptosporidiosis in the north central United States, *Proc. Am. Assoc. Vet. Lab. Diagn.*, 22, 131, 1979.
90. **Bergeland, M. E. and Henry, S. C.,** Infectious diarrheas of young pigs. *Vet. Clin. North Am. Large Anim. Pract.*, 4, 389, 1982.
91. **Berk, R. N., Wall, S. D., McArdle, C. B., McCutchan, J. A., Clemett, A. R., Rosenblum, J. S., Premkumer, A., and Megibow, A. J.,** Cryptosporidiosis of the stomach and small intestine in patients with AIDS, *Am. J. Radiol.*, 143, 549, 1984.

92. **Berkowitz, C. D. and Seidel, J. S.,** Spontaneous resolution of cryptosporidiosis in a child with acquired immunodeficiency syndrome, *Am. J. Dis. Child.*, 139, 967, 1985.
93. **Berkowitz, C. D.,** AIDS and parasitic infections, including *Pneumocystis carinii* and cryptosporidiosis, *Pediatr. Clin. N. Am.*, 32, 933, 1985.
93a. **Berkowitz, F. E., Vallabh, W., Buqwana, A., and Heney, C.,** Cryptosporidiosis in black South African children, *S. Afr. Med. J.*, 74, 272, 1988.
94. **Bermudez, A. J., Ley, D. H., Levy, M. G., Ficken, M. D., Guy, J. S., and Gerig, T. M.,** Intestinal and bursal cryptosporidiosis in turkeys following inoculation with *Cryptosporidium* sp. isolated from commercial poults, *Avian Dis.*, 32, 445, 1988.
95. **Berthier, M., Lacroix, C., Agius, G., Nivet, H., Boufassa, S., and Jacquemin, J. C., Hoppeler,** Haemolytic uraemic syndrome associated with cryptosporidium oocysts, *Eur. J. Pediatr.*, 147, 213, 1988.
96. **Bertram, U. and Lampert, F.,** Intestinale Kryptosporidiose bei Verdacht auf erworbene Immunschwäche, *Dtsch. Med. Wochenschr.*, 111, 587, 1986.
97. **Biggs, B.-A., Megna, R., Wickremesinghe, S., and Dwyer, B.,** Human infection with *Cryptosporidium* spp.: results of a 24-month survey, *Med. J. Aust.*, 147, 175, 1987.
98. **Bird, R. G. and Smith, M. D.,** Cryptosporidiosis in man: parasite life cycle and fine structural pathology, *J. Pathol.*, 132, 217, 1980.
99. **Bishburg, E., Sunderam, G., Reichman, L. B., and Kapila, R.,** Central nervous system tuberculosis with the acquired immunodeficiency syndrome and its related complex, *Ann. Int. Med.*, 105, 210, 1986.
100. **Bissenden, J. G.,** Cryptosporidium and diarrhoea, *Br. Med. J.*, 293, 287, 1986.
101. **Blacklow, N. R. and Wolfson, J. S.,** Case records of the Massachusetts General Hospital: case 39-1985, *N. Engl. J. Med.*, 313, 805, 1985.
102. **Blagburn, B. L. and Current, W. L.,** Accidental infection of a researcher with human *Cryptosporidium*, *J. Infect. Dis.*, 148, 772, 1983.
103. **Blagburn, B. L.,** Avian cryptosporidiosis, in *Proc. 1st Int. Workshop on Cryptosporidiosis*, Angus, K. W. and Blewett, D. A., Eds., September 7 to 8, 1988, Edinburgh, 1989, 27.
104. **Blagburn, B. L., Lindsay, D. S., Giambrone, J. J., Sundermann, C. A., and Hoerr, F. J.,** Experimental cryptosporidiosis in broiler chickens, *Poult. Sci.*, 66, 442, 1987.
105. **Blagburn, B. L., Lindsay, D. S., Hoerr, F. J., Giambrone, J. J., and Sundermann, C. A.,** Avian cryptosporidiosis: an emerging disease problem?, *Poult. Dig.*, 436, 1986.
105a. **Blagburn, B. L. and Lindsay, D. S.,** Unpublished data.
105b. **Blagburn, B.,** Unpublished data.
106. **Blakey, J.,** Cryptosporidium in stools, *Med. J. Aust.*, 141, 686, 1984.
107. **Blanchard, J. L., Baskin, G. B., Murphey-Corb, M., and Martin, L. N.,** Disseminated cryptosporidiosis in simian immunodeficiency virus/delta-infected rhesus monkeys, *Vet. Pathol.*, 24, 454, 1987.
108. **Blewett, D. A.,** Cryptosporidiosis in red deer, in *The Management and Health of Farmed Deer*, Reid, H. W., Ed., Kluwer Academic, Boston, 1988, 43.
109. **Blewett, D. A.,** Disinfection and oocysts, in *Proc. 1st Int. Workshop on Cryptosporidiosis*, Angus, K. W. and Blewett, D. A., Eds.,September 7 to 8, 1988, Edinburgh, 1989, 107.
110. **Blewett, D. A.,** Quantitative techniques in *Cryptosporidium* research, in *Proc. 1st Int. Workshop on Cryptosporidiosis*, Angus, K. W. and Blewett, D. A., Eds., September 7 to 8, 1988, Edinburgh, 1989, 85.
111. **Blumberg, R. S., Kelsey, P., Perrone, T., Dickersin, R., Laquaglia, M., and Ferruci, J.,** Cytomegalovirus- and Cryptosporidium-associated acalculous gangrenous cholecystitis, *Am. J. Med.*, 76, 1118, 1984.
112. **Boch, V. J., Göbel, E., Heine, J., Brändler, U., and Schloemer, L.,** Kryptosporidien-Infektion bei Haustieren, *Berl. Muench. Tieraerztl. Wochenschr.*, 95, 361, 1982.
113. **Bogaerts, J., Lepage, P., Rouvroy, D., and Vandepitte, J.,** *Cryptosporidium* spp., a frequent cause of diarrhea in Central Africa, *J. Clin. Microbiol.*, 20, 874, 1984.
114. **Bogaerts, J., Lepage, P., Rouvroy, D., Van Goethem, C., Nsengumuremye, F., Mohamed, O., Habyalimana, J. B., and Vandepitte, J.,** Cryptosporidiosis in Rwanda. Clinical and epidemiological features, *Ann. Soc. Belge Med. Trop.*, 67, 157, 1987.
115. **Boonpucknavig, S. and Bunyaratvej, S.,** Detection of smooth muscle antibodies in patients with cryptosporidiosis, *J. Clin. Lab. Immunol.*, 27, 153, 1988.
116. **Bossen, A. N. and Britt, E. M.,** Cryptosporidiosis in immunocompetent patients, *N. Engl. J. Med.*, 313, 1019, 1985.
117. **Boufassa-Ouzrout, S., Chermette, R., and Meissonier, E.,** *Cryptosporidiosis: A Cosmopolitan Disease in Animals and in Man*, 1st ed., Office International des Epizootiques, Technical Ser. No. 5, 1986.
118. **Boylan, T., Cunningham, M., and Reddacliff, G.,** Cryptosporidiosis in snakes in Taronga Zoo, *Thylacinus*, 10, 20, 1985.
119. **Brady, E., Margolis, M. L., and Korzenlowski, O. M.,** Pulmonary cryptosporidiosis in acquired immune deficiency syndrome, *J. Am. Med. Assoc.*, 252, 89, 1984.
120. **Brandonisio, O., Angarano, G., Altamura, M., Locaputo, S., Pastore, G., and Canestri-Trotti, G.,** Cryptosporidiosis in the acquired immunodeficiency syndrome: a case report, *Eur. J. Epidemiol.*, 3, 442, 1987.

121. **Brasitus, T. A.,** Parasites and malabsorption, *Am. J. Med.*, 67, 1058, 1979.
122. **Brasseur, P., Lemeteil, D., and Mallet, E.,** La cryptosporidiose chez l'enfant immunocompétent, *Presse Med.*, 16, 177, 1987.
123. **Brasseur, P., Lemeteil, D., and Ballet, J. J.,** Rat model for human cryptosporidiosis, *J. Clin. Microbiol.*, 26, 1037, 1988.
124. **Brett, R. W. and Ridgeway, J. W.,** Experience with chlorine dioxide in Southern Water Authority and Water Research Centre, *J. Inst. Water Eng. Sci.*, 35, 135, 1981.
125. **Bridger, J. C., Hall, G. A., and Brown, J. F.,** Characterization of a calici-like virus (Newbury agent) found in association with astrovirus in bovine diarrhea, *Infect. Immun.*, 43, 133, 1984.
126. **Britt, D. P. and Al-Ghawaby, M.A.,** *Cryptosporidium* in a child in Kuwait, *Ann. Trop. Med. Parasitol.*, 82, 407, 1988.
127. **Bronsdon, M. A.,** Rapid dimethyl sulfoxide-modified acid-fast stain of *Cryptosporidium* oocysts in stool specimens, *J. Clin. Microbiol.*, 19, 952, 1984.
128. **Brownstein, D. G., Strandberg, J. D., Montali, R. J., Bush, M., and Fortner, J.,** *Cryptosporidium* in snakes with hypertrophic gastritis, *Vet. Pathol.*, 14, 606, 1977.
129. **Bryant, J. L., Stills, H. F., and Middleton, C. C.,** *Cryptosporidia* in squirrel monkeys (*Saimiri sciureus*), *Lab. Anim. Sci.*, 33, 482, 1983.
130. **Burchard, G. D.,** Klinische Bedeutung, Epidemiologie und Laboratoriumsdiagnostik enteraler Infektionen mit Kryptosporidien, *Immun. Infekt.*, 2, 51, 1986.
131. **Buxton, D.,** Diseases of deer: neonatal enteritis, in *Management and Diseases of Deer*, Alexander, T. L., Ed., Publ. Vet. Deer Soc., 1986, 97.
132. **Buxton, D.,** Diseases of deer: clostridial diseases, in *Management and Diseases of Deer*, Alexander, T. L., Ed., Publ. Vet. Deer Soc., 1986, 81.
133. **Campbell, I., Tzipori, S., Hutchison, G., and Angus, K. W.,** Effect of disinfectants on survival of cryptosporidium oocysts, *Vet. Rec.*, 111, 414, 1982.
134. **Campbell, P. N. and Current, W. L.,** Demonstration of serum antibodies to *Cryptosporidium* sp. in normal and immunodeficient humans with confirmed infections, *J. Clin. Microbiol.*, 18, 165, 1983.
135. **Canestri-Trotti, G., Gramenzi, F., Foglini, A., and Tognato, G.,** Prima segnalazione di protozoi del genere *Cryptosporidium* Tyzzer, 1907 in vitelli in Italia, *Atti Soc. Ital. Buitaria*, 14, 289, 1982.
135a. **Canestri-Trotti, G. and Quesada, A.,** Primo reperto di *Cryptosporidium* sp. in bufali italiani (*Bubalus bubalis*), *Atti Soc. Ital. Sci. Vet.*, 37, 737, 1983.
135b. **Canestri-Trotti, G., Quesada, A., and Visconti, S.,** Ricerche sulla fauna protozoaria intestinale del buffalo (*Bubalus bubalis*), *Atti Soc. Ital. Buitaria*, 16, 433, 1984.
135c. **Canestri-Trotti, G., Pampiglione, S., and Visconti, S.,** *Cryptosporidium* sp. e *Isospora suis* nel suino in Italia, *Parassitologia*, 26, 299, 1984.
135d. **Canestri-Trotti, G. and Visconti, S.,** Indagine parassitologica su protozoi intestinali in equini dell'esercito italiano, *Atti Soc. Ital. Sci. Vet.*, 39, 758, 1985.
135e. **Canestri-Trotti, G. and Arnone, B.,** *Cryptosporidium muris* nel gatto a Bologna, *Parassitologia*, 28, 212, 1986.
135f. **Canestri-Trotti, G., Pomilio, F., and Visconti, S.,** Indagine parassitologica sulla fauna protozoaria intestinale di vitelli allevati in provincia di Chieti, *Atti Soc. Ital. Buiatria*, 18, 473, 1986.
135g. **Canestri-Trotti, G.,** Criptosporidiosi: contributi di ricerca dell' istituto di malattie infettive, profilassi e polizia veterinaria dell' Università di Bologna, *G. Mal. Infett. Parassit.*, 39, 476, 1987.
135h. **Canestri-Trotti, G. and Giannetto, S.,** Ulteriori dati sulla diffusione di *Cryptosporidium muris* in Italia, *Atti Soc. Ital. Sci. Vet.*, 41, 1193, 1987.
135i. **Caprioli, A., Gentile, G., Baldassarri, L., Bisicchia, R., Romoli, E., and Donelli, G.,** *Cryptosporidium* as a common cause of childhood diarrhoea in Italy, *Epidem. Infect.*, 102, 537, 1989.
136. **Card, C. E., Perdrizet, J. A., Georgi, M. E., and Shin, S. J.,** Cryptosporidiosis associated with bacterial enteritis in a goat kid, *J. Am. Vet. Med. Assoc.*, 191, 69, 1987.
137. **Carlson, B. L. and Nielsen, S. W.,** Cryptosporidiosis in a raccoon, *J. Am. Vet. Med. Assoc.*, 181, 1405, 1982.
137a. **Carneiro, C. S., Tarnau, C., and Chermette, R.,** Essais de transmission experimentale de cryptosporidies d'origine equine aux souriceaux et aux poussins, *Rev. Ecuat. Hig. Med. Trop.*, 37, 37, 1987.
138. **Carstensen, H., Hansen, H. L., and Kristiansen, H. O.,** Cryptosporidiosis in children in southern Guinea-Bissau, *Pediatr. Infect. Dis.*, 5, 718, 1986.
139. **Carstensen, H., Hansen, H. L., Kristiansen, H. O., and Gomme, G.,** The epidemiology of cryptosporidiosis and other intestinal parasitoses in children in southern Guinea-Bissau, *Trans. R.. Soc. Trop. Med. Hyg.*, 81, 860, 1987.
140. **Carter, M. J. and Anzimlt, T.,** Cryptosporidium: an important cause of gastrointestinal disease in immuno-competent patients, *N. Z. Med. J.*, 99, 101, 1986.
141. **Carter, M. J.,** Cryptosporidium infestation, *N. Z. Med. J.*, 97, 743, 1984.
142. **Casemore, D. P. and Jackson, B.,** Sporadic cryptosporidiosis in children, *Lancet*, 2, 679, 1983.

143. **Casemore, D. P. and Jackson, F. B.,** Hypothesis: cryptosporidiosis in human beings is not primarily a zoonosis, *J. Infect.*, 9, 153, 1984.
144. **Casemore, D. P., Armstrong, M., and Jackson, B.,** Screening for Cryptosporidium in stools, *Lancet*, 1, 734, 1984.
145. **Casemore, D. P., Sands, R. L., and Curry, A.,** Cryptosporidium species a "new" human pathogen, *J. Clin. Pathol.*, 38, 1321, 1985.
146. **Casemore, D. P., Armstrong, M., and Sands, R. L.,** Laboratory diagnosis of cryptosporidiosis, *J. Clin. Pathol.*, 38, 1337, 1985.
147. **Casemore, D. P., Jessop, E. G., Douce, D., and Jackson, F. B.,** *Cryptosporidium* plus campylobacter: an outbreak in a semi-rural population, *J. Hyg.*, 96, 95, 1986.
148. **Casemore, D. P.,** The antibody response to Cryptosporidium: development of a serological test and its use in a study of immunologically normal persons, *J. Infect.*, 14, 125, 1987.
149. **Casemore, D. P.,** Timing of symptoms and oocyst excretion in human cryptosporidiosis, *N. Engl. J. Med.*, 317, 168, 1987.
150. **Casemore, D. P.,** The epidemiology of human cryptosporidiosis, in *1st Int. Workshop on Cryptosporidiosis*, Angus, K. W. and Blewett, D. A., Eds., September 7 to 8, 1988, Edinburgh, 1989, 65-78.
151. **Casemore, D.,** Cryptosporidiosis: another source, *Br. Med. J.*, 298, 750, 1989.
151a. **Casemore, D. P.,** Epidemiological aspects of human cryptosporidiosis, *Epidemiol. Infect.*, 104, 1, 1990.
152. **Cenac, J., Deluol, A. M., Matheron, S., Coulaud, J. P., and Savel, J.,** La cryptosporidiose. I. Une nouvelle protozoose intestinale humaine, *Ann. Biol. Clin. (Paris)*, 42, 389, 1984.
153. **Chandrasekar, P. H.,** "Cure" of chronic cryptosporidiosis during treatment with azidothymidine in a patient with the acquired immune deficiency syndrome, *Am. J. Med.*, 83, 187, 1987.
154. **Chang, O. J.,** Cryptosporidiosis asociada a diarrea aguda en Oxapampa, provincia de Oxapampa, departamento de Pasco, Perú, *Bol. Chil. Parasitol.*, 42, 39, 1987.
155. **Chermette, R., Polack, B., Boufassa, S., and Bariaud, F.,** Observations de cryptosporidies chez des bovins adultes, in *Cryptosporidiose du Jeune Ruminant*, Navetat, H. and Espinasse, J., Eds., Lyons, 1984, 19.
156. **Chermette, R., Polack, B., Boufassa, S., Bariaud, F., Tarnau, C., Couderc, O., and Creuzon, F.,** Cryptosporidies chez les animaux adultes en France: role epidemiologique, in Commun. Congr. Natl. Soc. France Parasitol., Rennes, France, March 16, 1984.
157. **Chermette, R., Boufassa S., Soule, C., Tarnau, C., Courderc, O., and Lengronne, D.,** La cryptosporidiose équine: une parasitose méconnue, *CEREOPA Études et Réalisations Pédagogigues sûr le Cheval*, 13 éme Journée d'étude, Paris, CEREOPA, 1987, 81.
158. **Chermette, R. and Boufassa-Ouzrout, S.,** *Cryptosporidiosis: a Cosmopolitan Disease in Animals and in Man* (Tech. Ser. No. 5), 2nd ed., Office International des Épizooties, Paris, 1988, 1.
158a. **Chermette, R., Tarnau, C., Boufassa-Ouzrout, S., and Couderc, O.,** Survey on equine cryptosporidiosis in Normandy, *Coccidia and Intestinal Coccidiomorphs*, 5th Int. Coccidiosis Conf., Tours, France, October 17-20, 1989, INRA, 1989, 493.
158b. **Chermette, R. and Blondel, S.,** Cryptosporidiose des carnivores domestiques, résultats préliminaires en France, *Bull. Soc. Fr. Parasitol.*, 7, 31, 1989.
159. **Chiampi, N. P., Sundberg, R. D., Klompus, J. P., and Wilson, A. J.,** Cryptosporidial enteritis and pneumocystis pneumonia in a homosexual man, *Hum. Pathol.*, 14, 734, 1983.
159a. **Chrisp, C. E., Reid, W. C., Rush, H. G., Suckow, M. A., Bush, A. and Thomann, M. J.,** Cryptosporidiosis in guinea pigs: an animal model, *Infect. Immun.*, 56, 674, 1990.
160. **Chng, H. H., Shaw, D., Klesius, P., and Saxson, A.,** Inability of oral bovine transfer factor to eradicate cryptosporidial infection in a patient with congenital dysgammaglobulinemia, *Clin. Immunol. Immunopathol.*, 50, 402, 1989.
161. **Clarke, J. J.,** A study of coccidia met with in mice, *J. Micr. Sci.*, 37, 277, 1894.
161a. **Clotet, B., Sirera, G., Cofan, F., Monterola, J. M., Tortosa, F., and Foz, M.,** Efficacy of the somatostatin analogue (SMS-201-995), Sandostatin, for cryptosporidial diarrhoea in patients with AIDS, *AIDS*, 3, 857, 1989.
162. **Cockrell, B. Y., Valerio M. G., and Garner, F. M.,** Cryptosporidiosis in the intestines of rhesus monkeys (*Macaca mulatta*), *Lab. Anim. Sci.*, 24, 881, 1974.
163. **Cockerill, F. R., Hurley, D. V., Malagelada, J. R., La Russo, N. F., Edson, W. F., Katzmann, J., Banks, P. M., Wiltsie, J. C., Davis, J. P., Lack, E. E., Ishak, K. G., and Van Scoy, R. E.,** Polymicrobial cholongitis and Kaposi's sarcoma in blood products transfusion related acquired immune deficiency syndrome, *Am. J. Med.*, 80, 1237, 1986.
164. **Coelho, K. I. R., Maeda, S. A., and Marques, M. E. A.,** Intestinal cryptosporidiosis. Association with *Pneumocystis carinii*, cytomegalovirus and *Candida* sp. infections, *Rev. Inst. Med. Trop. Sao Paulo*, 29, 323, 1987.
165. **Cogswell, J. J. and Richardson, E. A.,** Acute self limited colitis, *Br. Med. J.*, 289, 561, 1984.
166. **Cohen, J. D., Ruhlig, L., Jayich, S. A., Tong, M. J., Lechago, J., and Snape, W. J.,** Cryptosporidium in acquired immunodeficiency syndrome, *Dig. Dis. Sci.*, 29, 773, 1984.

167. **Coleman, S.U., Klei, T.R., French, D.D., Chapman, M.R., and Corstvet, R.E.**, Prevalence of *Cryptosporidium* sp in equids in Louisiana, *Am. J. Vet. Res.*, 50, 575, 1989.

168. **Collier, A. C., Miller, R. A., and Meyers, J. D.**, Cryptosporidiosis after marrow transplantation: person-to-person transmission and treatment with spiramycin, *Ann. Intern. Med.*, 101, 205, 1984.

169. **Collignon, P.**, Staining of atypical oocysts from patients with cryptosporidiosis, *Lancet*, 1, 1494, 1987.

170. **Collins, G. H.**, Cryptosporidiosis: a "new" disease? *Med. J. Aust.*, 140, 516, 1984.

171. **Combee, C. L., Collinge, M. L., and Britt, E. M.**, Cryptosporidiosis in a hospital-associated day care center, *Pediatr. Infect. Dis.*, 5, 528, 1986.

172. **Connolly, G. M. and Gazzard, B. G.**, Toxic megacolon in cryptosporidiosis, *Postgrad. Med. J.*, 63, 1103, 1987.

173. **Connolly, G. M., Dryden, M. S., Shanson, D. C., and Gazzard, B. G.**, Cryptosporidial diarrhoea in AIDS and its treatment, *Gut*, 29, 593, 1988.

174. **Contrepois, M. and Vallet, A.**, Cryptosporidiose et diarrhée néo-natale en élevage bovin, *Point. Vet.*, 16, 235, 1984.

175. **Cook, D. J., Kelton, J. G., Stanisz, A. M., and Collins, S. M.**, Somatostatin treatment for cryptosporidial diarrhea in a patient with the acquired immunodeficiency syndrome (AIDS), *Ann. Intern. Med.*, 108, 708, 1988.

175a. **Cook, R. A. and Stetter, M.**, Cryptosporidiosis in a Mexican Rosey Boa, *J. Zoo. Anim. Med.*, 19, 60, 1988.

176. **Cooper, D. A., Wodak, A., Marriot, D. J. E., Harkness, J. L., Ralston, M., Hill, A., and Penny, R.**, Cryptosporidiosis in the acquired immune deficiency syndrome, *Pathology*, 16, 455, 1984.

177. **Corbett-Feeney, G.**, Cryptosporidium among children with acute diarrhoea in the west of Ireland, *J. Infect.*, 14, 79, 1987.

178. **Craun, G. F.**, Surface water supplies and health, *J. Am. Water Works Assoc.*, 80, 40, 1988.

179. **Crawford, F. G., Vermund, S. H., Ma, J. Y., and Deckelbaum, R. J.**, Asymptomatic cryptosporidiosis in a New York City day care center, *Pediatr. Infect. Dis. J.*, 7, 806, 1988.

180. **Crawford, F. G. and Vermund, S.H.**, Human cryptosporidiosis, *CRC Crit. Rev. Microbiol.*, 16, 113, 1988.

180a. **Crawshaw, G. J. and Mehren, K. G.**, Cryptosporidiosis in zoo and wild animals, in *Erkrankungen der Zootiere, Verhandlungsbericht des 29, Int. Symp. über die Erkrankungen der Zootiere von 20*, Ippen, R. and Schröder. H. D., Eds., Akademie-Verlag, Berlin, 1987, 353.

180b. **Cristino, J. A. G. M.**, Criptosporidiosise. A propósito dos primeiros casos de criptosporidiose identificados no Laboratório de Bacteriologia do Hospital de Santa Maria em Lisboa, *Argent. Inst. Bacteriol. Cam. Pestana*, 15, 31, 1987.

180c. **Cristino, J. A. G. M., Carvalho, I. P., and Salgado, M. J.**, An outbreak of cryptosporidiosis in a hospital day-care centre, *Epidem. Infect.*, 101, 355, 1988.

181. **Cross, J. H., Alcantara, A., Alquiza, L., Zaraspe, G., and Ranoa, C.**, Cryptosporidiosis in Philippine children, *Southeast Asian J. Trop. Med. Public Health*, 16, 257, 1985.

182. **Cross, R. F. and Moorhead, P. D.**, a rapid staining technic for cryptosporidia, *Mod. Vet. Pract.*, 65, 307, 1984.

183. **Cruickshank, R., Ashdown, L., and Croese, J.**, Human cryptosporidiosis in north Queensland, *Aust. N. Z. J. Med.*, 18, 582, 1988.

184. **Cruz, J. R., Cano, F., Caceres, P., Chew, F., and Pareja, G.**, Infection and diarrhea caused by *Cryptosporidium* sp. among Guatemalan infants, *J. Clin. Microbiol.*, 26, 88, 1988.

185. **Cubertafond, S., Deruaz, D., Mojon, M., and Hermier, M.**, Place des cryptosporidies dans les diarrhees de l'enfant: enquete lyonnaise, *Ann. Biol. Clin. (Paris)*, 45, 92, 1988.

186. **Current, W. L., Reese, N. C., Ernst, J. V., Bailey, W. S., Heyman, M. B., and Weinstein, W. M.**, Human cryptosporidiosis in immunocompetent and immunodeficient persons. Studies of an outbreak and experimental transmission, *N. Engl. J. Med.*, 308, 1252, 1983.

187. **Current, W. L.**, Human cryptosporidiosis, *N. Engl. J. Med.*, 309, 1326, 1983.

188. **Current, W. L. and Long, P. L.**, Development of human and calf *Cryptosporidium* in chicken embryos, *J. Infect. Dis.*, 148, 1108, 1983.

189. **Current, W. L. and Haynes, T. B.**, Complete development of *Cryptosporidium* in cell culture, *Science*, 224, 603, 1984.

190. **Current, W. L.**, Cryptosporidiosis, *J. Am. Vet. Med. Assoc.*, 187, 1334, 1985.

191. **Current, W. L.**, Cryptosporidiosis: a protozoologists view of an emerging zoonosis, *Microecol. Ther.*, 15, 165, 1985.

192. **Current, W. L.**, *Cryptosporidium*: its biology and potential for environmental transmission, *CRC Crit. Rev. Environ. Control*, 17, 21, 1986.

193. **Current, W. L.**, *Cryptosporidium* sp. in chickens: parasite life cycle and aspects of acquired immunity, in *Research in Avian Coccidiosis*, Proc. Georgia Coccidiosis Conf., McDougald, L. R., Joyner, L. P., and Long, P. L., Eds., November 19 to 21, 1985, Athens, GA, 1986, 124.

194. **Current, W. L., Upton, S. J., and Haynes, T. B.**, The life cycle of *Cryptosporidium baileyi* n. sp. (Apicomplexa, Cryptosporidiidae) infecting chickens, *J. Protozool.*, 33, 289, 1986.

195. **Current, W. L. and Reese, N. C.**, A comparison of endogenous development of three isolates of *Cryptosporidium* in suckling mice, *J. Protozool.*, 33, 98, 1986.

196. **Current, W. L. and Snyder, D. B.,** Development of and serologic evaluation of acquired immunity to *Cryptosporidium baileyi* by broiler chickens, *Poult. Sci.*, 67, 720, 1988.
197. **Current, W. L.,** The biology of *Cryptosporidium.*, *ASM News*, 54, 605, 1988.
198. **Current, W. L.,** Cryptosporidium and cryptosporidiosis, in *Proc. 1st Int. Workshop on Cryptosporidiosis*, Angus, K. W. and Blewett, D. A., Eds., September 7 to 8, 1988, Edinburgh, 1989, 1.
199. **Current, W. L. and Blck, P. H.,** The immunobiology of *Cryptosporidium* spp., *Pathol. Immun. Pathol. Res.*, 8, 141, 1989.
200. **D'Antonio, R. G., Winn, R. E., Taylor, J. P., Gustafson, T. L., Current, W. L., Rhodes, M. M., Gary, G. W., and Zajac, R. A.,** A waterborne outbreak of cryptosporidiosis in normal hosts, *Ann. Intern. Med.*, 103, 886, 1985.
201. **Dale, B. A. S., Gordon, G. Thomson, R., and Urquhart, R.,** Perinatal infection with cryptosporidium, *Lancet*, 1, 1042, 1987.
202. **Dan, S., Cristescu, P., and Neagoe, V.,** Identificarea genului Cryptosporidium în România, *Rev. Crest. Anim.*, 33, 42, 1983.
203. **Dan, S., Radu, S., Seicaru, A., and Cristescu, P.,** Date privind infestatita naturala si experimentala cu criptosporidii las pasari si animale de laborator, in Semin. Progrese in Terapia si Combaterea Zoonozelor Parazitare, November 6 to 7, 1987, Cluj-Napoca, 1987, 131.
204. **Dan, S., Ciocnitu, A. M., Begnescu, R., Cristescu, P., Barboi, G., Pirvulescu, M., Seicaru, A., Ma Zanet, M., Turcu, D., and Diaconu, S.,** Aspecke epizootologice si morfo-clinice in infestatia cu criptosporidii la vitei, Si societa de medicina veterinara cluj, in Semin. Progrese in Terapi si Combaterea Zoonozelor Parazitare, November 6 to 7, 1987, Cluj-Napoca, 1987, 138.
205. **Das, P., Pal, S., Dutta, D., Bhattacharya, M. K., and Pal, S. C.,** Cryptosporidiosis in Bengali children with acute diarrhoea, *Trans. R. Soc. Trop. Med. Hyg.*, 81, 241, 1987.
206. **Datry, A., Rozenbaum, W., Rosenheim, M., Danis, M., Duflo, B., and Gentilini, M.,** Parasitoses opportunistes et SIDA: méthodes de diagnostic, *Feuill. Biol.*, 26, 41, 1985.
206a. **Datry, A., Danis, M., and Gentilini, M.,** Développement complet de Cryptosporidium en culture cellulaire: applications, *Med./Sci.*, 5, 762, 1989.
207. **Davis, A. J. and Jenkins, S. J.,** Cryptosporidiosis and proliferative ileitis in a hamster, *Vet. Pathol.*, 23, 632, 1986.
208. **Davis, J. J., Heyman, M. B., Ferrell, L., Kerner, J., Kerlan, R., and Thaler, M. M.,** Sclerosing cholangitis associated with chronic cryptosporidiosis in a child with a congenital immunodeficiency disorder, *Am. J. Gastroentrol.*, 82, 1196, 1987.
209. **Davis, J. J. and Heyman, M. B.,** Cryptosporidiosis and sinusitis in an immunodeficient adolescent, *J. Infect. Dis.*, 158, 649, 1988.
210. **DeHovitz, J. A., Pape, J. W., Boncy, M., and Johnson, W. D.,** Clinical manifestations and therapy of *Isospora belli* infection in patients with the acquired immunodeficiency syndrome, *N. Engl. J. Med.*, 315, 87, 1986.
211. **De las Heras, M., Garciá de Jalón, J. A., Balaguer, L., and Badiola, J. J.,** Diarreas en corderos y cabritos asociadias a criptosporidios, *Med. Vet.*, 4, 273, 1987.
212. **De Leeuw, P. W., Moerman, A., Pol, J. M. A., Talmon, F. P., and van Zijderveld, F. G.,** Epidemiological observations on cryptosporidiosis in dairy herds in The Netherlands, in Proc. 13th World Congress on Diseases of Cattle, Durban, Republic of South Africa, 1984, 104.
213. **Deluol, A. M., Cenac, J., Matheron, S., Marche, C., and Savel, J.,** La cryptosporidiose. II. Diagnostic biologique, *Ann. Biol. Clin. (Paris)*, 42, 399, 1984.
214. **De Mol, P., Mukashema, S., Bogaerts, J., Hemelhof, W., and Butzler, J. P.,** Cryptosporidium related to measles diarrhoea in Rwanda, *Lancet*, 2, 42, 1984.
215. **DeRycke, J., Bernard, S., LaPorte, J., Naciri, M., Popoff, M. R., and Rodolakis, A.,** Prevalence of various enteropathogens in the feces of diarrheic and healthy calves, *Ann. Rech. Vet.*, 17, 159, 1986.
216. **Desgrandchamps, D. and Munzinger, J.,** Infektiöse Gastroenteritis beim immunkompetenten Kind. Bedeutung von Cryptosporidium spp. und Aeromonas spp, *Schweiz. Med. Wochenschr.*, 119, 276, 1989.
217. **Dhillon, A. S., Thacker, H. L., Dietzel, A. V., and Winterfield, R. W.,** Respiratory cryptosporidiosis in broiler chickens, *Avian Dis.*, 25, 747, 1981.
218. **Dias, R. M. D. S., Mangini, A. C. S., Torres, D. M. A. G. V., Corréa, M. O. A., Lupetti, N., Corréa, F. M. A., and Chieffi, P. P.,** Cryptosporidiosis among patients with acquired immunodeficiency syndrome (AIDS) in the county of São Paulo, Brazil, *Rev. Inst. Med. Trop. Sao Paulo*, 30, 310, 1988.
219. **Diers, J. and McCallister, G. L.,** Occurrence of *Cryptosporidium* in home daycare centers in west-central Colorado, *J. Parasitol.*, 75, 637, 1989.
220. **Dillehay, D. L., Boosinger, T. R., and MacKenzie, S.,** Gastric cryptosporidiosis in a chameleon, *J. Am. Vet. Med. Assoc.*, 189, 1139, 1986.
221. **Dobbins, W. O., III and Weinstein, W. M.,** Electron microscopy of the intestine and rectum in acquired immunodeficiency syndrome, *Gastroenterology*, 88, 738, 1985.
221a. **Dominguez, S. and Almarza, A.,** Primer hallazgo de Cryptosporidium en un canino en la Republica Argentina, *Pet's Ciencia*, 4, 244, 1988.

222. **Doster, A. R., Mahaffey, E. A., and McClearen, J. R.**, Cryptosporidia in the cloacal coprodeum of red-lored parrots (*Amazona autumnalis*), *Avian Dis.*, 23, 654, 1979.

223. **Driscoll, M. S., Thomas, V. L., and Sanford, B. A.**, *Cryptosporidium* infection in day-care centers, *Drug Intell. Clin. Pharm.*, 22, 636, 1988.

224. **Dryjanski, J., Gold, J. W. M., Ritchie, M. T., Kurtz, R. C., Lim, S. L., and Armstrong, D.**, Cryptosporidiosis. Case report in a health team worker, *Am. J. Med.*, 80, 751, 1986.

225. **Dubey, J. P., Speer, C. A., and Fayer, R.**, *Sarcocystosis of Animals and Man*, CRC Press, Boca Raton, FL, 1989, 93.

226. **Ducatelle, R., Maenhout, D., Charlier, G., Miry, C., Coussement, W., and Hoorens, J.**, Cryptosporidiosis in goats and in Mouflon sheep, *Vlaams Diergeneeskd. Tijdschr.*, 52, 7, 1983.

227. **DuPont, H. L.**, Cryptosporidiosis and the healthy host, *N. Eng. J. Med.*, 312, 1319, 1985.

228. **Duszynski, D. W.**, Two new coccidia (Protozoa: Eimeriidae) from Costa Rican lizards with a review of the *Eimeria* from lizards, *J. Protozool.*, 16, 581, 1969.

229. **Eckert, J.**, Kryptosporidiose, *Internist*, 25, 242, 1984.

230. **Edelman, M. J. and Oldfield, E. C.**, Severe cryptosporidiosis in an immunocompetent host, *Arch. Int. Med.*, 148, 1873, 1988.

231. **Elsser, K. A., Moricz, M., and Proctor, E. M.**, *Cryptosporidium* infections: a laboratory survey, *CMAJ*, 135, 211, 1986.

232. **Ericsson, C. D. and Dupont, H. L.**, *Cryptosporidium* and diarrhea, *Lancet*, 2, 914, 1983.

233. **Ernest, J. A., Blagburn, B. L., Lindsay, D. S., and Current, W. L.**, Infection dynamics of *Cryptosporidium parvum* (Apicomplexa: Cryptosporiidae) in neonatal mice (*Mus musculus*), *J. Parasitol.*, 72, 796, 1986.

234. **Fayer, R. and Leek, R. G.**, The effects of reducing conditions, medium, pH, temperature, and time on in vitro excystation of *Cryptosporidium*, *J. Protozool.*, 31, 567, 1984.

235. **Fayer, R., Ernst, J. V., Miller, R. G., and Leek, R. G.**, Factors contributing to clinical illness in calves experimentally infected with a bovine isolate of *Cryptosporidium*, *Proc. Helminthol. Soc. Wash.*, 52, 64, 1985.

236. **Fayer, R. and Ungar, B. L. P.**, *Cryptosporidium* spp. and cryptosporidiosis, *Microbiol. Rev.*, 50, 458, 1986.

237. **Fayer, R., Klesius, P. H., and Andrews, C.**, Efficacy of bovine transfer factor to protect neonatal calves against experimentally induced clinical cryptosporidiosis, *J. Parasitol.*, 73, 1061, 1987.

238. **Fayer, R., Perryman, L. E., and Riggs, M. W.**, Hyperimmune bovine colostrum neutralizes *Cryptosporidium* sporozoites and protects mice against oocyst challenge, *J. Parasitol.*, 75, 151, 1989.

239. **Fayer, R., Andrews, C., Ungar, B. L. P. and Blagburn, B.** Efficacy of hyperimmune bovine colostrum for prophylaxis of cryptosporidiosis in neonatal calves. *J. Parasitol.*, 75, 393, 1989.

239a. **Fayer, R., and Chrisp, P.**, Unpublished data.

240. **Feng I Chu (Qu Fengyi)**, Two cases of human infection with *Cryptosporidium* reported in the Nanging area (in Chinese), *Jiang. Su. Med.*, 87, 412, 1987.

241. **Fenwick, B. W.**, Cryptosporidiosis in a neonatal gazella, *J. Am. Vet. Med. Assoc.*, 183, 1331, 1983.

242. **Fernández, A., Gómez-Villamandos, S. C., Carrasco, L., Perea, A., Quezada, M., and Gómez, M. A.**, Brote diarréico en potros asociado a criptosporidios, *Med. Vet.*, 5, 311, 1988.

242a. **Fernández, A., Quezada, M. Gómez, M. A., Navarro, J. A., Rodríquez, J., and Sierra, M. A.**, Cryptosporidiosis in chickens from southern Spain, *Avian Dis.*, 34, 224, 1990.

243. **Fiedler, H. H.**, Zur Verbreitung von Kryptosporidien unter norddeutschen Rinderbeständen, *Tieraerztl. Umsch.*, 40, 526, 1985.

244. **Fine, D. L., Landon, J. C., Pienta, R. J., Kubicek, M. T., Valerio, M. G., Loeb, W. F., and Chopra, H. C.**, Responses of infant rhesus monkeys to inoculation with Mason-Pfizer monkey virus materials, *J. Natl. Cancer Inst.*, 54, 651, 1975.

245. **Fischer, M. C. and Agger, W. A.**, Cryptosporidiosis, *Am. Fam. Phys.*, 36, 201, 1987.

245a. **Fischer, O.**, Kryptosporidioza telat v obdobi mlecne vyzivy, *Vet. Med. (Prague)*, 27, 465, 1982.

246. **Fischer, O.**, Ekonomický význam kokcidií rodu *Cryptosporidium* pro odchov telat, *Vet. Med. (Prague)*, 29, 419, 1984.

247. **Fischer, O.**, Attempted therapy and prophylaxis of cryptosporidiosis in calves by administration of sulphadimidine, *Acta Vet. (Brno)*, 52, 183, 1983.

248. **Fitzgerald, W.**, An outbreak of cryptosporidiosis in neonatal calves — a regime for treatment and prevention, *Ir. Vet. News*, 10, 37, 1988.

249. **Flegg, P. J.**, *Cryptosporidium* in travellers from Pakistan, *Trans. R. Soc. Trop. Med. Hyg.*, 81, 171, 1987.

250. **Fletcher, A., Sims, T. A., and Talbot, I. C.**, Cryptosporidial enteritis without general or selective immune deficiency, *Br. Med. J.*, 285, 22, 1982.

251. **Fletcher, O. J., Munnell, J. F., and Page, R. K.**, Cryptosporidiosis of the bursa of Fabricius of chickens, *Avian Dis.*, 19, 630, 1975.

252. **Forgacs, P., Tarshis, A., Ma, P., Federman, M., Mele, L., Silverman, M. L., and Shea, J. A.**, Intestinal and bronchial cryptosporidiosis in an immunodeficient homosexual man, *Ann. Intern. Med.*, 99, 793, 1983.

253. **Freidank, H. and Kist, M.**, Cryptosporidia in immunocompetent patients with gastroenteritis, *Eur. J. Clin. Microbiol.*, 6, 56, 1987.

254. **Fukushima, K. and Helman, R. G.**, Cryptosporidiosis in a pup with distemper, *Vet. Pathol.*, 21, 247, 1984.
255. **Funk, R. S.**, Implications of cryptosporidiosis in Emerald Tree boas, *Corallus caninus*, in *Proc. 10th and 11th Int. Herpetological Symp. on Captive Propagation and Husbandry*, Rosenburg, M. J., Ed., Zoological Consortium, June 17 to 20, 1987, Chicago, 1988, 139.
256. **Gajadhar, A. A., Caron, J. P., and Allen, J. R.**, Cryptosporidiosis in two foals, *Can. Vet. J.*, 26, 132, 1985.
257. **Galbraith, N. S.**, Cryptosporidiosis: another source, *Br. Med. J.*, 298, 276, 1989.
258. **Gallaher, M. M., Herndon, J. L., Nims, L. J., Sterling, C. R., Grabowski, D. J., and Hull, H. F.**, Cryptosporidiosis and surface water, *Am. J. Public Health*, 79, 39, 1989.
259. **Garcia, L. S., Bruckner, D. A., Brewer, T. C., and Shimizu, R. Y.**, Techniques for the recovery and identification of *Cryptosporidium* oocysts from stool specimens, *J. Clin. Microbiol.*, 18, 185, 1983.
260. **Garcia, L. S., Brewer, R. C., and Bruckner, D. A.**, Fluorescence detection of *Cryptosporidium* oocysts in human fecal specimens by using monoclonal antibodies, *J. Clin. Microbiol.*, 25, 119, 1987.
261. **Garcia, L. S. and Current, W. L.**, Cryptosporidiosis: clinical features and diagnosis, *CRC Crit. Rev. Clin. Lab. Sci.*, 27, 439, 1989.
262. **Garcia, L. S., Brewer, T. C., and Bruckner, D. A.**, Incidence of *Cryptosporidium* in all patients submitting stool specimens for ova and parasite examination: monoclonal antibody IFA method, *Diagn. Microbiol. Infect. Dis.*, 11, 25, 1989.
263. **García-Rodríguez, J. A., Martín Sánchez, A. M., Canut, A., and Moraleda, J. M.**, Criptosporidiosis en un paciente con SIDA no perteneciente a grupo de riesgo, *Rev. Clin. Esp.*, 182, 86, 1988.
264. **Garcia-Rodgriguez, J. A., Martin Sánchez, A. M., Canut-Blasco, A., Cedeno Montano, J., and Heras de Pedro, M. I.**, The incidence of cryptosporidiosis in children: a one-year prospective survey in a general hospital in Spain, *Eur. J. Epidemiol.*, 5, 70, 1989.
266. **Gardiner, C. H. and Imes, G. D.**, *Cryptosporidium* sp. in the kidney of a black-throated finch, *J. Am. Vet. Med. Assoc.*, 185, 1401, 1984.
267. **Gari-Toussaint, M., Marty, P., Le Fichoux, Y., Pesce, A., and Saint-Paul, M. C.**, Cryptosporidiose intestinale et pulmonaire temporairement asymptomatique chez un sideen, *Med. Malad. Infect.*, 11, 843, 1988.
268. **Garone, M. A., Winston, B. J., and Lewis, J. H.**, Cryptosporidiosis of the stomach, *Am. J. Gastroenterol.*, 81, 465, 1986.
269. **Garza, D., Hopfer, R. L., Eichelberger, C., Eisenbach, S., and Fainstein, V.**, Fecal staining methods for screening *Cryptosporidium* oocysts, *J. Med. Technol.*, 1, 560, 1984.
270. **Geldreich, E. E.**, Potable water: new directions in microbial regulations, *Am. Soc. Microbiol. News*, 52, 530, 1986.
270a. **Genchi, C. and Leoni, A.**, La Criptosporidiosi, patologia, emergente dell' allevamento intesivo del govino, *Obiett. Vet.*, 4, 17, 1984.
271. **Gentile, G., Baldassarri, L., Caprioli, A., Donelli, G., Vinditti, M., Avvisati, G., and Martino. P.**, Colonic vascular invasion as a possible route of extraintestinal cryptosporidiosis, *Am. J. Med.*, 82, 574, 1987.
272. **Gerstoft, J., Holten-Andersen, W., Blom, J., and Nielsen, J. O.**, Cryptosporidium enterocolitis in homosexual men with AIDS, *Scand. J. Infect. Dis.*, 16, 385, 1984.
273. **Gharagozlou, M. J. and Khodashenas, M.**, Cryptosporidiosis in a native rooster with a chronic proliferative enteritis, *Arch. Vet.*, 17, 129, 1985.
274. **Gibson, J. A., Hill, M. W. M., and Huber, M. J.**, Cryptosporidiosis in Arabian foals with severe combined immunodeficiency, *Aust. Vet. J.*, 60, 378, 1983.
275. **Gibson, S. V. and Wagner, J. E.**, Cryptosporidiosis in guinea pigs: a retrospective study, *J. Am. Vet. Med. Assoc.*, 189, 1033, 1986.
276. **Gillespie, D.**, Cryptosporidiosis in reptiles, *Curr. Vet. Ther.*, 9, 748, 1986.
277. **Gjerde, B.**, Circumstantial evidence of an extracellular (epicellular) localization of developmental stages of Cryptosporidium, *Acta Vet. Scand.*, 27, 629, 1986.
278. **Glaze, W. H.**, Drinking-water treatment with ozone, *Environ. Sci. Technol.*, 21, 224, 1987.
278a. **Glazebrook, J. S. and Campbell, R. S. F.**, Diseases of barramundi (*Lates calcarifer*) in Australia: a review, in *Management of Wild and Cultured Sea Bass/Barramundi (Lates calcarifer)*, Copeland, J. W. and Grey, D. L., Eds., Australian Centre for International Agricultural Research, Canberra, 204, 1987.
279. **Glisson, J. R., Brown, T. P., Brugh, M., Page, R. K., Kleven, S. H., and Davis, R. B.**, Sinusitis in turkeys associated with respiratory cryptosporidiosis, *Avian Dis.*, 28, 783, 1984.
280. **Goebel, E. and Braendler, U.**, Ultrastructure of microgametogenesis, microgametes and gametogamy of *Cryptosporidium* sp. in the small intestine of mice, *Protistologica*, 18, 331, 1982.
281. **Göbel, E.**, Diagnose und Therapie der akuten Kryptosporidiose beim Kalb, *Tieraerztl. Umsch.*, 42, 863, 1987.
282. **Godshalk, C. P., MacCoy, D. M., Patterson, J. S., and McKiernan, B. C.**, Gastric hypertrophy associated with cryptosporidiosis in a snake, *J. Am. Vet. Med. Assoc.*, 189, 1126, 1986.
283. **Gomez, E., Alonso, M., Blandino, T., Frías, M. T., and Merino, N.**, Cryptosporidium spp en terneros de una recria de la provincia Habana, *Rev. Salud Anim.*, 4, 65, 1982.
284. **Gómez, E., Blandino, T., Alonso, M., Rodríguez D. J., and Abreu, R.**, *Cryptosporidium* spp.: inoculación experimental en ratones y ratas de laboratorio, *Rev. Salud Anim.*, 9, 273, 1987.

285. **Goodstein, R. F., Colombo, C. S., Illfelder, M. A., and Skaggs, R. E.**, Bronchial and gastrointestinal cryptosporidiosis in AIDS, *J. A. O. A.*, 89, 195, 1989.
286. **Goodwin, M. A., Latimer, K. S., Brown, W. L., Steffens, P. W., Martin, P. W., Resurreccion, R. S., Smeltzer, M. A., and Dickson, T. G.**, Respiratory cryptosporidiosis in chickens, *Poult. Sci.*, 67, 1684, 1988.
287. **Goodwin, M. A. and Brown, J.**, Histologic incidence and distribution of Cryptosporidium sp. infection in chickens: 68 cases in 1986, *Avian Dis.*, 32, 365, 1988.
288. **Goodwin, M. A., Steffens, W. L., Russell, I. D., and Brown, J.**, Diarrhea associated with intestinal cryptosporidiosis in turkeys, *Avian Dis.*, 32, 63, 1988.
289. **Goodwin, M. A.**, Small-intestinal cryptosporidiosis in a chicken, *Avian Dis.*, 32, 844, 1988.
290. **Goodwin, M. A. and Brown, J.**, Light-microscopic lesions associated with naturally occurring bursal cryptosporidiosis in chickens, *Avian Dis.*, 33, 74, 1989.
290a. **Goodwin, M. A. and Brown, J.**, A geographic locus of respiratory cryptosporidiosis in Georgia broilers, *Avian Dis.*, 33, 368, 1989.
290b. **Goodwin, M. A.**, Cryptosporidiosis in birds — a review, *Avian Pathol.*, 18, 365, 1989.
290c. **Goodwin, M. A. and Krabill, V. A.**, Diarrhea associated with small-intestinal cryptosporidiosis in a budgerigar and in a cockatiel, *Avian Dis.*, 33, 829, 1989.
290d. **Goodwin, M. A. and Brown, J.**, Intestinal cryptosporidiosis in chickens, *Avian Dis.*, 33, 770, 1989.
290e. **Goodwin, M. A. and Barsnat, J. A.**, Intractable diarrhea associated intestinal cryptosporidiosis in a domestic cat also infected with feline leukemia virus, *J. Am. Anim. Hosp. Assoc.*, in press.
290f. **Goodwin, M. A., Brown, J., and Fletcher, O. J.**, The relationship of *Cryptosporidium* sp infection of the Bursa of Fabricius, intestinal tract, and respiratory systems of chickens in Georgia, 1974—1988, *Avian Dis.*, in press.
290g. **Goodwin, M. A. and Brown, J.**, Effect of *Cryptosporidium baileyi* on broilers infected at 26 days of age, *Avian Dis.*, 34, in press.
291. **Gorham, S. L., Mallinson, E. T., Synder, D. B., and Odor, E. M.**, Cryptosporidia in the bursa of Fabricius — a correlation with mortality rates in broiler chickens, *Avian Pathol.*, 16, 205, 1987.
291a. **Gorman, T., Alcaíno, H., and Santelices, J.**, *Cryptosporidium* y otras coccidias intestinales en terneros de lechería. Region metropolitana, Chile, *Arch. Med. Vet.*, 21, 29, 1989.
292. **Gouet, Ph.**, Pathologie digestive du chevreau nouveauń in *Les maladies de la chevre, Niort, France, October 9 to 11, 1984, Les colloques de l'INRA*, 1984, 433.
293. **Grabowski, D. J., Powers, K. M., Knott, J. A., Tanuz, M. V., Nims, L. J., Savitt-Kring, M. I., Lauren, C. M., Stevenson, B. I., and Hull, H. F.**, Epidemiologic Notes and Reports, Cryptosporidiosis — New Mexico, 1986, *Morbid. Mortal.*, 36, 561, 1987.
294. **Gray, J.**, Cryptosporidiosis, *J. Infect.*, 10, 1, 1985.
295. **Greene, H. J. and Bakheit, H. A.**, A study of the aetiology, epidemiology and control of calf diarrhoea in Ireland, *Ir. Vet. J.*, 38, 63, 1984.
295a. **Greenberg, R. E., Mir, R., Bank, S., and Siegal, F. P.**, Resolution of intestinal cryptosporidiosis after treatment of AIDS with AZT, *Gastroenterology*, 97, 1327, 1989.
296. **Greyson-Fleg, R. T., Jones, B., Fishman, E.K., Hamilton, S. R., and Siegelman, F. F.**, Computed tomography findings in gastrointestinal involvement by opportunistic organisms in acquired immune deficiency syndrome, *J. Comp. Tomogr.*, 10, 175, 1986.
297. **Grime, J. S.**, Investigation of an outbreak of cryptosporidiosis in Hull, *Commun. Med.*, 9, 310, 1987.
298. **Griner, L. A.**, *Pathology of Zoo Animals*, Zoological Society of San Diego, San Diego, California, 1983, 1.
299. **Gross, T. L., Wheat, J., Bartlett, M., and O'Connor, K. W.**, AIDS and multiple system involvement with Cryptosporidium, *Am. J. Gastroenterol.*, 81, 456, 1986.
300. **Guarda, L. A., Stein, S. A., Cleary, K. A., and Ordóñez, N. G.**, Human cryptosporidiosis in the acquired immune deficiency syndrome, *Arch. Pathol. Lab. Med.*, 107, 562, 1983.
301. **Guarda, L. A., Luna, M. A., Smith, L., Mansell, P. W. A., Gyorkey, F., and Roca, A. N.**, Acquired immune deficiency syndrome: postmortem findings, *Am. J. Clin. Pathol.*, 81, 549, 1984.
302. **Guderian, R. H., Sandoval, V. C. A., and Mackenzie, C. D.**, Cryptosporidiosis in Ecuadorian children with acute diarrhoea, *J. Trop. Pediatr.*, 32, 290, 1986.
303. **Günther, H.**, Kryptosporidien beim Kalb — Bedeutung, Nachweis, Bekämpfung, *Monatsh. Veterinaermed.*, 38, 653, 1983.
304. **Günther, H.**, Bekämpfung der bovinen Kryptosporidiose, *Monatsh. Veterinaermed.*, 39, 730, 1984.
305. **Guy, J. S., Levy, M. G., Ley, D. H., Barnes, H. J., and Gerig, T. M.**, Experimental reproduction of enteritis in bobwhite quail (*Colinus virginianus*) with *Cryptosporidium* and Reovirus, *Avian Dis.*, 31, 713, 1987.
306. **Guy, J. S., Levy, M. G., Ley, D. H., Barnes, H. J., and Gerig, T. M.**, Interaction of reovirus and *Cryptosporidium baileyi* in experimentally infected chickens, *Avian Dis.*, 32, 381, 1988.
307. **Hall, G. A., Reynolds, D. J., Chanter, N., Morgan, J. H., Parsons, K. R., Debney, T. G., Bland, A. P., and Bridger, J. C.**, Dysentery caused by *Escherichia coli* (S102-9) in calves: natural and experimental disease, *Vet. Pathol.*, 22, 156, 1985.
308. **Hall, G. A., Reynolds, D. J., Parsons, K. R., Bland, A. P., and Morgan, J. H.**, Pathology of calves with diarrhoea in southern Britain, *Res. Vet. Sci.*, 45, 240, 1988.

309. **Hamilton, W. J.,** Management of deer: structure of the industry. The stratification of the industry, in *Management and Diseases of Deer,* Alexander, T. L., Ed., Publ. Vet. Deer Soc., 1986, 10.
310. **Hamilton, W. J.,** Management of deer: structure of the industry. Land resources and related systems of deer farming, in *Management and Diseases of Deer,* Alexander, T. L., Ed., Publ. Vet. Deer Soc., 1986, 7.
311. **Hamoudi, A. C., Qualman, S. J., Marcon, M. J., Hribar, M., McClung, H. J., Murray, R. D., and Cannon, H. J.,** Do regional variations in prevalence of cyptosporidiosis occur? The central Ohio experience, *Am. J. Public Health,* 78, 273, 1988.
312. **Hampton, J. C. and Rosario, B.,** The attachment of protozoan parasites to intestinal epithelial cells of the mouse, *J. Parasitol.,* 52, 939, 1966.
313. **Hannah, J. and Riordan, T.,** Case to case spread of cryptosporidiosis; evidence from a day nursery outbreak, *Public Health,* 102, 539, 1988.
314. **Harari, M. D., West, B., and Dwyer, B.,** Cryptosporidium as a cause of laryngotracheitis in an infant, *Lancet,* 1, 1207, 1986.
315. **Harp, J. A., Wannemuehler, M. W., Woodmansee, D. B., and Moon, H. W.,** Susceptibility of germfree or antibiotic-treated adult mice to *Cryptosporidium parvum, Infect. Immun.,* 56, 2006, 1988.
315a. **Harp, J. A., Woodmansee, D. B., and Moon, H. W.,** Effect of colostral antibody on susceptibility of calves to *Cryptosporidium parvum* infection, *Am. J. Vet. Res.,* 50, 2117, 1989.
316. **Harris, L. F. and Kinzer, G. M.,** Cryptosporidiosis, *Ala. Med.,* 54, 18, 1985.
317. **Hart, C. A., Baxby, D., and Blundell, N.,** Gastro-enteritis due to Cryptosporidium: a prospective survey in a children's hospital, *J. Infect.,* 9, 264, 1984.
318. **Hart, C. A. and Baxby, D.,** Cryptosporidiosis in immunocompetent patients, *N. Engl. J. Med.,* 313, 1018, 1985.
319. **Hart, C. A. and Baxby, D.,** Management of cryptosporidiosis, *J. Antimicrob. Chemother.,* 15, 3, 1985.
319a. **Hatkin, J. M., Lindsay, D. S., Giambrone, J. J., Hoerr, F. J., and Blagburn, B. L.,** Experimental biliary cryptosporidiosis in broiler chickens, *Avian Dis.,* 34, 1990, in press.
320. **Hawkins, S. P., Thomas, R. P., and Tesdate, C.,** Acute pancreatitis: a new finding in cryptosporidium enteritis, *Br. Med. J.,* 294, 483, 1987.
321. **Hay, E. M., Winfield, J., and McKendrick, M. W.,** Reactive arthritis associated with cryptosporidium enteritis, *Br. Med. J.,* 295, 248, 1987.
322. **Hayes, E. B., Matte, T. D., O'Brien, T. R., McKinley, T. W., Logsdon, G. S., Rose, J. B., Ungar, B. L. P., Word, D. M., Pinsky, P. F., Cummings, M. L., Wilson, M. A., Long, E. G., Hurwitz, E. S., and Juranek, D. D.,** Large community outbreak of cryptosporidiosis due to contamination of a filtered public water supply, *N. Eng. J. Med.,* 320, 1372, 1989.
323. **Heidl, M., Hering, L., and Scholz, H.,** Kryptosporidiose bei Kindern, *Infection,* 14, 173, 1986.
324. **Heijbel, H., Slaine, K., Seigel, B., Wall, P., McNabb, S. J. N., Gibbons, W., and Istre, G. R.,** Outbreak of diarrhea in a day care center with spread to household members: the role of *Cryptosporidium, Pediatr. Infect. Dis. J.,* 6, 532, 1987.
324a. **Heijmans, J. F.,** Cryptosporidiose bij slachtkuikens in Nederland, *Tijdschr. Diergeneeskd.,* 577, 1988.
325. **Heine, J., Moon, H. W., Woodmansee, D. B., and Pohlenz, J. F. L.,** Experimental tracheal and conjunctival infections with *Cryptosporidium* sp. in pigs, *Vet. Parasitol.,* 17, 17, 1984.
326. **Heine, J. and Boch, J.,** Kryptosporidien-Infektionen beim Kalb. Nachweis, Vorkommen und experimentelle Übertragung, *Berl. Muench. Tieraerztl. Wochenschr.,* 94, 289, 1981.
327. **Heine, J.,** Eine einfache Nachweismethode für Kryptosporidien im Kot, *Zentralbl. Veterinaermed. Reihe B,* 29, 324, 1982.
328. **Heine, J., Moon, H. W., and Woodmansee, D. B.,** Persistent *Cryptosporidium* infection in congenitally athymic (nude) mice, *Infect. Immun.,* 43, 856, 1984.
329. **Heine, J., Pohlenz, J. F. L., Moon, H. W., and Woode, G. N.,** Enteric lesions and diarrhea in gnotobiotic calves monoinfected with *Cryptosporidium* species, *J. Infect. Dis.,* 150, 768, 1984.
331. **Heldstab, A., Berger, R., and Bestetti, G.,** Demonstration of virus, Chlamydiae and Cryptosporidae in snake gastroenteritis, *Herpetopathologia,* 1, 101, 1989.
332. **Henkel, G.,** Kryptosporidiose des Menschen, *Muench. Med. Wochenschr.,* 125, 1099, 1983.
333. **Henricksen, S. A. and Krogh, H. V.,** A note on bovine infections with Cryptosporidia in Denmark, *Nord. Vet. Med.,* 32, 501, 1980.
334. **Henricksen, S. A. and Pohlenz, J. F. L.,** Staining of cryptosporidia by a modified Ziehl-Neelsen technique, *Acta Vet. Scand.,* 22, 594, 1981.
335. **Henriksen, S. A. and Krogh, H. V.,** Bovine cryptosporidiosis in Denmark. I. Prevalence, age distribution, and seasonal variation, *Nord. Vet. Med.,* 37, 34, 1985.
336. **Henriksen, S. A.,** Epidemiology of cryptosporidiosis in calves, in *1st Int. Workshop on Cryptosporidiosis,* Angus, K. W. and Blewett, D. A., Eds., September 7 to 8, 1988, Edinburgh, 1989, 79.
337. **Henrickson, R. V., Maul, D. H., Osborn, K. G., Sever, J. I., Madden, D. L., Ellingsworth, L. R., Anderson, J. H., Lowenstine, L. J., and Gardner, M. B.,** Epidemic of acquired immunodeficiency in rhesus monkeys, *Lancet,* 1, 388, 1983.

338. **Henry, M. C., De Clercq, D., Lokombe, B., Kayembe, K., Kapita, B., Mamba, K., Mbendi, N., and Mazebo, P.**, Parasitological observations of chronic diarrhoea in suspected AIDS adult patients in Kinshasa (Zaire), *Trans. Roy. Soc. Trop. Med. Hyg.*, 80, 309, 1986.
339. **Hering, L., Heidl, M., and Dörffel, J.**, Nachweis von Kryptosporidien bei Enteritis — Kasuistische Mitteilung, *Klin. Med.*, 41, 597, 1986.
340. **Heuschele, W. P., Oosterhuis, J., Janssen, D., Robinson, P. T., Ensley, P. K., Meier, E., Olson, T., Anderson, M. P., and Benirschke, K.**, Cryptosporidial infections in captive wild animals, *J. Wild. Dis.*, 22, 493, 1986.
341. **Heyman, M. B., Shigekuni, L. K., and Ammann, A. J.**, Separation of Cryptosporidium oocysts from fecal debris by density gradient centrifugation and glass bead columns, *J. Clin. Microbiol.*, 23, 789, 1986.
342. **Heyworth, M. F. and Owen, R. L.**, Gastrointestinal aspects of the acquired immunodeficiency syndrome, *Surv. Digest. Dis.*, 3, 197, 1985.
343. **Hiepe, Th., Jungmann, R., Plath, H., and Shuster, R.**, Untersuchungen über Vorkommen, Nachweis und Krankheitsbild der Kryptosporidien-Infektion neugeborener Schaflämmer., *Monatsh. Veterinaermed.*, 40, 524, 1985.
344. **Hiepe, Th., Jungmann, R., and Roffeis, R. P.**, Vorkommen, Verlauf, Nachweis und Bekämpfung der Kryptosporidiose unter den Bedingungen der Kälber-Intensivhaltung, *Monatsh. Veterinaermed.*, 43, 470, 1988.
345. **Hill, B. D.**, Immune responses in *Cryptosporidium* infections. in *1st Int. Workshop on Cryptosporidiosis*, Angus, K. W. and Blewett, D. A., Eds., September 7 to 8, 1988, Edinburgh, 1989, 97.
345a. **Hill, B. D., Blewett, D. A., Dawson, A. M., and Wright, S.**, Analysis of the kinetics, isotype and specificity of serum and coproantibody in lambs infected with *Cryptosporidium parvum*, *Res. Vet. Sci.*, 48, 76, 1990.
346. **Hillyer, G. V. and Climent, C.**, Acquired immunodeficiency syndrome (AIDS) and parasitic diseases in Puerto Rico, *Bol. Assoc. Med. P. R.*, 80, 312, 1988.
347. **Hinnant, K. L., Rotterdam H. Z., Bell, E. T., and Tapper, M. L.**, Cytomegalovirus infection of the alimentary tract: a clinicopathological correlation, *Am. J. Gastroenterol.*, 81, 944, 1986.
348. **Hinnant, K., Swartz, A., Rotterdam, H., and Rudski, C.**, Cytomegaloviral and cryptosporidial cholecystitis in two patients with AIDS, *Am. J. Surg. Pathol.*, 13, 57, 1989.
348a. **Hira, P. R., Al-Ali, F., Zaki, M., Saleh, Q., Sharda, D., and Behbehani, K.**, Human cryptosporidiosis in the Arabian Gulf: first report of infections in children in Kuwait, *J. Trop. Med. Hyg.*, 92, 249, 1989.
349. **Hoerr, F. J., Ranck, F. M., and Hastings, T. F.**, Respiratory cryptosporidiosis in turkeys, *J. Am. Vet. Med. Assoc.*, 173, 1591, 1978.
350. **Hoerr, F. J., Current, W. L., and Haynes, T. B.**, Fatal cryptosporidiosis in quail, *Avian Dis.*, 30, 421, 1986.
351. **Højlyng, N., Molbak, K., and Jepsen, S.**, Cryptosporidiosis in Liberian children, *Lancet*, 1, 734, 1984.
353. **Højlyng, N., Molbak, K, and Jepsen, S.**, Cryptosporidiosis in human beings is not primarily a zoonosis, *J. Infect.*, 11, 270, 1985.
354. **Højlyng, N., Mølbak, K., and Jepsen, S.**, *Cryptosporidium* spp., a frequent cause of diarrhea in Liberian children, *J. Clin. Microbiol.*, 23, 1109, 1986.
355. **Højlyng, N., Holten-Andersen, W., and Jepsen, S.**, Cryptosporidiosis: a case of airborne transmission, *Lancet*, 2, 271, 1987.
356. **Højlyng, N. and Jensen, B. N.**, Respiratory cryptosporidiosis in HIV-positive patients, *Lancet*, 1, 590, 1988.
356a. **Holland, R. E., Herdt, T. H., and Refsal, K. R.**, Pulmonary excretion of H_2 in calves with *Cryptosporidium*-induced malabsorption, *Dig. Dis. Sci.*, 34, 1399, 1989.
357. **Holley, H. P. and Thiers, B. H.**, Cryptosporidiosis in a patient receiving immunosuppressive therapy: possible activation of latent infection, *Digest. Dis. Sci.*, 31, 1004, 1986.
358. **Holley, H. P. and Dover, C.**, *Cryptosporidium*: a common cause of parasitic diarrhea in otherwise healthy individuals, *J. Infect. Dis.*, 153, 365, 1986.
359. **Holten-Anderson, W., Gerstoft, J., Henriksen, S. A., and Pedersen, N. S.**, Prevalence of Cryptosporidium among patients with acute enteric infection, *J. Infect.*, 9, 277, 1984.
360. **Hoover, D. M., Hoerr, F. J., Carlton, W. W., Hinsman, E. J., and Ferguson, H. W.**, Enteric cryptosporidiosis in a naso tang, *Naso lituratus* Bloch and Schneider, *J. Fish Dis.*, 4, 425, 1981.
361. **Horen, W. P.**, Detection of *Cryptosporidium* in human fecal specimens, *J. Parasitol.*, 69, 622, 1983.
362. **House, J. A.**, Economic impact of rotavirus and other neonatal disease agents of animals, *J. Am. Vet. Med. Assoc.*, 173, 573, 1978.
363. **Howerth, E. W.**, Bovine cryptosporidiosis, *J. S. Afr. Vet. Assoc.*, 52, 251, 1981.
364. **Hunt, D. A., Shannon, R., Palmer, S. R., and Jephcott, A. E.**, Cryptosporidiosis in an urban community, *Br. Med. J.*, 289, 814, 1984.
365. **Hunt, R. D., Blake, B. J., Chalifoux, L. V., Sehgal, P. K., King, N. W., and Letvin, N. L.**, Transmission of naturally occurring lymphoma in macaque monkeys, *Proc. Natl. Acad. Sci. U.S.A.*, 80, 5085, 1983.
366. **Hussain Qadri, S. M. and Khalil, S. H.**, Intestinal parasites: incidence and etiology in over 1,000 patients at King Faisal Specialist Hospital in Riyadh, *Ann. Sudan. Med.*, 7, 207, 1987.
367. **Inman, L. R. and Takeuchi, A.**, Spontaneous cryptosporidiosis in an adult female rabbit, *Vet. Pathol.*, 16, 89, 1979.

368. **Isaac-Renton, J. L., Fogel, D., Stibbs, H. H., and Ongerth, J. E.,** Giardia and Cryptosporidium in drinking water, *Lancet*, 1, 973, 1987.

369. **Isaacs, D., Hunt, G. H., Phillips, A. D., Price, E. H., Raafat, F., and Walker-Smith, J. A.,** Cryptosporidiosis in immunocompetent children, *J. Clin. Pathol.*, 38, 76, 1985.

370. **Iseki, M.,** *Cryptosporidium felis* sp. n. (Protozoa: Eimeriorina) from the domestic cat, *Jpn. J. Parasitol.*, 28, 285, 1979.

371. **Iseki, M.,** Two species of *Cryptosporidium* naturally infecting house rats, *Rattus-norvegicus*, *Jpn. J. Parasitol.*, 35, 521, 1986.

372. **Iseki, M., Maekawa, T., Moriya, K., Uni, S., and Takada, S.,** Infectivity of *Cryptosporidium muris* (strain RN 66) in various laboratory animals, *Parasitol. Res.*, 75, 218, 1989.

373. **Iskander, A. R., Tawfeek, A., and Farid, A. F.,** Cryptosporidial infection among buffalo calves in Egypt, *Indian J. Anim. Sci.*, 57, 1057, 1987.

374. **Itakura, C., Goryo, M., and Umemura, T.,** Cryptosporidial infection in chickens, *Avian Pathol.*, 13, 487, 1984.

375. **Itakura, C., Nakamura, H., Umemura, T., and Goryo, M.,** Ultrastructure of cryptosporidial life cycle in chicken host cells, *Avian Pathol.*, 14, 237, 1985.

376. **Izquierdo, J., Antúnez, I., Calderón, M. T., Pérez Giraldo, C., and Muñoz Sanz, A.,** Diarrea por Cryptosporidium y neoplasia colónica, *Rev. Clin. Esp.*, 182, 393, 1987.

377. **Jackson, B. and Casemore, D. P.,** Cryptosporidium and diarrhoea, *Br. Med. J.*, 293, 85, 1986.

377a. **Jacobson, E.,** Personal communication.

378. **Jakubowski, W. and Ericksen, T. H.,** Methods for detection of *Giardia* cysts in water supplies, in Proc. Symp. Waterborne Transmission of Giardiasis, September 18 to 20, 1978, EPA 600/9-79-001, 1979, 193.

379. **Jakubowski, W.,** Detection of *Giardia* cysts in drinking water: State of the art, in *Giardia and Giardiasis*, Erlandsen, S. L. and Meyer, E. A., Eds., Plenum Press, New York, 1984, 263.

380. **Janda, J., Machala, V., Lebl, J., Šmidáková, E., Bláhová, K., Nevoral, J., and Pavlásek, I.,** Střevní kryptosporidióza u 14 letého děvčete s malabsorpčním syndromem a nejasným získanym defektem imunity, *Cs. Pediatr.*, 42, 149, 1987.

381. **Janoff, E. N. and Reller, L. B.,** *Cryptosporidium* species, a protean protozoan, *J. Clin. Microbiol.*, 25, 967, 1987.

381a. **Jantanavivat, C.,** Apparent inability of *Cryptosporidium baileyi* of chickens to infect poultry handlers in Bangkok, *Trans. R. Soc. Trop. Med. Hyg.*, 83, 651, 1989.

382. **Jarroll, E. L., Bingham, A. K., and Meyer, E. A.,** Effect of chlorine on *Giardia lamblia* cyst viability, *Appl. Environ. Microbiol.*, 41, 483, 1981.

383. **Jerrett, I. V. and Snodgrass, D. R.,** Cryptosporidia associated with outbreaks of neonatal calf diarrhoea, *Aust. Vet. J.*, 57, 434, 1981.

384. **Jervis, H. R., Merrill, T. G., and Sprinz, H.,** Coccidiosis in the guinea pig small intestine due to a *Cryptosporidium*, *Am. J. Vet. Res.*, 27, 408, 1966.

385. **Jirouš, J., Cyprichová, V., and Vaništa, J.,** A case of human cryptosporidiosis in Czechoslovakia, *J. Hyg. Epidemiol. Microbiol. Immunol.*, 30, 103, 1986.

386. **Jokipii, L. and Jokipii, A. M. M.,** Timing of symptoms and oocyst excretion in human cryptosporidiosis, *N. Engl. J. Med.*, 315, 1643, 1986.

387. **Jokipii, L., Pohjola, S., and Jokipii, A. M. M.,** Cryptosporidium: a frequent finding in patients with gastrointestinal symptoms, *Lancet*, 2, 358, 1983.

388. **Jokipii, A., Hemilä, M., and Jokipii, L.,** Prospective study of acquisition of *Cryptosporidium*, *Giardia lamblia*, and gastrointestinal illness, *Lancet*, 2, 487, 1985.

389. **Jokipii, L., Pohjola, S., and Jokipii, A. M. M.,** Cryptosporidiosis associated with traveling and giardiasis, *Gastroenterology*, 89, 838, 1985.

390. **Jokipii, L. and Jokipii, A. M. M.,** Timing of symptoms and oocyst excretion in human cryptosporidiosis, *N. Engl. J. Med.*, 317, 169, 1987.

391. **Jonas, C., Deprez, C., De Maubeuge, J., Taelman, H., Panzer, J. M., and Deltenre, M.,** Cryptosporidium in patient with acquired immunodeficiency syndrome, *Lancet*, 2, 964, 1983.

393. **Jungmann, R. and Hiepe, T.,** Vorkommen und Intravitaldiagnostik der Kryptosporidiose bei neugeboronen Kälbern (Kuruzmitteilung), *Monatsh. Veterinaermed.*, 38, 299, 1983.

394. **Just, G., Neisel, F., Helm, E. B., Brodt, H. R., Stümer, S., and Stille, W.,** Kryptosporidien-Infektionen bei AIDS, *Dtsch. Med. Wochenschr.*, 112, 378, 1987.

395. **Kageruka, P., Brandt, J. R. A., Taelman, H., and Jonas, C.,** Modified Köster staining method for the diagnosis of cryptosporidiosis, *Am. Soc. Beige Méd. Trop.*, 64, 171, 1984.

396. **Kahn, D. G., Garfinkle, J. M., Klonoff, D. C., Pembrook, L. J., and Morrow, D. J.,** Cryptosporidial and cytomegaloviral hepatitis and cholecystitis, *Arch. Pathol. Lab. Med.*, 111, 879, 1987.

397. **Kansouzkou-Kanakaudif, A., Danielides, B., and Arzanitidou-Zayonas, T.,** Cryptosporidial enteritis, *Acta Microbiol. Hell.*, 29, 369, 1984 (in Greek).

399. **Katz, M. D., Erstad, B. L., and Rose, C.**, Treatment of severe Cryptosporidium-related diarrhea with octreotide in a patient with AIDS, *Drug Intell. Clin. Pharm.*, 22, 134, 1988.
400. **Katzenelson, E., Kletter, B., and Shuval, H. I.**, Inactivation kinetics of viruses and bacteria in water by use of ozone, *J. Am. Water Works Assoc.*, 66, 725, 1974.
401. **Kavin, H., Jonas, R. B., Chowdhury, L., and Kabins, S.**, Acalculous cholecystitis and cytomegalovirus infection in the acquired immunodeficiency syndrome, *Ann. Intern Med.*, 104, 53, 1986.
402. **Kazlow, P. G., Shah, K., Benkov, K. J., Dische, R., and LeLeiko, N. S.**, Esophagial cryptosporidiosis in a child with acquired immune deficiency syndrome, *Gastroenterology*, 91, 1301, 1986.
403. **Kennedy, G. A., Kreitner, G. L., and Strafuss, A. C.**, Cryptosporidiosis in three pigs, *J. Am. Vet. Med. Assoc.*, 170, 348, 1977.
404. **Keren, G., Barzilai, A., Barzilay, Z., Goldschmied-Reouven, A., Bogokowsky, B., and Rubinstein, E.**, Life-threatening cryptosporidiosis in immunocompetent infants, *Eur. J. Pediatr.*, 146, 187, 1987.
405. **Kern, P., Toy, J., and Dietrich, M.**, Preliminary clinical observations with recombinant interleukin-2 in patients with AIDS or LAS, *Blut*, 50, 1, 1985.
406. **Kern, W., Mayer, S., Kreuzer, P., and Vanek, E.**, Low prevalence of intestinal cryptosporidiosis among immunocompetent and immunocompromised patients with and without diarrhoea in southern Germany, *Infection*, 15, 440, 1987.
407. **Khan, Z. A., Namnyak, S. S., Al Jama, A. A., and Madan, I.**, Prevalence of cryptosporidiosis in Dammam and Alkhobar, Saudi Arabia, *Ann. Trop. Pediatr.*, 8, 170, 1988.
408. **Kibbler, C. C., Smith, A., Hamilton-Dutoit, S. J., Milburn, H., Pattinson, J. K., and Prentice, H. G.**, Pulmonary cryptosporidiosis occurring in a bone marrow transplant patient, *Scand. J. Infect. Dis.*, 19, 581, 1987.
409. **Kilani, R. T. and Sekla, L.**, Purification of *Cryptosporidium* oocysts and sporozoites by cesium chloride and percoll gradients, *Am. J. Trop. Med. Hyg.*, 36, 505, 1987.
410. **Kim, C. W.**, Chemotherapeutic effect of arprinocid in experimental cryptosporidiosis, *J. Parasitol.*, 73, 663, 1987.
411. **Kim, C. W.**, *Cryptosporidium* sp.: experimental infection in Syrian golden hamsters, *Exp. Parasitol.*, 63, 243, 1987.
412. **King, N. W., Hunt, R. D., and Letvin, N. L.**, Histopathologic changes in macaque monkeys with an acquired immune deficiency syndrome (AIDS), *Am. J. Pathol.*, 113, 382, 1983.
413. **King, N. W.**, Simian models of acquired immunodeficiency syndrome (AIDS): a review, *Vet. Pathol.*, 23, 345, 1986.
414. **Kirkpatrick, C. E. and Farrell, J. P.**, Cryptosporidiosis, *Compend. Contin. Educ. Pract. Vet.*, 6, S154, 1984.
415. **Kirkpatrick, C. E.**, *Cryptosporidium* infection as a cause of calf diarrhea, *Vet. Clin. North Am. Food Anim. Pract.*, 1, 515, 1985.
416. **Kirkpatrick, C. E.**, Animal reservoirs of *Cryptosporidium* spp. and *Isospora belli*, *J. Infect Dis.*, 158, 909, 1988.
417. **Kirkpatrick, C. E., DiPietro, J. A., Todd, K. S., Jr., and Baszcer, T. V.**, An outbreak of cryptosporidiosis in a group of neonatal ponies, to be published.
418. **Klesius, P. H., Haynes, T. B., and Malo, L. K.**, Infectivity of *Cryptosporidium* sp isolated from wild mice for calves and mice, *J. Am. Vet. Med. Assoc.*, 189, 192, 1986.
419. **Koch, K. L., Shankey, T. V., Weinstein, G. S., Dye, R. E., Abt, A. B., Current, W. L., and Eyster, M. E.**, Cryptosporidiosis in a patient with hemophilia, common variable hypogammaglobulinemia and the acquired immunodeficiency syndrome, *Ann. Intern. Med.*, 99, 337, 1983.
420. **Koch, K. L., Phillips, D. J., Aber, R. C., and Current, W.**, Cryptosporidiosis in hospital personnel: evidence for person-to-person transmission, *Ann. Intern. Med.*, 102, 593, 1985.
421. **Koch, K. L., Phillips, D. J., Aber, R. C., and Current, W.L.**, Kryptosporidiose beim Pflegepersonal als Folge beruflicher Exponierung, *Verdauungskrankheiten J.*, 6, 148, 1988.
422. **Kocoshis, S. A., Cibull, M. L., Davis, T. E., Hinton, J. T., Seip, M., and Banwell, J. G.**, Intestinal and pulmonary cryptosporidiosis in an infant with severe combined immune deficiency, *J. Pediatr. Gastroenterol. Nutr.*, 3, 149, 1984.
422a. **Koffi-Akoua, G., Ferly-Therizol, M., Kouassi-Beugre, M. T., Konan, A., Timite, A. M., Assi Adou, J., and Assale, G.**, Cryptosporidies et *Candida* dans les diarrhées infantiles a Abidjan, *Bull. Soc. Pathol. Exot.*, 82, 451, 1989.
422b. **Korich, D. G., Mead, J. R., Madore, M. S., and Sinclair, N. A., and Sterling, C. R.**, Effects of ozone, chlorine dioxide, chlorine and monochloramine on *Cryptosporidium* oocyst viability, *Appl. Environ. Microbiol.*, 56, 1423, 1990.
423. **Kořinek, J. and Chroust, K.**, Dynamics of the incidence of cryptosporidia in calves, *Acta Vet. (Brno)*, 57, 39, 1988.
424. **Körner, E. and Winkelmann, J.**, Health aspects of deer farming, in *The Management and Health of Farmed Deer*, Reid, H. W., Ed., Kluwer Academic, Boston, 1988, 89.
425. **Korsholm, H. and Henriksen, S. A.**, Infection with Cryptosporidium in roe deer (*Capreolus capreolus* L.). A preliminary report, *Nord. Vet. Med.*, 36, 266, 1984.

426. **Koudela, B., Vítovec, J., Dat, D. T., and Lan, P. D.,** Preliminary communication on cryptosporidiosis of pigs in Viet-Nam, *Folia Parasitol. (Prague)*, 33, 301, 1986.
426a. **Koudela, B., Vítovec, J., Štěrba, J. and Miláček, P.,** An unusual localization of developmental stages of *Cryptospordium parvum* Tyzzer, 1912 in the cells of small intestine of a gnotobiotic piglet, *Folia Parasitol. (Prague)*, 36, 219, 1989.
427. **Kovatch, R. M. and White, J. D.,** Cryptosporidiosis in two juvenile rhesus monkeys, *Vet. Pathol.*, 9, 426, 1972.
428. **Kozakiewicz, B., and Maszewska, I.,** Ekstensywność inwazji Cryptosporidium sp. u cielqt z objawami biegunki w gospodarstwach wielkostadnych, *Med. Weter.*, 44, 404, 1988.
429. **Krause, M., Burger, H. R., Vollenweider, R., and Anabitarte, M.,** Pathologisch-anatomische Befunde beim erworbenen Immundefektsyndrom (AIDS), *VASA*, 13, 99, 1984.
430. **Krogh, H. V. and Henriksen, S. A.,** Bovine cryptosporidiosis in Denmark. II. Cryptosporidia associated with neonatal calf diarrhea, *Nord. Vet. Med.*, 37, 42, 1985.
431. **Kwaga, J. K. P., Umoh, J. U., and Odoba, M. B.,** *Cryptosporidium* infections in humans with gastroenteritis in Zaria, Nigeria, *Epidem. Infect.*, 101, 93, 1988.
432. **Kyle, R.,** *A Feast in the Wild*, KUDU Publishing in Wildlife Management, Kidlington, Oxon, England, 1987, 1.
433. **Lacroix, C., Berthier, M., Agius, G., Nivet, H., Bouffassa, S., Hoppeler, A., and Jacquemin, J. L.,** Syndrome hemolytique et uremique associe a des oocystes de *Cryptosporidium* dans les selles, *Bull. Soc. Fr. Parasitol.*, 5, 53, 1987.
434. **Lacroix, C., Berthier, M., Agius, G., Bonneau, D., Pallu, B., and Jacquemin, J. L.,** *Cryptosporidium* oocysts in immunocompetent children: epidemiologic investigations in the day-care centers of Poitiers, France, *Eur. J. Epidemiol.*, 3, 381, 1987.
435. **Lähdevirta, J., Jokipi, A. M. M., Sammalkoysi, K., and Jokipii, L.,** Perinatal infection with Cryptosporidium and failure to thrive, *Lancet*, 1, 48, 1987.
436. **Lambertus, M. W. and Mathisen, G. E.,** *Mycobacterium marinum* infection in a patient with cryptosporidiosis and the acquired immunodeficiency syndrome, *Cutis*, 41, 38, 1988.
437. **Landsberg, J. H. and Paperna, I.,** Ultrastructural study of the coccidian *Cryptosporidium* sp. from stomachs of juvenile cichlid fish, *Dis. Aquat. Org.*, 2, 13, 1986.
438. **Landsberg, J. H. and Paperna, I.,** Intestinal infections by *Eimeria* (S. L.) *vanasi* n. sp. (Eimeriidae, Apicomplexa, Protozoa) in cichlid fish, *Ann. Parasitol. Hum. Comp.*, 62, 283, 1987.
439. **Landsverk, T.,** Cryptosporidiosis and the follicle-associated epithelium over the ileal Peyer's patch in calves, *Res. Vet. Sci.*, 42, 299, 1987.
440. **Lasser, K. H., Lewin, K. J., and Ryning, F. W.,** Cryptosporidial enteritis in a patient with congenital hypogammaglobulinemia, *Hum. Pathol.*, 10, 234, 1979.
441. **Latimer, K. S., Goodwin, M. A., and Davis, M. K.,** Rapid cytologic diagnosis of respiratory cryptosporidiosis in chickens, *Avian Dis.*, 32, 826, 1988.
442. **Laughon, B. E., Druckman, D. A., Vernon, A., Quinn, T. C., Polk, B. F., Modlin, J. F., Yolken, R. H., and Bartlett, J. G.,** Prevalence of enteric pathogens in homosexual men with and without acquired immunodeficiency syndrome, *Gastroenterology*, 94, 984, 1988.
443. **Laxer, M. A., Alcantara, A. K., Javato-Laxer, M., Cui, M. D., Leano, R. A., Bautista, S., Ramilo, N., and Gavina, T. L.,** *Cryptosporidium* from Palawan, Republic of the Philippines, *Philipp. J. Microbiol. Infect. Dis.*, 17, 1, 1988.
443a. **Laxer, M. A., Alcantara, A. K., Javato-Laxer, M., Menorca, D. M., Fernando, M. T., and Ranoa, C. P.,** Immune response to cryptosporidiosis in Philippine children, *Am. J. Trop. Med. Hyg.*, 42, 131, 1990.
444. **Lazo, A., Barriga, O. O., Redman, D. R., and Bech-Nielsen, S.,** Identification by transfer blot of antigens reactive in the enzyme-linked immunosorbent assay (ELISA) in rabbits immunized and a calf infected with *Cryptosporidium* sp., *Vet. Parasitol.*, 21, 151, 1986.
445. **Leach, G.,** *Energy and Food Production*, IPC Science and Technology Press, Guildford, Surrey, England, 1976, 21.
446. **Leahy, J. G., Rubin, A. J., and Sproul, O. J.,** Inactivation of *Giardia muris* cysts by free chlorine, *Appl. Environ. Microbiol.*, 53, 1448, 1987.
447. **Le Charpentier, Y., Galian, A., Messing, B., Andreani, T., Modegliani, R., Lavergne, A., Hoang, C., Puel, F., and Houin, R.,** Diagnostic ultrastructural d'une infection intestinale humaine, a *Cryptosporidium* sp., *Ann. Pathol.*, 2, 336, 1982.
448. **LeChevallier, M. W., Trok, T., Burns, M., and Lee, R. G.,** Immunofluorescence techniques for detection of parasites in drinking water, presented at Am. Water Works Assoc. Water Quality Technical Conf., St. Louis, November 13 to 17, 1988.
449. **Leek, R. G. and Fayer, R.,** Prevalence of *Cryptosporidium* infections and their relation to diarrhea in calves on 12 dairy farms in Maryland, *Proc. Helminthol. Soc. Wash.*, 51, 360, 1984.
450. **Lefkowitch, J. H., Krumholz, S., Feng-Chen, K. C., Griffin, P., Despommier, D., and Brasitus, T. A.,** Cryptosporidiosis of the human small intestine: a light and electron microscopic study, *Hum. Pathol.*, 15, 746, 1984.

451. **Lengronne, D., Regnier, G., Veau, P., Chermette, R., Boufassa, S., and Soulé, C.**, Cryptosporidiose chez les poulains diarrhóiques, *Point Vet.*, 17, 528, 1985.
452. **Lerner, C. W. and Tapper, M. L.**, Opportunistic infection complicating acquired immune deficiency syndrome. Clinical features of 25 cases, *Medicine*, 63, 155, 1984.
453. **Letvin, N. L., Eaton, K. A., Aldrich, W. R., Sehgal, P. K., Blake, B. J., Schlossman, S. F., King, N. W., and Hunt, R. D.**, Acquired immunodeficiency syndrome in a colony of macaque monkeys, *Proc. Natl. Acad. Sci. U.S.A.*, 80, 2718, 1983.
454. **Letvin, N. L., King, N. W., Daniel, M. D., Aldrich, W. R., Blake, B. J., and Hunt, R. D.**, Experimental transmission of macaque AIDS by means of inoculation of macaque lymphoma tissue, *Lancet*, 2, 599, 1983.
455. **Letvin, N. L., Daniel, M. D., Sehgal, P. K., Desrosiers, R. C., Hunt, R. D., Waldron, L. M., MacKey, J. J., Schmidt, D. K., Chalifoux, L. V., and King, N. W.**, Induction of AIDS-like disease in macaques monkeys with T-cell tropic retrovirus STLV-III, *Science*, 230, 71, 1985.
456. **Levine, N. D. and Tadros, W.**, Named species and hosts of *Sarcocystis* (Protozoa: Apicomplexa: Sarcocystidae), *Syst. Parasitol.*, 2, 41, 1980.
457. **Levine, N. D.**, Some corrections of coccidian (Apicomplexa: Protozoa) nomenclature, *J. Parasitol.*, 66, 830, 1980.
458. **Levine, N. D.**, Taxonomy and review of the coccidian genus *Cryptosporidium* (Protozoa, Apicomplexa), *J. Protozool.*, 31, 94, 1984.
459. **Levine, N. D.**, The genera *Cryptosporidium* and *Epieimeria* in the coccidian family Cryptosporidiidae (Protozoa, Apicomplexa), *Trans. Am. Microsc. Soc.*, 103, 205, 1984.
460. **Levine, N. D.**, Phylum II. Apicomplexa Levine 1970, in *Illustrated Guide to the Protozoa*, Lee, J. J., Hutner, S. H., and Bovee, E. C., Eds., Society of Protozoologists, Lawrence, KS, 1985, 322.
461. **Levine, N. D.**, The taxonomy of *Sarcocystis* (Protozoa, Apicomplexa) species, *J. Parasitol.*, 72, 372, 1986.
462. **Levine, J. F., Levy, M. G., Walker, R. L., and Crittenden, S.**, Cryptosporidiosis in veterinary students, *J. Am. Vet. Med. Assoc.*, 193, 1413, 1988.
463. **Levy, M. G., Ley, D. H., Barnes, H. J., Gerig, T. M., and Corbett, W. T.**, Experimental cryptosporidiosis and infectious bursal disease virus infeciton of specific-pathogen-free chickens, *Avian Dis.*, 32, 803, 1988.
464. **Lewis, I. J., Hart, C. A., and Baxby, D.**, Diarrhoea due to *Cryptosporidium* in acute lymphoblastic leukemia, *Arch. Dis. Child.*, 60, 60, 1985.
465. **Ley, D. H., Levy, M. G., Hunter, L., Corbett, W., and Barnes, H. J.**, Cryptosporidia-positive rates of avian necropsy accessions determined by examination of auramine O-stained fecal smears, *Avian Dis.*, 32, 108, 1988.
466. **Liebler, E. M., Pohlenz, J. F., and Woodmansee, D. B.**, Experimental intrauterine infection of adult BALB/c mice with *Cryptosporidium* sp., *Infect. Immun.*, 54, 255, 1986.
467. **Lin, J. A., Shieh, H. K., and Wang, J. S.**, Studies on the avian cryptosporidiosis: pathological and electron microscopic observation, *Taiwan J. Vet. Anim. Husb.*, 44, 43, 1984.
468. **Lindsay, D. S., Blagburn, B. L., and Sundermann, C. A.**, Host specificity of *Cryptosporidium* sp. isolated from chickens, *J. Parasitol.*, 72, 565, 1986.
469. **Lindsay, D. S., Blagburn, B. L., Sundermann, C. A., Hoerr, F. J., and Ernest, J. A.**, Experimental *Cryptosporidium* infections in chickens: oocyst structure and tissue specificity, *J. Am. Vet. Res.*, 47, 876, 1986.
470. **Lindsay, D. S. and Blagburn, B. L.**, *Cryptosporidium* sp. infections in chickens, produced by intra-cloacal inoculation of oocysts, *J. Parasitol.*, 74, 615, 1986.
471. **Lindsay, D. S., Blagburn, B. L., Sundermann, C. A., and Ernest, J. A.**, Chemoprophylaxis of cryptosporidiosis in chickens, using halofuginone, salinomycin, lasalocid, or monensin, *Am. J. Vet. Res.*, 48, 354, 1987.
472. **Lindsay, D. S., Blagburn, B. L., Hoerr, F. J., and Giambrone, J. J.**, Experimental *Cryptosporidium baileyi* infections in chickens and turkeys produced by ocular inoculation of oocysts, *Avian Dis.*, 31, 355, 1987.
473. **Lindsay, D. S., Blagburn, B. L., and Hoerr, F. J.**, Experimentally induced infections in turkeys with *Cryptosporidium baileyi* isolated from chickens, *Am. J. Vet. Res.*, 48, 104, 1987.
474. **Lindsay, D. S., Blagburn, B. L., and Ernest, J. A.**, Experimental *Cryptosporidium parvum* infections in chickens, *J. Parasitol.*, 73, 242, 1987.
475. **Lindsay, D. S., Blagburn, B. L., Sunderman, C. A., Hoerr, F. J., and Giambrone, J. J.**, *Cryptosporidium baileyi*: effects of intra-abdominal and intravenous inoculation of oocysts on infectivity and site of development in broiler chickens, *Avian Dis.*, 31, 841, 1987.
476. **Lindsay, D. S., Hendrix, C. M., and Blagburn, B. L.**, Experimental *Cryptosporidium parvum* infections in opossums (*Didelphis virginiana*), *J. Wildl. Dis.*, 24, 157, 1988.
477. **Lindsay, D. S., Sundermann, C. A., and Blagburn, B. L.**, Cultivation of *Cryptosporidium baileyi*: studies with cell cultures, avian embryos, and pathogenicity of chicken embryo-passaged oocysts, *J. Parasitol.*, 74, 288, 1988.
478. **Lindsay, D. S., Blagburn, B. L., Sundermann, C. A., and Giambrone, J. J.**, Effect of broiler chicken age on susceptibility to experimentally induced *Cryptosporidium baileyi* infection, *Am. J. Vet. Res.*, 49, 1412, 1988.
479. **Lindsay, D. S., Blagburn, B. L., Sundermann, C. A., and Hoerr, F. J.**, Experimental infections in domestic ducks with *Cryptosporidum baileyi* isolated from chickens, *Avian Dis.*, 33, 69, 1989.

480. **Lindsay, D. S., Blagburn, B. L., and Sundermann, C. A.**, Morphometric comparison of the oocysts of *Cryptosporidium meleagridis* and *Cryptosporidium baileyi* from birds, *Proc. Helminthol. Soc. Wash.*, 56, 91, 1989.

480a. **Lindsay, D. S. and Blagburn, B. L.**, Unpublished data.

481. **Links, I. J.**, Cryptosporidial infection of piglets, *Aust. Vet. J.*, 58, 60, 1982.

481a. **Loening, W. E. K., Coovadia, Y. M., and Van Den Ende, J.**, Etiological factors of infantile diarrhea; a community-based study, *Ann. Trop. Paediatr.*, 9, 248, 1989.

482. **Logsdon, G. S. and Hoff, J. C.**, Barriers to the transmission of waterborne disease, in *Waterborne Diseases in the United States*, Craun, G. G., Ed., CRC Press, Boca Raton, FL, 1986, 255.

483. **Logsdon, G. S., Juranek, D., Mason, L., Ongerth, J. E., Rose, J. B., Sterling, C. R., and Ungar, B. L. P.**, Roundtable: Cryptosporidium, *J. Am. Water Works Assoc.*, 80, 14, 1988.

484. **Loose, J. H., Sedergran, D. J., and Cooper, H. S.**, Identification of *Cryptosporidium* in paraffin-embedded tissue sections with the use of a monoclonal antibody, *Am. J. Clin. Pathol.*, 91, 206, 1989.

485. **Lopez, J. W., Allen, S. D., Mitchell, J., and Quinn, M.**, Rotavirus and *Cryptosporidium* shedding in dairy calf feces and its relationship to colostrum immune transfer, *J. Dairy Sci.*, 71, 1288, 1988.

486. **Lopez-Brea, M., Garcia Picazo, L., Del Rey, M. C.**, Cryptosporidium in children and adults in Madrid, *J. Infect.*, 13, 314, 1986.

487. **Lopez-Brea, M., Garcia-Picazo, L., Del Ray, M., and Jimenez, M. L.**, Cryptosporidium in stool specimens in Madrid, *Trans. R. Soc. Trop. Med. Hyg.*, 79, 422, 1985.

488. **Louie, E., Borkowsky, W., Klesius, P. H., Haynes, T. B., Gordon, S., Bonk, S., and Lawrence, H. S.**, Treatment of cryptosporidiosis with oral bovine transfer factor, *Clin. Immunol. Immunopathol.*, 44, 329, 1987.

489. **Loureiro, E. C. B., Linhares, A. C., and Mata, L.**, Acute diarrhoea associated with *Cryptosporidium* sp. in Belém, Brazil (preliminary report), *Rev. Inst. Med. Trop. Sao Paulo*, 28, 138, 1986.

490. **Luft, B. J., Payne, D., Woodmansee, D., and Kim, C. W.**, Characterization of the *Cryptosporidium* antigens from sporulated oocysts of *Cryptosporidium parvum*, *Infect. Immun.*, 55, 2436, 1987.

491. **Lukešová, D., Trojáková, H., Celerová, L.**, Experimentální invaze laboratorních zvířat kokcidiemi z rodu *Cryptosporidium*, *Vet. Med. (Prague)*, 61, 365, 1988.

492. **Lumb, R., Erlich, S., and Davidson, G. P.**, Cryptosporidia detection, *Med. J. Aust.*, 142, 329, 1985.

493. **Lumb, R., Smith, K., O'Donoghue P. J., and Lanser, J. A.**, Ultrastructure of the attachment of *Cryptosporidium* sporozoites to tissue culture cells, *Parasitol. Res.*, 74, 531, 1988.

494. **Lumb, R., Lanser, J. A., and O'Donoghue, P. J.**, Electrophoretic and immunoblot analysis of *Cryptosporidium* oocysts, *Immunol. Cell Biol.*, 66, 369, 1988.

495. **Ma, P. and Soave, R.**, Three-step stool examination for cryptosporidiosis in 10 homosexual men with protracted watery diarrhea, *J. Infect. Dis.*, 147, 824, 1983.

496. **Ma, P.**, Cryptosporidium and the enteropathy of immune deficiency, *J. Pediatr. Gastroenterol. Nutr.*, 3, 488, 1984.

497. **Ma, P., Villanueva, T. G., Kaufman, D., and Gillooley, J. F.**, Respiratory cryptosporidiosis in the acquired immune deficiency syndrome. Use of modified cold Kinyoun and Hemacolor stains for rapid diagnoses, *J. Am. Med. Assoc.*, 252, 1298, 1984.

498. **Ma, P.**, Laboratory diagnosis of coccidiosis, in *Microbiology — 1984*, Leive, L., and Schlessinger, D., Eds., American Society for Microbiology, Washington, D C, 1984, 224.

499. **Ma, P., Kaufman, D. L., Helmick, C. G., D'Souza, A. J., and Navin, T. R.**, Cryptosporidiosis in tourists returning from the Caribbean, *N. Engl. J. Med.*, 312, 647, 1985.

500. **Ma, P.**, Cryptosporidiosis and immune enteropathy: a review, in *Current Clinical Topics in Infectious Diseases*, Vol. 8, Remington, J. S. and Swartz, M. N., Eds., McGraw-Hill, New York, 1987, 99.

501. **MacFarlane, D. E. and Horner-Bryce, J.**, Cryptosporidiosis in wellnourished and malnourished children, *Acta Paediatr. Scand.*, 76, 474, 1987.

502. **Madore, M. S., Rose, J. B., Gerba, C. P., Arrowood, M. J., and Sterling, C. R.**, Occurrence of *Cryptosporidium* oocysts in sewage effluents and selected surface waters, *J. Parasitol.*, 73, 702, 1987.

502a. **Maekawa, T.**, Studies on *Cryptosporidium muris* parasitizing the murine stomach, *J. Osaka City Med. Cent.*, 37, 679, 1988 (in Japanese).

503. **Magnasco, E. J.**, Cryptosporidium: agente asociado a la diarrea neonatal de los terneros. *Vet. Argent.*, 1, 342, 1984.

504. **Mahajan, R. C., Gaguly, N. K., Sehgal, R., and Malla, N.**, Parasitic diarrhoea, *Indian J. Pediatr.*, 55, 829, 1988.

505. **Mai Nguyen, X.**, Intestinale Cryptosporidiose: eine seltene Durchfallserkrankung beim Menschen, *Schweiz. Med. Wochenschr.*, 115, 1205, 1985.

505a. **Mai, N. X.**, Cryptosporidial diarrhoea in children, *Infection*, 15, 444, 1987.

505b. **Mair, T. S., Taylor, F. G. R., Harbour, D. A., and Pearson, G. R.**, Concurrent *Cryptosporidium* and coronavirus infections in an Arabian foal with combined immunodeficiency syndrome, *Vet. Rec.*, 126, 127, 1990.

506. **Malla, N., Sehgal, R., Ganguly, N. K., and Mahajan, R. C.,** Cryptosporidiosis in children in Chandigarh, *Indian J. Med. Res.*, 86, 722, 1987.

506a. **Malla, N., Sehgal, R., Ganguly, N. K., and Mahajan, R. C.,** Cryptosporidiosis — the Indian scene, *Indian J. Pediatr.*, 56, 6, 1989.

507. **Malebranche, R., Arnoux, E., Guérin, J. M., Pierre, G. O., Laroche, A. C., Péan-Guichard, C., Elie, R., Morisset, P. H., Spira, T., Mandeville, R., Drotman, P., Seemayer, T., and Dupuy, J. M.,** Acquired immunodeficiency syndrome with severe gastrointestinal manifestations in Haiti, *Lancet*, 2, 873, 1983.

508. **Mallolas, J., Latorre, X., Gatell, J. M., Miró, J. M., Francés, A., Mariscal, D., and Soriano, E.,** Análisis clínico de los primeros 145 casos consecutivos de síndrome de inmunodeficiencia adquirida, *Med. Clin. (Barcelona)*, 91, 721, 1988.

509. **Manivel, C., Filipovich, A., and Snover, D. C.,** Cryptosporidiosis as a cause of diarrhea following bone marrow transplantation, *Dis. Colon Rectum*, 28, 741, 1985.

510. **Mann, E. D. and Sekla, L. H.,** Cryptosporidiosis in Manitoba, *Can. Med. Assoc. J.*, 135, 1072, 1986.

511. **Mann, E. D., Sekla, L. H., and Eibisch, G.,** *Cryptosporidium* antibodies in Manitoba cattle: a pilot study using an indirect fluorescent antibody procedure, *Can. Vet. J.*, 28, 126, 1987.

512. **Mann, E. D., Sekla, L. H., Nayar, G. P. S., and Koschik, C.,** Infection with *Cryptosporidium* spp. in humans and cattle in Manitoba, *Can. J. Vet. Res.*, 50, 174, 1986.

513. **Marcial, M. A. and Madara, J. L.,** *Cryptosporidium*: cellular localization, structural analysis of absorptive cell-parasite membrane-membrane interactions in guinea pigs, and suggestion of protozoan transport by M cells, *Gastroenterology*, 90, 583, 1986.

514. **Margulis, S. J., Honig, C. L., Soave, R., Govoni, A. F., Mouradian, J. A., and Jacobson, I. M.,** Biliary tract obstruction in the acquired immunodeficiency syndrome, *Ann. Intern. Med.*, 105, 207, 1986.

515. **Markus, M. B. and Bush, J. B.,** Staining of coccidial oocysts, *Vet. Rec.*, 121, 329, 1987.

516. **Marquart, K. H., Müller, H. A. G., Sailer, J., and Moser, R.,** Slim disease (AIDS), *Lancet*, 2, 1186, 1985.

517. **Marshall, A. R., Al-Jumaili, I. J., Fenwick, G. A., Bint, A. J., and Record, C. O.,** Cryptosporidiosis in patients at a large teaching hospital, *J. Clin. Microbiol.*, 25, 172, 1987.

518. **Martino, P., Gentile, G., Caprioli, A., Baldassarri, L., Donelli, G., Arcese, W., Fenu, S., Micozzi, A., Venditti, M., and Mandelli, F.,** Hospital-acquired cryptosporidiosis in a bone marrow transplantation unit, *J. Infect. Dis.*, 158, 647, 1988.

519. **Mason, R. W. and Hartley, W. J.,** Respiratory cryptosporidiosis in a peacock chick, *Avian Dis.*, 24, 771, 1980.

520. **Mason, R. W., Hartley, W. J., and Tilt, L.,** Intestinal cryptosporidiosis in a kid goat, *Aust. Vet. J.*, 57, 386, 1981.

521. **Mason, R. W.,** Conjunctival cryptosporidiosis in a duck, *Avian Dis.*, 30, 598, 1986.

522. **Mata, L., Simhon, A., Urrutia, J. J., Kronmal, R. A., Fernández, R., and García, B.,** Epidemiology of rotaviruses in a cohort of 45 Guatamalan Mayan Indian children observed from birth to the age of three years, *J. Infect. Dis.*, 148, 452, 1983.

523. **Mata, L., Bolaños, H., Pizarro, D., and Vives, M.,** Criptosporidiosis en niños de Costa Rica: estudio transveral y longitudinal, *Rev. Biol. Trop.*, 32, 129, 1984.

524. **Mata, L., Bolaños, H., Pizarro, D., and Vives, M.,** Cryptosporidiosis in children from some highland Costa Rican rural and urban areas, *Am. J. Trop. Med. Hyg.*, 33, 24, 1984.

525. **Mata, L.,** *Cryptosporidium* and other protozoa in diarrheal disease in less developed countries, *Pediatr. Infect. Dis.*, 5, 117, 1986.

526. **Mathan, M. M., Venkatesan, S., George, R., Mathew, M., and Mathan, V. I.,** Cryptosporidium and diarrhoea in southern Indian children, *Lancet*, 2, 1172, 1985.

527. **Matheron, S., Cenec, J., Deluol, A. M., Savel, J., Souala, F., Coulaud, J. P., Saimot, A. G., and Charmot, G.,** Cryptosporidiose: a propos de 11 observations, *Bull. Soc. Pathol. Exot.*, 77, 449, 1984.

528. **Matovelo, J. A., Landsverk, T., and Posada, G. A.,** Cryptosporidiosis in Tanzanian goat kids. Scanning and transmission electron microscopic observations, *Acta Vet. Scand.*, 25, 322, 1984.

529. **Matuschka, F.-R.,** Reptiles as intermediate and/or final hosts for sarcosporidia, *Parasitol. Res.*, 73, 22, 1987.

530. **Maurage, C., Naciri, M., and Arnaud-Battandier, F.,** Cryptosporidiose: enquête épidémiologique, prospective, chez des enfants hospitalisés atteints de diarrhée, *Arch. Fr. Pediatr.*, 44, 177, 1987.

531. **McClure, H. M., Anderson, D. C., Fultz, P. N. Ansari, A. A., Lockwood, E., and Brodie, A.,** Spectrum of disease in macaque monkeys chronically infected with SIV/SMM, *Vet. Immunol. Immunopathol.*, 21, 13, 1989.

532. **McColl, D. and Mooney, T. J.,** Cryptosporidia in Tasmania, *Med. J. Aust.*, 22, 900, 1984.

533. **McDonald, V. and Barer, M. E.,** Growth of *Cryptosporidium* in six cell lines, in *Proc. 1st Int. Workshop on Cryptosporidiosis*, Angus, K. W. and Blewett, D. A., Eds., Animal Disease Research Association, September 7 to 8, 1988, Edinburgh, 1989, 119.

534. **McKenzie, R. A., Green, P. E., Hartley, W. J., and Pollitt, C. C.,** *Cryptosporidium* in a red-bellied black snake (*Pseudechis porphyriacus*), *Aust. Vet. J.*, 54, 365, 1978.

535. **McLauchlin, J., Casemore, D. P., Harrison, T. G., Gerson, P. J., Samuel, D., and Taylor, A. G.,** Identification of Cryptosporidium oocysts by monoclonal antibody, *Lancet*, 1, 51, 1987.

535a. **McMeeking, A., Borkowsky, W., Klesius, P. H., Bonk, S., Holzman, R. S., and Lawrence, H. S.**, A controlled trial of bovine dialzyable leukocyte extract for cryptosporidiosis in patients with AIDS, *J. Infect. Dis.*, 161, 108, 1990.

536. **McNabb, S. J. N., Hensel, D. M., Welch, D. F., Heijbel, H., McKee, G. L., and Istre, G. R.**, Comparison of sedimentation and flotation techniques for identification of *Cryptosporidium* sp. oocysts in a large outbreak of human diarrhea, *J. Clin. Microbiol.*, 22, 587, 1985.

537. **Mead, G. M., Sweetenham, J. W., Ewins, D. L., Furlong, M., and Lowes, J. A.**, Intestinal cryptosporidiosis: a complication of cancer treatment, *Cancer Treat. Rep.*, 70, 769, 1986.

538. **Mead, J. R., Arrowood, M. J., and Sterling, C. R.**, Antigens of *Cryptosporidium* sporozoites recognized by immune sera of infected animals and humans, *J. Parasitol.*, 74, 135, 1988.

539. **Mead, J. R., Arrowood, M. J., Current, W. L., and Sterling, C. R.**, Field inversion gel electrophoretic separation of *Cryptosporidium* spp. chromosome-sized DNA, *J. Parasitol.*, 74, 366, 1988.

539a. **Mead, J. R., Humphreys, R. C., Sammons, D. W., and Sterling, C. R.**, Identification of isolate-specific sporozoite proteins of *Cryptosporidium parvum* by two-dimensional gel electrophoresis, *Infect. Immun.*, 58, 1990, in press.

540. **Meisel, J. L., Perera, D. R., Meligro, C., and Rubin, C. E.**, Overwhelming watery diarrhea associated with a Cryptosporidium in an immunosuppressed patient, *Gastroenterology*, 70, 1156, 1976.

543. **Meuten, D. J., Van Kruiningen, H. J., and Lein, D. H.**, Cryptosporidiosis in a calf, *J. Am. Vet. Med. Assoc.*, 165, 914, 1974.

544. **Meyer, E. A.**, Waterborne *Giardia* and *Cryptosporidium, Parasitol. Today*, 4, 200, 1988.

545. **Miláčvek, P. and Vítovec, J.**, Differential staining of cryptosporidia by aniline-carbol-methyl violet and tartrazine in smears from faeces and scrapings of intestinal mucosa, *Folia Parasitol. (Prague)*, 32, 50, 1985.

546. **Miller, R. A., Holmberg, R. E., Jr., and Clausen, C. R.**, Life-threatening diarrhea caused by *Cryptosporidium* in a child undergoing therapy for acute lymphocytic leukemia, *J. Pediatr.*, 103, 256, 1983.

547. **Miller, R. A., Wasserheit, J. N., Kirihara, J., and Coyle, M. B.**, Detection of *Cryptosporidium* oocysts in sputum during screening for mycobacteria, *J. Clin. Microbiol.*, 20, 1192, 1984.

547a. **Miller, R. A., Bronsdon, M. A., and Morton, W. R.**, Experimental cryptosporidiosis in a primate model, *J. Infect. Dis.*, 161, 312, 1990.

548. **Miller, N. M. and van den Ende, J.**, Seasonal prevalence of *Cryptosporidium*-associated diarrhoea in young children, *SAMJ*, 70, 636-637, 1986.

548a. **Miline, R.**, Parasite in farm waste threatens water supplies, *New Sci.*, 1675, 22, 1989.

549. **Modigliani, R., Bories, C., Le Charpentier, Y., Salmeron, M., Messing, B., Galian, A., Rambaud, J. C., Lavergne, A., Cochand-Priollet, B., and Desportes, I.**, Diarrhoea and malabsorption in acquired immune deficiency syndrome: a study of four cases with special emphasis on opportunistic protozoan infestations, *Gut*, 26, 179, 1985.

550. **Möller, K.**, Endoparasitenbefall bei Sauen unter dem Einfluβ der sich wandelnden Haltungsformen in rheinland-pfälzischen Ferkelerzeugerbetrieben, *Tieraerztl. Umsch.*, 39, 525, 1984.

551. **Molnár, K. and Baska, F.**, Light and electron microscopic studies on *Epieimeria anguillae* (Léger & Hollande, 1922), a coccidium parasitizing the European eel, *Anguilla anguilla* L., *J. Fish Dis.*, 9, 99, 1986.

552. **Molnár, K.**, Occurrence of two new *Goussia* species in the intestine of the sterlet (*Acipenser ruthenus*), *Acta Vet. Hung.*, 34, 169, 1986.

553. **Montessori, G. A. and Bischoff, L.**, Cryptosporidiosis: a cause of summer diarrhea in children, *Can. Med. Assoc. J.*, 132, 1285, 1985.

554. **Monticello, T. M., Levy, M. G., Bunch, S. E., and Fairley, R. A.**, Cryptosporidiosis in a feline leukemia virus-positive cat, *J. Am. Vet. Med. Assoc.*, 191, 705, 1987.

555. **Moon, H. W., McClurkin, A. W., Isaacson, R. E., Pohlenz, J., Skartvedt, S. M., Gillette, K. G., and Baetz, A. L.**, Pathogenic relationships of rotavirus, *Escherichia coli* and other agents in mixed infections in calves, *J. Am. Vet. Med. Assoc.*, 173, 577, 1978.

556. **Moon, H. W. and Bemrick, W. J.**, Fecal transmission of calf cryptosporidia between calves and pigs, *Vet. Pathol.*, 18, 248, 1981.

557. **Moon, H. W., Woode, G. N., and Ahrens, F. A.**, Attempted chemoprophylaxis of cryptosporidiosis in calves, *Vet. Rec.*, 110, 181, 1982.

558. **Moon, H. W., Schwartz, A., Welch, M. J., McCann, P. P., and Runnels, P. L.**, Experimental fecal transmission of human cryptosporidia to pigs and attempted treatment with an ornithine decarboxylase inhibitor, *Vet. Pathol.*, 19, 700, 1982.

559. **Moon, H. W., Pohlenz, J. F. L., Woodmansee, D. B., Woode, G. N., Heine, J., Abel, S., and Jarvinen, J. A.**, Intestinal cryptosporidiosis: pathogenesis and immunity, *Microecol. Ther.*, 15, 103, 1985.

560. **Moon, H. W. and Woodmansee, D. B.**, Cryptosporidiosis, *J. Am. Vet. Med. Assoc.*, 189, 643, 1986.

561. **Moon, H. W., Woodmansee, D. B., Harp, J. A., Abel, S., and Ungar, B. L. P.**, Lacteal immunity to enteric cryptosporidiosis in mice: immune dams do not protect their suckling pups, *Infect. Immun.*, 56, 649, 1988.

562. **Moore, J. A., Blagburn, B. L., and Lindsay, D. S.**, Cryptosporidiosis in animals including humans, *Compend. Contin. Educ. Pract. Vet.*, 10, 275, 1988.

563. **Morch, J.**, Prevalence of Cryptosporidium in broiler chickens in Denmark, *Isr. J. Vet. Med.*, 42, 11, 1986.

564. **Morin, M., Larivière, S., and Lallier, R.**, Pathological and microbiological observations made on spontaneous cases of acute neonatal calf diarrhea, *Can. J. Comp. Med.*, 40, 228, 1976.

565. **Morin, M., Larivière, S., Lallier, R., Begin, M., Roy, R., and Ethier, R.**, Neonatal calf diarrhea: pathology and microbiology of spontaneous cases in dairy herds and incidence of the enteropathogenes implicated as etiological agents, in *Proc. 2nd Int. Symp. on Neonatal Calf Diarrhea*, 1978, 347.

566. **Morteo, C. G., Estrella, S. G., and de Aluja, A. S.**, Criptosporidiosis en bovinos lactantes (histopatología, microscopía electrónica de transmisión y de barrido), *Vet. Mex.*, 14, 12, 1983.

566a. **Moriya, K.**, Ultrastructural observations on oocysts, sporozoites and oocyst residuum of *Cryptosporidium muris* (strain RN66), *J. Osaka City Med. Cent.*, 38, 177, 1989 (in Japanese).

567. **Moskovitz, B. L., Stanton, T. L. and Kusmierek, J. J. E.**, Spiramycin therapy for cryptosporidial diarrhoea in immunocompromised patients, *J. Antimicrob. Chemother. Suppl. B.*, 22, 189, 1988.

568. **Moss, G., Braunstein, F. M., and Newkirk, R. E.**, Postoperative enteral hyperalimentation for cryptosporidial acute cholecystitis associated with AIDS and enteritis, *J. Am. Coll. Nutr.*, 6, 351, 1987.

569. **Müller, H. E.**, Kryptosporidiose — Erreger, Epidemiologie, Klinik und Nachweis, *Dtsch. Med. Wochenschr.*, 111, 146, 1986.

570. **Musial, C. E., Arrowood, M. J., Sterling, C. R., and Gerba, C. P.**, Detection of *Cryptosporidium* in water by using polypropylene cartridge filters, *Appl. Environ. Microbiol.*, 53, 687, 1987.

571. **Myers, L. L., Firehammer, B. D., Border, M. M., and Shoop, D. S.**, Prevalence of enteric pathogens in the feces of healthy beef calves, *Am. J. Vet. Res.*, 45, 1544, 1984.

572. **Nachamkin, I., Jones, A., and Hasyn, H.**, Routine parasitological examination for *Cryptosporidium*, *J. Infect. Dis.*, 154, 369, 1986.

573. **Naciri, M. and Yvore, P.**, La cryptosporidiose de bovins, *Recl. Med. Vet.*, 159, 221, 1983.

575. **Naciri, M., Yvore, P., de Boissieu, C., and Esnault, E.**, Multiplication de *Cryptosporidium muris* (Tyzzer 1907) in vitro entretien d'une souche sur oeufs embryonnós, *Recl. Med. Vet.*, 162, 51, 1986.

576. **Naciri, M.**, Cryptosporidiose: nouveautós bibliographiques et observations personnelles, *Bull. G.T.V.*, 39, 1987.

576a. **Naciri, M. and Mazzella, O.**, Association cryptosporidiose et maladie de Marek chez des poulets nains, *Rec. Med. Vet.*, 164, 311, 1988.

576b. **Naciri, M.**, Animal and human cryptosporidiosis: opportunist infection? Pathogenicity of the genus *Cryptosporidium*, *in Coccidia and Intestinal Coccidiomorphs*, Proc. 5th Int. Coccidiosis Conf., Tours, France, October 17 to 20, 1989. INRA, 1989, 51.

576c. **Naciri, M., Mazzella, O., and Coudert, F.**, Interactions between *Cryptosporidia* and pathogenic viruses, in *Coccidia and Intestinal Coccidiomorphs*, Proc. 5th Int. Coccidiosis Conf., Tours, France, October 17 to 20, 1989, INRA, 1989, 235.

576d. **Naciri, M., Yvore, P., Banting, A. de L., and Gregoire, S.**, Induced and natural cryptosporidiosis in conventional calves, to be published.

576e. **Naciri, M. and Yvore, P.**, Treatment of cryptosporidiosis in lambs using halofugnone lactate. Abstract in *Proc. 1st Int. Workshop on Cryptosporidiosis*, Agus, K. W. and Blewitt, D. A. Eds., September 7 to 8, 1988, Edinburgh, 1989, 120.

576f. **Naciri, M., Mazella, O., and Coudert, F.**, Interactions cryptosporidies-virus-sauvage ou vaccinal de la maladie de marek chez le poulet, Rec. Med. Vet., 165, 383, 1989.

576g. **Naciri, M. and Yvore, P.**, Efficacite du lactate d'halofuginone dans le traitment de la cryptosporidiose chez l'angneau, *Rec. Med. Vet.*, 165, 823, 1989.

577. **Nagy, B., Antal, Á., and Rátz, F.**, A borjak cryptosporidiosisának magyarországi elöfordulása, *Magy. Állatorv. Lapja*, 34, 585, 1979.

578. **Nagy, B., Antal, Á., and Lakner, J.**, Significance of intestinal cryptosporidiosis in calf diarrhea, in *Proc. 2nd Int. Symp. Veterinary Laboratory Diagnosticians*, Vol. 3, Lucerne, Switzerland, 1980, 431.

579. **Nagy, B. and Pohlenz, J.**, Die bovine Kryptosporidiose: Diagnose und Therapie, *Tieraerztl. Prax.*, 10, 163, 1982.

580. **Nagy, B., Bozsó, M., Pálfi, V., Nagy, Gy., and Sahiby, M.A.**, Studies on cryptosporidial infections of goat kids, in *Les Maladies da la Chevre*, October 9 to 11, 1984, Niort, France, *Les colloques de l'INRA*, 1984, 443.

581. **Nagy, B., Csontos, L., Pálfi, V., Nagy, G., and Bozsó, M.**, Polyätiologische diagnostische Erfahrungen bei Kälberdurchfällen im ersten Lebensmonat, *Wien. Tieraerztl. Monatschr.*, 73, 181, 1986.

581a. **Nagy, B., Jankó, M., and Molnár, P.**, Cryptosporidium — ürítés hasmenéses borjak és kecskegidák, valamint gondozoik körében, *Magy. Allatorv. Lapja*, 44, 389, 1989.

582. **Nakamura, K. and Abe, F.**, Respiratory (especially pulmonary) and urinary infections of *Cryptosporidium* in layer chickens, *Avian Pathol.*, 17, 703, 1988.

583. **Naqash, H. U. and Wattal, C.**, Cryptosporidiosis in Kashmir, *Trans. R. Soc. Trop. Med. Hyg.*, 82, 499, 1988.

584. **Naranjo, J. G., Miranda, E. S., Pacheco, L. L., Sack, B., Verastegui, M. P.**, Cryptosporidiosis en el Perú, *Rev. Gast. Peru*, 5, 24, 1985.

585. **Navin, T. R. and Juranek, D. D.**, Cryptosporidiosis: clinical, epidemiologic and parasitologic review, *Rev. Infect. Dis.*, 6, 313, 1984.
586. **Navin, T. R.**, Cryptosporidiosis in humans: review of recent epidemiologic studies, *Eur. J. Epidemiol.*, 1, 77, 1985.
587. **Navin, T. R. and Hardy, A. M.**, Cryptosporidiosis in patients with AIDS, *J. Infect. Dis.*, 155, 150, 1987.
590. **Nichols, G. and Thom, B. T.**, Screening for *Cryptosporidium* in stools, *Lancet*, 1, 735, 1984.
590a. **Nicolas, J. A., Dubost, G., Gayaud, C., and Noel, F.**, Importance des cryptosporidies dans les diarrées néonatales du veau, *Point Vet.*, 16, 350, 1984.
591. **Nime, F. A., Burek, J. D., Page, D. L., Holscher, M. A., and Yardley, J. H.**, Acute enterocolitis in a human being infected with the protozoan Cryptosporidium, *Gastroenterology*, 70, 592, 1976.
592. **Nishikawa, H., Takashe, K., and Yamada, S.**, Four cases of cryptosporidial parasitization in chickens, *Jpn. J. Vet. Med. Assoc.*, 37, 667, 1984 (in Japanese).
593. **Nobel, T. A., Kuttin, E., Yakobson, B., and Perl, S.**, First diagnosis of bovine cryptosporidiosis in Israel, *Refu. Vet.*, 39, 10, 1982.
594. **Nooruddin, M. and Sarma, D. K.**, Role of Cryptosporidium in calf dairrhoea, *Livestock Advisor (India)*, 12, 49, 1987.
595. **Nozais, J. P., Datry, A., Karabinis, A., Rozenbaum, W., and Gentilini, M.**, La cryptosporidiose humaine. A propos de 4 cas, *Bull. Soc. Pathol. Exot. Fil.*, 77, 441, 1984.
596. **O'Donoghue, P. J.**, *Cryptosporidium* infections in man, animals, birds and fish, *Aust. Vet. J.*, 62, 253, 1985.
597. **O'Donoghue, P. J., Tham, V. L., de Saram, W. G., Paull, K. L., and McDermott, S.**, *Cryptosporidium* infections in birds and mammals and attempted cross-transmission studies, *Vet. Parasitol.*, 26, 1, 1987.
598. **Ogassawara, S., Benassi, S., Larsson, C. E., and Hagiwara, M. K.**, *Cryptosporidium curyi* sp. n. in the feces of cats in the the city of São Paulo, Brazil, *Rev. Microbiol, Sao Paulo*, 17, 346, 1986.
599. **Ogimoto, K., Inamoto, T., Soga, T., and Itakura, C.**, Experimental infection of chickens with Cryptosporidium, *Zentrabl. Bakteriol. Hyg. A*, 264, 343, 1987.
600. **Oh, S. H., Jaffe, N., Fainstein, V., and Pickering, V.**, Cryptosporidiosis and anticancer chemotherapy, *J. Pediatr.*, 104, 963, 1984.
601. **Ojeh, C. K.**, An outbreak of neonatal calf diarrhoea involving *Cryptosporidium*, *Trop. Vet.*, 4, 71, 1986.
602. **Ongerth, J. E. and Stibbs, H. H.**, Identification of *Cryptosporidium* oocysts in river water, *Appl. Environ. Microbiol.*, 53, 672, 1987.
603. **Ongerth, J. E. and Stibbs, H. H.**, Prevalence of *Cryptosporidium* infection in dairy calves in western Washington, *Am. J. Vet. Res.*, 50, 1069, 1989.
603a. **Orenstein, J. M.**, Personal communication, George Washington University Medical School.
604. **Orr, J. P.**, *Cryptosporidium* infection associated with proliferative enteritis (wet tail) in Syrian hamsters, *Can. Vet. J.*, 29, 843, 1988.
605. **Orr, M. B., Mackintosh, C. G., and Suttie, J. M.**, Cryptosporidiosis in deer calves, *N. Z. Vet. J.*, 33, 151, 1985.
605a. **Ortolani, E. L.**, Study of some epidemiological aspects of cryptosporidiosis in calves in the state of São Paulo, Brazil, Abstract in *Proc. 1st Int. Workshop on Cryptosporidiosis*, September 7 to 8, 1988, Edinburgh, 1989, 122.
606. **Osborn, K. G., Prahalada, S., Lowenstine, L. J., Gardner, M. B., Maul, D. H., and Henrickson, R. V.**, The pathology of an epizootic of acquired immunodeficiency in rhesus macaques, *Am. J. Pathol.*, 114, 94, 1984.
606a. **Palkovič, L., and Maroušek, V.**, The pathogenity of *cryptosporidium parvum* Tyzzer, 1912 and *C. bailey* Current, Upton et Haynes, 1986 for chickens, *Folia Parasitol. (Prague)*, 36, 209, 1989.
606b. **Palkovič, L. and Pecka, Z.**, Natural cryptosporidiosis in domestic geese, *Anser anser* L., in Czechoslovakia, *Folia Parasitol. (Prague)*, 36, 126, 1989.
607. **Panciera, R. J., Thomassen, R. W., and Garner, F. M.**, Cryptosporidial infection in a calf, *Vet. Pathol.*, 8, 479, 1971.
608. **Papadopoulou, C., Xylouri, E., Zisides, N.**, Cryptosporidial infection in broiler chickens in Greece, *Avian Dis.*, 32, 842, 1988.
608a. **Papadopoulu, C., Xylouri, E., Mantizios, A., Spyropoulos, G., and Stoforos, S.**, Cryptosporidiosis in farm animals in Greece, Abstract in *Proc. 1st Int. Workshop on Cryptosporidiosis*, Angus, K. W. and Blewitt, D. A., Eds., September 7 to 8, 1988, Edinburgh, 1989, 123.
609. **Pape, J. W., Levine, E., Beaulieu, M. E., Marshall, F., Verdier, R., and Johnson, W. D., Jr.**, Cryptosporidiosis in Haitian children, *Am. J. Trop. Med. Hyg.*, 36, 333, 1987.
610. **Paperna, I.**, Scanning electron microscopy of the coccidian parasite *Cryptosporidium* sp from cichlic fishes, *Dis. Aquat. Org.*, 3, 231, 1987.
611. **Parker, R. J., Scott, C. D. B., and Jeboult, J. W.**, Cryptosporidial diarrhoea, *Med. J. Aust.*, 143, 426, 1985.
612. **Pashkin, P.I., Lakovinkova, E. V., Losket, V. I., Novinsky, K. K., and Sidorenko, N. Y.**, Epizootiology of cryptosporidiosis in livestock forms of Leningrad districts, in Invasive Diseases of Agricultural Animals, Sbornik of papers of the Leningrad Veterinary Institute, Leningrad, 1988, 60.
613. **Pavlásek, I.**, First record of Cryptosporidium sp. in calves in Czechoslovakia, *Folia Parasitol. (Prague)*, 28, 187, 1981.

614. **Pavlásek, I.**, First detection of Cryptosporidium sp. oocysts in calf faeces by flotation method, *Folia Parasitol. (Prague)*, 29, 115, 1982.

615. **Pavlásek, I.**, Dynamics of the release of oocysts of Cryptosporidium sp. in spontaneously infected calves, *Folia Parasitol. (Prague)*, 29, 295, 1982.

616. **Pavlásek, I. and Mareš, J.**, Vliv jednorázové dezinfekce farmy na průběh kryptosporidiózních infekcí telat, *Vet. Med. (Prague)*, 28, 449, 1983.

617. **Pavlasek, I., Zikmund, B., and Klíma, F.**, Vliv ruzného zpusobu ustájení telat po narození na výskyt *Cryptosporidium* sp., *Vet. Med. (Prague)*, 28, 31, 1983.

618. **Pavlásek, I.**, Cryptosporidium sp. in *Cyprinus carpio* Linné, 1758 in Czechoslovakia, *Folia Parasitol. (Prague)*, 30, 248, 1983.

619. **Pavlásek, I.**, Experimental infection of cat and chicken with Cryptosporidium sp. oocysts isolated from a calf, *Folia Parasitol. (Prague)*, 30, 121, 1983.

620. **Pavlásek, I. and Nikitin, V. F.**, Finding of Cryptosporidium sp in calves in the USSR, *Folia Parasitol. (Prague)*, 30, 4, 1983.

621. **Pavlásek, I.**, Účinek dezinfecnich prostředků na infekceschopnost oocyst Cryptosporidium sp., *Cesk. Epidemiol. Mikrobiol. Immunol.*, 33, 97, 1984.

622. **Pavlásek, I.**, First record of developmental stages of Cryptosporidium sp. in various organs of experimentally infected mice and spontaneously infected calves, *Folia Parasitol. (Prague)*, 31, 191, 1984.

623. **Pavlásek, I.**, První nález spontánní kryptosporidiózní infekce kočky domácí v ČSSR, *Veterinarstvi*, 35, 125, 1985.

624. **Pavlásek, I. and Palkovic, L.**, Spontánní kryptosporidiózní infekce kuřat, *Veterinarstvi*, 36, 131, 1986.

625. **Pavlasek, I. and Nikitin, V. F.**, Finding of coccidia of the genus cryptosporidium in the organs of calf excretory system, *Folia Parasitol.*, 34, 197, 1987.

626. **Pavlásek, I.**, Studies on the life cycle on Cryptosporidium coccidia in experimentally infected chickens, *Folia Parasitol.*, 34, 193, 1987.

626a. **Pavlásek, I., Kopachka, M., Nikitin, V. F., Kozlova, G. A., and Korovikov, S. E.**, Cryptosporidiosis in boiler fowls [In Russian], *Veterinariya*, 2, 39, 1989.

626b. **Pavlásek, I. and Kozakiewicz, B.**, Occurrence and localization of Cryptosporidium sp. in spontaneously infected pheasants (*Phasianus colchicus* L.), *Folia Parasitol. (Prague)*, 36, 287, 1989.

627. **Payne, P., Lancaster, L. A., Heinzman, M., and McCutchan, J. A.**, Identification of Cryptosporidium in patients with acquired immunodeficiency syndrome, *N. Engl. J. Med.*, 309, 613, 1983.

628. **Pearson, G. R. and Logan, E. F.**, Demonstration of cryptosporidia in the small intestine of a calf by light, transmission electron and scanning electron microscopy, *Vet. Rec.*, 103, 212, 1978.

629. **Pearson, G. R., Logan, E. F., and McNulty, M. S.**, Distribution of cryptosporidia within the gastrointestinal tract of young calves, *Res. Vet. Sci.*, 33, 228, 1982.

630. **Pearson, G. R. and Logan, E. F.**, Scanning and transmission electron microscopic observations on the host-parasite relationship in intestinal cryptosporidiosis of neonatal calves, *Res. Vet. Sci.*, 34, 149, 1983.

631. **Pearson, G. R. and Logan, E. F.**, The pathology of neonatal enteritis in calves with observations on *E. coli*, rotavirus and *Cryptosporidium*, *Ann. Rech. Vet.*, 14, 422, 1983.

632. **Peeters, J. E., van Opdenbosch, E., and Glorieux, B.**, Demonstration of cryptosporidia in calf faeces: a comparative study, *Vlaams Diergeneeskd. Tijdschr.*, 51, 513, 1982.

633. **Peeters, J. E., Geeroms, R., Carman, R. J., and Wilkins, T. D.**, Significance of *Clostridium spiroforme* in the enteritis-complex of commercial rabbits, *Vet. Microbiol.*, 12, 25, 1986.

633a. **Peeters, J. E., Charlier, G. J., and Dussart P.**, Pouvoir pathogene de Cryptosporidium sp. chez les lapereaux avant et apres sevrage. *4ómes Journóes de la Recherche Cunicole en France*, Ed. ITAVI, Paris, Vol. II, communication n° 37, 1 to 9, 1986.

633b. **Peeters, J. E.**, Etiology and pathology of diarrhoea in weanling rabbits, in *Rabbit Production Systems Including Welfare*, A seminar in the community programme for the coordination of agricultural research, Auxilia, T., Ed., Commission of the European Communities, Luxemburg, 1986, 127.

634. **Peeters, J. E., Ares Mazás, E., Masschelein, W. J., Villacorta Martinez de Maturana, I., and Debacker, E.**, Effect of disinfection of drinking water with ozone or chlorine dioxide on survival of *Cryptosporidium parvum* oocysts, *Appl. Environ. Microbiol.*, 55, 1519, 1989.

634a. **Perera, J.**, Cryptosporidium associated childhood diarrhoea in Sri Lanka — a preliminary study, *Ceylon Med. J.*, 33, 101, 1988.

635. **Perez-Schael, I., Boher, Y., Mata, L., Perez, M., and Tapia, F. J.**, Cryptosporidiosis in Venezuelan children with acute diarrhea, *Am. J. Trop. Med. Hyg.*, 34, 721, 1985.

635a. **Perryman, L. E., Riggs, M. W., Mason, P. H., and Fayer, R.**, Kinetics of *Cryptosporidium parvum* sporozoite neutralization by monoclonal antibodies, immune serum, and immune bovine colostrum, *Infect. Immun.*, 58, 257, 1990.

636. **Petras, R. E., Carey, W. D., and Alanis, A.**, Cryptosporidial enteritis in a homosexual male with an acquired immunodeficiency syndrome, *Cleveland Clin. Q.*, 50, 41, 1983.

637. **Peters, C. S. and Kathpalia, S. B., Chittom-Swiatlo, A. L., Kallick, C. A., Sherer, R., and Kocka, F. E.,** *Isospora belli* and *Cryptosporidium* sp. from a patient not suspected of having acquired immunodeficiency syndrome, *Diagn. Microbiol. Infect. Dis.*, 8, 197, 1987.
637a. **Pherwani, A. V., Bhave, S. Y., Bijm, A. M., and Desai, A. G.,** Prevalence of Cryptosporidium in children with acute diarrhoea, *Indian J. Pediatr.*, 56, 133, 1989.
638. **Pitlik, S. D., Fainstein, V., Garza, D., Guarda, L., Bolivar, R., Rios, A., Hopfer, R. L., and Mansell, P. A.,** Human cryptosporidiosis: spectrum of disease. Report of six cases and review of the literature, *Arch. Intern. Med.*, 143, 2269, 1983.
639. **Pitlik, S. D., Fainstein, V., Rios, A., Guarda, L., Mansell, P. W. A., and Hersh, E. M.,** Cryptosporidial cholecystitis, *N. Engl. J. Med.*, 308, 967, 1983.
640. **Pivont, P., Meunier, J., Lefevre, F., Baudouin, P., Bughin, J., and Antoine, H.,** Fréquence des cryptosporidies dans les matières fécales des veaux d'une clientèle vétérinaire, *Ann. Med. Vet.*, 128, 369, 1984.
641. **Pohjola, S.,** Survey of cryptosporidiosis in feces of normal healthy dogs, *Nord. Vet. Med.*, 36, 189, 1984.
642. **Pohjola, S.,** Negative staining method with nigrosin for the detection of cryptosporidial oocysts: a comparative study, *Res. Vet. Sci.*, 36, 217, 1984.
643. **Pohjola, S., Jokipii, L., and Jokipii, A. M. M.,** Dimethylsulphoxide-Ziehl-Neelsen staining technique for detection of cryptosporidial oocysts, *Vet. Rec.*, 115, 442, 1984.
644. **Pohjola, S., Oksanen, H., Jokipii, L., and Jokipii, A. M. M.,** Outbreak of cryptosporidiosis among veterinary students, *Scand. J. Infect. Dis.*, 18, 173, 1986.
645. **Pohjola, S., Oksanen, H., Neuvonen, E., Veijaleinen, P., and Henriksson, K.,** Certain enteropathogens in calves of Finnish dairy herds with recurrent outbreaks of diarrhea, *Prev. Vet. Med.*, 3, 547, 1986.
646. **Pohjola, S., Neuvonen, E., Niskanen, A., and Rantama, A.,** Rapid immunoassay for detection of Cryptosporidium oocysts, *Acta Vet. Scand.*, 27, 71, 1986.
647. **Pohjola, S. and Lindberg, L.-A.,** Experimental cryptosporidiosis in mice, calves and chicken, *Acta Vet. Scand.*, 27, 80, 1986.
648. **Pohjola, S., Jokipii, A. M. M., and Jokipii, L.,** Sporadic cryptosporidiosis in a rural population is asymptomatic and associated with contact to cattle, *Acta Vet. Scand.*, 27, 91, 1986.
649. **Pohlenz, J., Moon, H. W., Cheville, N. F., and Bemrick, W. J.,** Cryptosporidiosis as a probable factor in neonatal diarrhea of calves, *J. Am. Vet. Med. Assoc.*, 172, 452, 1978.
651. **Pohlenz, J., Bemrick, W. J., Moon, H. W., and Cheville, N. F.,** Bovine cryptosporidiosis: a transmission and scanning electron microscopic study of some stages in the life cycle and of the host parasite relationship, *Vet. Pathol.*, 15, 417, 1978.
652. **Pohlenz, V. J.,** Die Kryptosporidiose bei Menschen und Tieren, *Dtsch. Tieraerztl. Wochenschr.*, 94, 67, 1987.
653. **Polack, B., Chermette, R., Savey, M., and Bussieras, J.,** Les cryptosporidies en France. Techniques usuelles d'identification et résultats préliminaires d'enquêtes épidémiologiques, *Point. Vet.*, 15, 41, 1983.
654. **Polack, B. and Perrin, G.,** La cryptosporidiose du chevreau, *Bull. Group Tech. Vet. Fr.*, 3, 45, 1987.
654a. **Pongs, P.,** Kryptosporidien-Infektion biem Kalb Behandlungsversuch mit Lasalocid-Na unter Praxisbedingungen, *Tieraerztl. Umsch.*, 44, 100, 1989.
655. **Poonacha, K. B. and Pippin, C.,** Intestinal cryptosporidiosis in a cat, *Vet. Pathol.*, 19, 708, 1982.
655a. **Poonacha, K. B. and Tuttle, P. A.,** Intestinal cryptosporidiosis in two thoroughbred foals, *Equine Pract.*, 11, 6, 1989.
656. **Portnoy, D., Whiteside, M. E., Buckley, E., and MacLeod, C. L.,** Treatment of intestinal cryptosporidiosis with spiramycin, *Ann. Intern. Med.*, 101, 202, 1984.
657. **Portús, M., Serra, T., Botet, J., and Gállego, J.,** Cryptosporidiosis humana en España, *Med. Clin.*, 84, 462, 1985.
658. **Pospischil, A., Stiglmair-Herb, M. T., von Hegel, G., and Wiesner, H.,** Abomasal cryptosporidiosis in mountain gazelles, *Vet. Rec.*, 121, 379, 1987.
659. **Powell, H. S., Holscher, M. A., and Heath, J. E.,** Bovine cryptosporidiosis (a case report), *Vet. Med. Small Anim. Clin.*, 71, 205, 1976.
660. **Prado, V., Brinck, P. M., and Martinez, T. M. J.,** Enteritis por Cryptosporidium en un pacinte leucémico, *Rev. Chil. Pediatr.*, 56, 251, 1985.
661. **Proctor, S. J. and Kemp, R. L.,** *Cryptosporidium anserinum* sp. n. (Sporozoa) in a domestic goose *Anser. anser* L. from Iowa, *J. Protozool.*, 21, 664, 1974.
662. **Qadripur, V. S. A. and Klose, L.,** Cryptosporidium enteritidis Verseuchungsausmaß in Südniedersachsen Untersuchung am Beispiel Urtikaria, *Dermatol. Monatschr.*, 171, 438, 1985.
663. **Radu, S. and Dan, S.,** Identificarea cryptosporidiilor la puii broiler de curcă si de găină în România, *Rev. Crest. Anim.*, No. 6, 55, 1985.
664. **Rahaman, A. S. M. H., Sanyal, S. C., Al-Mahmud, K. A., Sobhan, A., Hossain, K. S., and Anderson, B. C.,** Cryptosporidiosis in calves and their handlers in Bangladesh, *Lancet*, 2, 221, 1984.
664a. **Rahman, M., Shahid, N. S., Rahman, H., Sack, D. A., Rahman, N., and Hossain, S.,** Cryptosporidiosis: a cause of diarrhea in Bangladesh, *Am. J. Trop. Med. Hyg.*, 42, 127, 1990.

665. **Ramsden, K. and Freeth, M.**, Cryptosporidial infection presenting as an acute appendicitis, *Histopathology*, 14, 209, 1989.

666. **Ranck, F. M., Jr. and Hoerr, F. J.**, Cryptosporidia in the respiratory tract of turkeys, *Avian Dis.*, 31, 389, 1987.

667. **Randall, C. J.**, Cryptosporidiosis of the Bursa of Fabricius and trachea in broilers, *Avian Pathol.*, 11, 95, 1982.

668. **Randall, C. J.**, Renal and nasal cryptosporidiosis in a junglefowl (*Gallus sonneratii*), *Vet. Rec.*, 119, 130, 1986.

669. **Randall, C. J.**, Conjunctivitis in pheasants associated with cryptosporidial infection, *Vet. Rec.*, 118, 211, 1986.

670. **Ratnam, S., Paddock, J. McDonald, E., Whitty, D., Jong, M., and Cooper, R.**, Occurrence of *Cryptosporidium* oocysts in fecal samples submitted for routine microbiological examination, *J. Clin. Microbiol.*, 22, 402, 1985.

671. **Reduker, D. W., Speer, C. A., and Blixt, J. A.**, Ultrastructure of *Cryptosporidium parvum* oocysts and excysting sporozoites as revealed by high resolution scanning electron microscopy, *J. Protozool.*, 32, 708, 1985.

672. **Reduker, D. W., Speer, C. A., and Blixt, J. A.**, Ultrastructural changes in the oocyst wall during excystation of *Cryptosporidium parvum* (Apicomplexa; Eucoccidiorida), *Can. J. Zool.*, 63, 1892, 1985.

673. **Reduker, D. W. and Speer, C. A.**, Factors influencing excystation in *Cryptosporidium* oocysts from cattle, *J. Parasitol.*, 71, 112, 1985.

674. **Reese, N. C., Current, W. L., Ernst, J. V., and Bailey, W. S.**, Cryptosporidiosis of man and calf: a case report and results of experimental infections in mice and rats, *Am. J. Trop. Med. Hyg.*, 31, 226, 1982.

675. **Rehg, J. E., Lawton, G. W., and Pakes, S. P.**, *Cryptosporidium cuniculus* in the rabbit (*Oryctolagus cuniculus*), *Lab. Anim. Sci.*, 29, 656, 1979.

676. **Rehg, J. E., Hancock, M. L., and Woodmansee, D. B.**, Characterization of cyclophosphamide-rat model of cryptosporidiosis, *Infect. Immun.*, 55, 2669, 1987.

677. **Rehg, J. E., Gigliotti, F. and Stokes, D. C.**, Cryptosporidiosis in ferrets, *Lab. Anim. Sci.*, 38, 155, 1988.

678. **Rehg, J., Hancock, M. L., and Woodmansee, D. B.**, Characterization of a dexamethasone-treated rat model of cryptosporidial infection, *J. Infect. Dis.*, 158, 1406, 1988.

679. **Rehg, J. E., Hancock, M. L., Woodmansee, D. B.**, Anticryptosporidial activity of sulfadimethoxine, *Antimicrob. Agents Chemother.*, 32, 1907, 1988.

679a. **Reif, J. S., Wimmer, L., Smith, J. A. Dargatz, D. A., and Cheney, J. M.**, Human cryptosporidiosis associated with an epizootic in calves, *Am. J. Public Health*, 79, 1528, 1989.

680. **Reinemeyer, C. R., Kline, R. C., and Stauffer, G. D.**, Absence of cryptosporidium oocysts in faeces of neonatal foals, *Equine Vet. J.*, 16, 217, 1984.

681. **Reinthaler, F. F., Hermentin, K., Mascher, F., Klem, G., and Sixl, W.**, Cryptosporidiosis in Ogun State, south-west Nigeria, *Trop. Med. Parasitol.*, 38, 51, 1987.

682. **Reinthaler, F. F., Link, G., Klem, G., Mascher, F., Sixl, W.**, Cryptosporidiosis in children with diarrhoea from slum areas in San Salvador, *Ann. Trop. Med. Parasitol.*, 82, 209, 1988.

682a. **René, E., Marche, C., Regnier, B., Saimot, A. G., Vilde, J. L., Perrone, C., Michon, C., Wolf, M., Chevalier, Th., Vallot, Th., Brun-Vesinet, F., Pangon, B., Deluol, A. M., Camus, F., Roze, C., Pignon, J. P., Mignon, M., and Bonfils, S.**, Intestinal infections in patients with acquired immunodeficiency syndrome: a prospective study in 132 patients, *Dig. Dis. Sci.*, 34, 773, 1989.

683. **Reynolds, D. J., Hall, G. A., Debney, T. G., and Parsons, K. R.**, Pathology of natural rotavirus infection in clinically normal calves, *Res. Vet. Sci.*, 38, 264, 1985.

684. **Reynolds, D. J., Morgan, J. H., Chanter, N., Jones, P. W., Bridger, J. C., Debney, T. G., and Bunch, K. J.**, Microbiology of calf diarrhoea in southern Britain, *Vet. Rec.*, 119, 34, 1986.

685. **Ribeiro, C. D. and Palmer, S. R.**, Family outbreak of cryptosporidiosis, *Br. Med. J.*, 292, 377, 1986.

686. **Rice, V. A., Andrews, F. N., Warwick, J. E., and Legates, J. E.**, *Breeding and Improvement of Farm Animals*, 6th ed., McGraw-Hill, New York, 1967, 1.

687. **Ridgeway, H. F. and Olson, B. H.**, Chlorine resistance patterns of bacteria from two drinking water distribution systems, *Appl. Environ. Microbiol.*, 44, 972, 1982.

688. **Riggs, J. L., Dupuis, K. W., Nakamura, K., and Spath, D. P.**, Detection of *Giardia lamblia* by immunofluorescence, *Appl. Environ. Microbiol.*, 45, 698, 1983.

689. **Riggs, M. W. and Perryman, L. E.**, Infectivity and neutralization of *Cryptosporidium parvum* sporozoites, *Infect. Immun.*, 55, 2081, 1987.

690. **Riggs, M. W., McGuire, T. C., Mason, P. H., and Perryman, L. E.**, Neutralization-sensitive epitopes are exposed on the surface of infectious *Cryptosporidium parvum* sporozoites, *J. Immunol.*, 143, 1340, 1989.

690a. **Riggs, M. W.**, Unpublished observations.

691. **Rio, B., Le Tourneau, A., Sobahni, I., Audouin, J., Bouvet, A., Diebold, J., and Zittoun, R.**, Cryptosporidiose au cours d'une greffe de moelle osseuse allogónique, *Presse Med.*, 15, 1414, 1986.

692. **Ritter, G. D., Ley, D. H., Levy, M., Guy, J., and Barnes, H. J.**, Intestinal cryptosporidiosis and reovirus isolation from Bobwhite quail (*Colinus virginianus*) with enteritis, *Avian Dis.*, 30, 603, 1985.

692a. **Rivero, D., Mendez, M., and Zuniga, W.**, Presencia de Cryptosporidium sp en terneros diarreicos de la Provincia de Holguín, *Rev. Cubana Cienc. Vet.*, 19, 317, 1989.

692b. **Roberts, W. G., Green, P. H., Ma, J., Carr, M., and Ginsberg, A. M.,** Prevalence of cryptosporidiosis in patients undergoing endoscopy: evidence for an asymptomatic carrier state, *Am. J. Med.*, 87, 537, 1989.

693. **Robinson, M., Hart, C. A., Baxby, D., Battin, M., Suliman, G. I., El Seed, A. M., and Coulters, J. B. S.,** *Cryptosporidium* as a cause of gastro-enteritis in Sudanese children, *Ann. Trop. Paediatr.*, 6, 155, 1986.

693a. **Rojo-Vázquez, F. A., Gass, A., Izquierdo, M., and Ortiz Menéndez, J. C.,** Estudios sobre la criptosporidiosis de los pequeños ruminants (ovinos y caprinos) en España, *Med. Vet.*, 4, 263, 1987.

694. **Rolston, K. V. I. and Fainstein, V.,** Cryptosporidiosis, *Eur. J. Clin. Microbiol.*, 5, 135, 1986.

694a. **Rolston, K. V. I., Fainstein, V., and Body, G. P.,** Intestinal cryptosporidiosis treated with eflornithine: a prospective study among patients with AIDS, *J. AIDS*, 2, 426, 1989.

695. **Roncoroni, A. J., Gomez, M. A., Mera, J., Cagnoni, P., and Michel, M. D.,** *Cryptosporidium* infection in renal transplacent patients, *J. Infect. Dis.*, 160, 559, 1989.

696. **Rose, J. B., Cifrino, A., Madore, M. S., Gerba, C. P., Sterling, C. R., and Arrowood, M. J.,** Detection of *Cryptosporidium* from wastewater and freshwater environments, *Am. Water Sci. Technol.*, 18, 233, 1986.

697. **Rose, J. B., Madore, M. S., Riggs, J. L., and Gerba, C. P.,** Detection of *Cryptosporidium* and *Giardia* in environmental waters, in *Proc. Water Quality Technology Conf.* (Portland, OR, November 16 to 20, 1986), American Water Works Association, Denver, 1987, 417.

698. **Rose, J. B., Darbin, H., and Gerba, C. P.,** Correlations of the protozoa, *Cryptosporidium* and *Giardia*, with water quality variables in a watershed, *Water Sci. Technol.*, 20, 271, 1988.

699. **Rose, J. B.,** Occurrence and significance of Cryptosporidium in water, *J. Am. Water Works Assoc.*, 80, 53, 1988.

699a. **Rose, J. B., Kayed, D., Madore, M. S., Gerba, C. P., Arrowood, M. J., Sterling, C. R., and Riggs, J. L.,** Methods for the recovery of *Giardia* and *Cryptosporidium* from environmental waters and their comparative occurrence, in *Advances in Giardia Research*, Wallis, P. M. and Hammond, B. R., Eds., University of Calgary Press, Calgary, 1988, 205.

699b. **Rose, J. B., Landeen, L. K., Riley, K. R., and Gerba, C. P.,** Evaluation of immunofluorescence techniques for detection of *Cryptosporidium* oocysts and *Giardia* cysts from environmental samples, *Appl. Environ. Microbiol.*, 55, 3189, 1989.

700. **Rose, M. E., Wakelin, D., Joysey, H. S., and Hesketh, P.,** Immunity to coccidiosis: T-cell control of infection with *Eimeria vermiformis* in mice does not require cooperation with inflammatory cells, *Parasite Immunol.*, 11, 231, 1989.

700a. **Rossanigo, E. C., Gialletti, L., Grelloni, V., Fioroni, A., and Rivero, V. B.,** Diagnosi di criptosporidiosi in alcuni allevamenti dell'Italia Centrale, *Riv. Zoot. Vet.*, 15, 9, 1987.

700b. **Rossi, P., Pozio, E., and Besse, M. G.,** Cryopreservation of *Cryptosporidium* sp. oocysts, *Trans. R. Soc. Trop. Med. Hyg.*, 84, 68, 1990.

700c. **Rossi, P., Pozio, E., Besse, M. G., Gomez Morales, M. A., and La Rosa, G.,** Experimental cryptosporidiosis in hamsters, *J. Clin. Microbiol.*, 28, 356, 1990.

701. **Roy, C., Engelbrecht, R. S., and Chian, E. S. K.,** Comparative inactivation of six enteroviruses by ozone, *J. Am. Water Works Assoc.*, 74, 660, 1982.

702. **Rózsa, J., Matyi, A., and Fíldes, J.,** Adatok a borjak cryptosporidiosisához, *Magy. Allatorv. Lapja*, 41, 455, 1986.

703. **Rubin, A. J., Engel, J. P., and Sproul, O. J.,** Disinfection of amoebic cysts in water with free chlorine, *J. Water Pollut. Control Fed.*, 55, 1174, 1983.

704. **Rush, B. A., Chapman, P. A., and Ineson, R. W.,** Cryptosporidium and drinking water, *Lancet*, 2, 632, 1987.

705. **Rusnak, J., Hadfield, T. L., Rhodes, M. M. and Gaines, J. K.,** Detection of *Cryptosporidium* oocysts in human fecal specimens by an indirect immunofluorescence assay with monoclonal antibody, *J. Clin. Microbiol.*, 27, 1135, 1989.

706. **Russell, R. G., Rosenkranz, S. L., Lee, L. A., Howard, H., DiGiacomo, R. F., Bronsdon, M. A., Blakley, G. A., Tsai, C-C., and Morton, W. R.,** Epidemiology and etiology of diarrhea in colony-born *Macaca nemestrina*, *Lab. Anim. Sci.*, 37, 309, 1987.

707. **Ryan, M. J., Sundberg, J. P., Sauerschell, R. J., and Todd, K. S.,** *Cryptosporidium* in a wild cottontail rabbit (*Sylvilagus floridanus*), *J. Wildl. Dis.*, 22, 267, 1986.

708. **Sáez-Llorens, X., Odio, C. M., Umaña, M. A., and Morales, M. V.,** Spiramycin *vs.* placebo for treatment of acute diarrhea caused by *Cryptosporidium*, *Pediatr. Infect. Dis. J.*, 8, 136, 1989.

709. **Sáez-Llorens, X.,** Spiramycin for treatment of *Cryptosporidium* enteritis, *J. Infect. Dis.*, 160, 342, 1989.

710. **Sallon, S., Deckelbaum, R. J., Schmid, I. I., Harlap, S., Baras, M., and Spira, D. T.,** *Cryptosporidium*, malnutrition, and chronic diarrhea in children, *Am. J. Dis. Child.*, 142, 312, 1988.

711. **Sánchez, P. A., Blandino, T., Alonso, M., and García, F. C.,** *Cryptosporidium* sp.: primer hallazgo en humano en Cuba, *Rev. Salud Anim.*, 8, 105, 1986.

712. **Sanford, S. A. and Josephson, G. K. A.,** Bovine crytosporidiosis: clinical and pathological findings in forty-two scouring neonatal calves, *Can. Vet. J.*, 23, 343, 1982.

713. **Sanford, S. E.,** Porcine neonatal coccidiosis: clinical, pathological, epidemiological, and diagnostic features, *Calif. Vet.*, 37, 26, 1983.

714. **Sanford, S. E.**, Enteric cryptosporidial infection in pigs: 184 cases (1981-1985), *J. Am. Vet. Med. Assoc.*, 190, 695, 1987.

715. **Sanjurjo, E., del Valle, M., Guevara, A., and Gálvez, M. D.**, Criptosporidiosis: reporte de un caso, *Rev. Cubana Med. Trop.*, 40, 24, 1988.

716. **Santa Cruz, G., Frau, G., Montaldo, C., and Piro, S.**, Criptosporidiosi: studio ultrastrutturale in un pazients affetto da AIDS, *Rass. Med. Sarda*, 90, 487, 1987.

717. **Saraswathi, K., Pandit, D. V., Deodhar L. P., and Bichile, L. S.**, Prevalence of cryptosporidia in patients with diarrhoea in Bombay, *Indian J. Med. Res.*, 87, 221, 1988.

718. **Sarfati, C., Derouin, F., and Beauvais, B.**, Cryptosporidies, *Isopora belli et Entamoeba histolytica*, *Bull. Soc. Fr. Parasitol.*, 5, 49, 1982.

719. **Sauch, J. F.**, Use of immunofluorescence and phase-contrast microscopy for detection and identification of *Giardia* cysts in water samples, *Appl. Environ. Microbiol.*, 50, 1434, 1985.

720. **Saxon, A. and Weinstein, W.**, Oral administration of bovine colostrum anti-cryptosporidia antibody fails to alter the course of human cryptosporidiosis, *J. Parasitol.*, 73, 413, 1987.

721. **Scaglia, M., Senaldi, G., Di Perri, G., and Mino, L.**, Unusual low-grade cryptosporidial enteritis in AIDS: a case report, *Infection*, 14, 43, 1986.

722. **Schmidt, U. and Nienhoff, H.**, Kryptosporidiose biem Schwein, *Dtsch. Tieraerztl. Wochenschr.*, 89, 437, 1982.

723. **Schmitz, J. A. and Smith, D. H.**, *Cryptosporidium* infection in a calf, *J. Am. Vet. Med. Assoc.*, 167, 731, 1975.

724. **Schneiderman, D. J., Cello, J. P., and Laing, F. C.**, Papillary stenosis and sclerosing cholangitis in the acquired immunodeficiency syndrome, *Ann. Intern. Med.*, 106, 546, 1987.

725. **Schulz, W.**, Die bovine Kryptosporidiose — Nachweis und Bedeutung, *Monatsh. Veterinaermed.*, 41, 330, 1986.

726. **Scott, C. D.**, Screening faecal smears for *Cryptosporidium* and *Isospora belli* using a modified Ziehl-Neelsen stain, *Aust. J. Med. Lab. Sci.*, 9, 80, 1988.

726a. **Sehgal, R., Mahajan, R. C., Thapa, B. R., and Gangly, N. K.**, Cryptosporidium causing chronic diarrhoea, *Indian J. Pediatr.*, 56, 129, 1989.

727. **Senaud, J., Mehlhorn, H., Heydorn, A. O., and Matuschka, F. R.**, Étude ultrastructurale des kystes de *Besnoitia besnoiti* chez le boeuf, *Protistologica*, 22, 287, 1986.

727a. **Sewankambo, N., Mugerwa, R. D., Goodgame, R., Carswell, J. W., Moody, A., Lloyd, G., and Lucas, S. B.**, Enteropathic AIDS in Uganda. An endoscopic, histological and microbiological study, *AIDS*, 1, 9, 1987.

728. **Shahid, N. S., Rahman, A. S. M. H., Anderson, B. C., Mata, L. J., and Sanyal, S. C.**, Cryptosporidiosis in Bangladesh, *Br. Med. J.*, 290, 114, 1985.

729. **Shahid, N. S., Rahman, A. S. M. H., and Sanyal, S. C.**, Cryptosporidium as a pathogen for diarrhoea in Bangladesh, *Trop. Geogr. Med.*, 39, 265, 1987.

730. **Sharp, D. G. and Leong, J.**, Inactivation of poliovirus I (Brunhilde) single particles by chlorine in water, *Appl. Environ. Microbiol.*, 40, 381, 1980.

731. **Shepherd, R. C., Reed, C. L., and Sinha, G. P.**, Shedding of oocysts of *Cryptosporidium* in immunocompetent patients, *J. Clin. Pathol.*, 41, 1104, 1988.

732. **Shepherd, R. C., Sinha, G. P., Reed, C. L., and Russell, F. E.**, Cryptosporidiosis in the west of Scotland, *Scot. Med. J.*, 33, 365, 1988.

733. **Shepherd, R. C., Smail, P.J., and Sinha, G. P.**, Reactive arthritis complicating cryptosporidial infection, *Arch. Dis. Child*, 64, 743, 1989.

734. **Sherwood, D., Angus, K. W., Snodgrass, D. R., and Tzipori, S.**, Experimental cryptosporidiosis in laboratory mice, *Infect. Immun.*, 38, 471, 1982.

735. **Sidorenko, N. V.**, Diagnosis of cryptosporidiosis: microscopic detection of oocysts, *Veterinarya, (Moscow)*, 2, 48, 1988 (in Russian).

736. **Simpson, J. W., Burnie, A. G., Miles, R. S., Scott, J. L., and Lindsay, D. I.**, Prevalence of giardia and cryptosporidium infection in dogs in Edinburgh, *Vet. Rec.*, 123, 445, 1988.

736a. **Singh, B. K., Gulati, A. K., Prasad, K. N., and Singla, P. N.**, *Cryptosporidium* infection in children with acute diarrhoea. *J. Com. Dis.*, 20, 136, 1988.

737. **Sinśki, E., Szklarczyk, J., Oralewska, B., Świątkowska, E. and Socha J.**, *Cryptosporidium* sp. infection in children with symptoms of gastroenteritis, *Acta Parasitol. Polon.*, 33, 295, 1988.

738. **Sisk, D. B., Gosser, H. S., and Styer, E. L.**, Intestinal cryptosporidiosis in two pups, *J. Am. Vet. Med. Assoc.*, 184, 835, 1984.

739. **Skeels, M. R., Sokolow, R., Hubbard, C. V., and Foster, L. R.**, Screening for coinfection with *Cryptosporidium* and *Giardia* in Oregon public health clinic patients, *J. Public Health.*, 76, 270, 1986.

740. **Slavin, D.**, *Cryptosporidium meleagridis* (sp. nov.), *J. Comp. Pathol.*, 65, 262, 1955.

741. **Sloper, K. S., Dourmashkin, R. R., Bird, R. B., Slavin, G., and Webster, A. D. B.**, Chronic malabsorption due to cryptosporidiosis in a child with immunoglobulin deficiency, *Gut*, 23, 80, 1982.

742. **Smith, G. and van den Ende, J.**, Cryptosporidiosis among black children in hospital in South Africa, *J. Infect.*, 13, 25, 1986.

743. **Smith, H. V., Girwood, R. W. A., Patterson, W. J., Hardie, R., Green, L. A., Benton, C., Tulloch, W., Sharp, J. C. M., and Forbes, G. I.**, Waterborne outbreak of cryptosporidiosis, *Lancet*, 2, 1484, 1988.
743a. **Smith, H. V., McDiarmid, A., Smith, A. L., Hinson, A. R., and Gilmour, R. A.**, An analysis of staining methods for the detection of *Cryptosporidium* spp. oocysts in water-related samples, *Parasitology*, 99, 323, 1989.
743b. **Smith, H. V. and Rose, J. B.**, Waterborne cryptosporidiosis, *Parasitol. Today*, 6, 8, 1990.
744. **Smith, P. D., Lane, H. C., Gill, V. J., Manischewitz, J. F., Quinnan, G. V., Fauci, A. S., and Masur, H.**, Intestinal infections in patients with the acquired immunodeficiency syndrome (AIDS). Etiology and response to therapy, *Ann. Intern. Med.*, 108, 328, 1988.
745. **Snodgrass, D. R., Angus, K. W., Gray, E. W., Keir, W. A., and Clerihew, L. W.**, Cryptosporidia associated with rotavirus and an *Escherichia coli* in an outbreak of calf scour, *Vet. Rec.*, 106, 458, 1980.
746. **Snodgrass, D. R., Angus, K. W., and Gray, E. W.**, Experimental cryptosporidiosis in germfree lambs, *J. Comp. Pathol.*, 94, 141, 1984.
747. **Snodgrass, D. R.**, Evaluation of a combined rotavirus and enterotoxigenic *Escherichia coli* vaccine in cattle, *Vet. Rec.*, 119, 39, 1986.
748. **Snyder, D. B., Current, W. L., Russek-Cohen, E., Gorham, S. L., Mallinson, E. T., Marquardt, W. W., and Savage, P. K.**, Serologic incidence of *Cryptosporidium* in Delmarva broiler flocks, *Poult. Sci.*, 67, 730, 1988.
749. **Snyder, D. E.**, Indirect immunofluorescent detection of oocysts of *Cryptosporidium parvum* in the feces of naturally infected raccoons (*Procyon lotor*), *J. Parasitol.*, 74, 1050, 1988.
750. **Snyder, S. P., England, J. J., and McChesney, A. E.**, Cryptosporidiosis in immunodeficient Arabian foals, *Vet. Pathol.*, 15, 12, 1978.
751. **Soave, R., Danner, R. L., Honig, C. L., Ma, P., Hart, C. C., Nash, T., and Roberts, R. B.**, Cryptosporidiosis in homosexual men, *Ann. Intern. Med.*, 100, 504, 1984.
752. **Soave, R. and Ma, P.**, Cryptosporidiosis. Traveler's diarrhea in two families, *Arch. Intern. Med.*, 145, 70, 1985.
753. **Soave, R. and Armstrong, D.**, *Cryptosporidium* and cryptosporidiosis, *Rev. Infect. Dis.*, 8, 1012, 1986.
754. **Soave, R. and Johnson, W. D., Jr.**, *Cryptosporidium* and *Isospora belli* infections, *J. Infect. Dis.*, 157, 225, 1988.
755. **Soave, R., Ruiz, J., Garcia-Saucedo, V., Garrocho, C., and Kean, B. H.**, Cryptosporidiosis in a rural community in central Mexico, *J. Infect. Dis.*, 59, 1160, 1989.
756. **Soave, R.**, Clinical cryptosporidiosis in man, in *Proc. 1st Int. Workshop on Cryptosporidiosis*, Angus, K. W. and Blewett, D. A., Eds., September 7 to 8, 1988, Edinburgh, 1989, 19.
756a. **Soave, R.**, Unpublished data.
757. **Sobieh, M., Tacal, J. V., Wilcke, B. W., Lawrence, W., and El-Ahraf, A.**, Investigation of cryptosporidial infection in calves in San Bernardino County, California, *J. Am. Vet. Med. Assoc.*, 191, 816, 1987.
758. **Soph, M., Lávička, M., and Zajíček, D.**, Histologická diagnostika kryptosporidiózy u telat; ověření výskytu u telat porážených MPZ Praha-Písnice, *Veterinarstvi*, 34, 107, 1984.
759. **Sorenson, S. K., Riggs, J. L., Dileanis, P. D., and Suk, T. J.**, Isolation and detection of *Giardia* cysts from water using direct immunofluorescence, *Water Res. Bull.*, 22, 843, 1986.
760. **Soulé, C., Plateau, E., Perret, C., Chermette, R., and Feton, M. M.**, Observation de cryptosporidies chez le poulain. Note préliminaire, *Rec. Med. Vet.*, 719, 1983.
761. **Soulen, M. C., Fishman, E. K., Scatarige, J. C., Hutchins, D., and Zerhouni, E. A.**, Cryptosporidiosis of the gastric antrum: detection using CT, *Radiology*, 159, 705, 1986.
762. **Speer, C. A. and Reduker, D. W.**, Oocyst age and excystation of *Cryptosporidium parvum*, *Can. J. Zool.*, 64, 1254, 1986.
763. **Spillmann, S. K., Eckert, J., Merk, W., and Frey, R.**, Zum Vorkommen von Cryptosporidien bei Kälbern in der Schweiz, *Schweiz. Arch. Tierheilkd.*, 128, 111, 1986.
764. **Steeb, S., Hagedorn, H.-J., and Krone, J.-R.**, Kryptosporidiose bei immunkompetenten Patienten. Epidemiologie und Klinik, *Dtsch. Med. Wochenschr.*, 112, 990, 1987.
765. **Steele, A. D., Bothma, M. T., and Fripp, P. J.**, Routine testing for rotavirus and *Cryptosporidium* in children with gastro-enteritis, *S. Afr. Med. Trop.*, 74, 425, 1988.
766. **Stehr-Green, J. K., Juranek, D. J., McCaig, L., Remsen, H. M., and Rains, C. S.**, Chickenpox and infection with cryptosporidiosis, *Am. J. Dis. Child.*, 140, 1213, 1986.
767. **Stehr-Green, J. K., McCaig, L., Remsen, H. M., Rains, C. S., Fox, M., and Juranek, D. D.**, Shedding of oocysts in immunocompetent individuals infected with *Cryptosporidium*, *Am. J. Trop. Med. Hyg.*, 36, 338, 1987.
768. **Stein, E., Boch, J., Heine, J., and Henkel, G.**, Der Verlauf natürlicher *Cryptosporidium*-Infektionen in vier Rinderzuchtbetrieben, *Berl. Muench. Tieraerztl. Wochenschr.*, 96, 222, 1983.
769. **Stemmermann, G. N., Hayashi, T., Glober, G. A., Oishi, N., and Frankel, R. I.**, Cryptosporidiosis. Report of a fatal case complicated by disseminated toxoplasmosis, *Am. J. Med.*, 69, 637, 1980.
770. **Štěrba, F. and Šulcová, I.**, Laboratory diagnosis of cryptosporidiosis in calves, *Veterinarstvi*, 38, 110, 1988 (in Russian).

771. **Štěrba, F.**, Histopathological findings in emergency slaughtered calves with spontaneous cryptosporidial infections in the ileum, *Veterinarstvi*, 38, 84, 1988 (in Russian).
772. **Sterling, C. R., Seegar, K., and Sinclair, N. A.**, *Cryptosporidium* as a causative agent of traveler's diarrhea, *J. Infect. Dis.*, 153, 380, 1986.
773. **Sterling, C. R. and Arrowood, M. J.**, Detection of *Cryptosporidium* sp. infections using a direct immunofluorescent assay, *Pediatr. Infect. Dis.*, 5, 139, 1986.
774. **Sterling, C. R., Arrowood, M. J., Marshall, M. M., and Stetzenbach, L. D.**, The detection of *Giardia* and *Cryptosporidium* from water sources using monoclonal antibodies, *Proc. Am. Water Works Assoc. Water Quality Technical Conf.*, 15, 271, 1987.
774a. **Sterling, C. R.**, *Cryptosporidium*: the potential for waterborne transmission, *Proc. Am. Water Works Assoc. Water Quality Technical Conf.*, 15, 885, 1987.
775. **Sterling, C. R., Kutob, R. M., Gizinski, M. J., Verastegui, M., and Stetzenbach, L.**, *Giardia* detection using monoclonal antibodies recognizing determinants of *in vitro* derived cysts, in *Advances in Giardia Research*, Wallis, P. M. and Hammond, B. R., Eds., University of Calgary Press, Calgary, 1988, 219.
775a. **Stetzenbach, L. D., Arrowood, M. J., Marshall, M. M., and Sterling, C. R.**, Monoclonal antibody based immunofluorescent assay for *Giardia* and *Cryptosporidium* detection in water samples, *Water Sci. Technol.*, 20, 193, 1988.
776. **Stibbs, H. H. and Ongerth, J. E.**, Immunofluorescence detection of *Cryptosporidium* oocysts in fecal smears, *J. Clin. Microbiol.*, 24, 517, 1986.
777. **Stine, K. C., Harris, J.-A., Lindsey, N. J., and Cho, C. T.**, Spontaneous remission of cryptosporidiosis in a child with acute lymphocytic leukemia, *Clin. Pediatr.*, 24, 722, 1985.
778. **Stuart, J. and Burgess, M.**, Acute abdominal distension following *Cryptosporidium* infection, *Aust. Paediatr. J.*, 23, 257, 1987.
779. **Stürchler, D., DiGiacomo, R. F., and Rausch., L.**, Parasitic infections in Yakima Indians, *Ann. Trop. Med. Parasitol.*, 81, 291, 1987.
781. **Sundberg, J. P., Hill, D., and Ryan, M. J.**, Cryptosporidiosis in a gray squirrel, *J. Am. Vet. Med. Assoc.*, 181, 1420, 1982.
782. **Sundermann, C. A., Lindsay, D. S., and Blagburn, B. L.**, In vitro excystation of *Cryptosporidium baileyi* from chickens, *J. Protozool.*, 34, 28, 1987.
783. **Sundermann, C. A., Lindsay, D. S., and Blagburn, B. L.**, Evaluation of disinfectants for ability to kill avian *Cryptosporidium* oocysts, *Compan. Anim. Pract.*, 2, 36, 1987.
783a. **Sundermann, C. A.**, Unpublished data.
784. **Sung, R. T. M., Raberin, H., Decousus, H., Perpoint, B., Boucheron, S., Queneau, P.**, Double parasitose opportuniste a Cryptosporidium et Toxoplasma chez un homme presentant un S.I.D.A., *Bull. Soc. Fr. Parasitol.*, 1, 33, 1985.
785. **Suteu, E., Chirila, F., and Cosma, V.**, Valoarea comparativa a unor metode coproscopice in diagnosticul infestatiei cu cryptosporidiue, Seminarul: Reproductia, Patologia Reproductiei si Bolile Neonatale la Animalele de Ferma, Cluj-Napoca, Romania, 1984, 24.
786. **Szabo, J. R. and Moore, K.**, Cryptosporidiosis in a snake, *Vet. Med.*, 79, 96, 1984.
787. **Tacal, J. V., Jr., Sobieh, M., and El-Ahraf, A.**, Cryptosporidium in market pigs in southern California, USA, *Vet. Rec.*, 120, 615, 1987.
788. **Tarazona-Lafarga, R. and Domínguez-Carmonia, M.**, Criptosporidiosis e inmunodepresion: comportamiento de ratones inmunodeprimidos e inmunocompetentes freuta a un *Cryptosporidium* de origen ovino, *Rev. Iber. Parasitol.*, 48, 357, 1988.
789. **Tarwid, J. N., Cawthorn, R. J., and Riddell, C.**, Cryptosporidiosis in the respiratory tract of turkeys in Saskatchewan, *Avian Dis.*, 29, 528, 1985.
790. **Tassarao, R., Weitz, V. J. C., Mercado, P. T. M. R., Espinoza, M. N., and Atias, M. A.**, Cryptosporidiosis humana: characteristicas clinicas de 11 casos, *Bol. Hosp. S. J. Dios*, 33, 237, 1986.
791. **Taylor, J. P., Perdue, J. N., Dingley, D, Gustafson, T. L., Patterson, M., and Reed, L. A.**, Cryptosporidiosis outbreak in a day-care center, *Am. J. Dis. Child.*, 139, 1023, 1985.
792. **Taylor, D. N. and Echeverria, P.**, When does Cryptosporidium cause diarrhoea?, *Lancet*, 2, 320, 1986.
793. **Tchobanoglous, G. and Schroeder, E. D.**, *Water Quality, Characteristics, Modeling, Modification*, Addison-Wesley, Reading, MA, 1985.
794. **Terzolo, H. R., Lawson, G. H. K., Angus, K. W., and Snodgrass, D. R.**, Enteric campylobacter infection in gnotobiotic calves and lambs, *Res. Vet. Sci.*, 43, 72, 1987.
795. **Tham, V. L., Kniesberg, S., and Dixon, B. R.**, Cryptosporidiosis in quails, *Avian Pathol.*, 11, 619, 1982.
796. **Thomson, M. A., Benson, J. W. T., and Wright, P. A.**, Two year study of cryptosporidium infection, *Arch. Dis. Child.*, 62, 559, 1987.
797. **Tikasingh, E. S., Smith, S., Poon-King, T., and Bratt, D.**, Cryptosporidium in man in Trinidad and Tobago, *Trans. R. Soc. Trop. Med. Hyg.*, 80, 998, 1986.

798. **Tilley, M. and Upton, S. J.**, Electrophoretic characterization of *Cryptosporidium parvum* (KSU-1 isolate) (Apicomplexa: Cryptosporidiidae), *Can. J. Zool.*, in press.

798a. **Tilley, M., Upton, S. J., Blagburn, B. L., and Anderson, B. C.**, Identification of outer oocyst wall proteins of three species of *Cryptosporidium* (Apicomplexa: Cryptosporidiidae) by ^{125}I surface labeling, *Infect. Immun.*, 58, 252, 1990.

799. **Townsend, K. G. and Lance, D. M.**, Cryptosporidiosis in calves, *N.Z. Vet. J.*, 35, 216, 1987.

801. **Treem, W. R.**, The visualization of 'yeast cells' in stools: a clue to the presence of *Cryptosporidium*, *Am. J. Dis. Child.*, 141, 1037, 1987.

802. **Triffit, M. J.**, Observations on two new species of coccidia parasitic in snakes, *Protozoology*, 1, 19, 1925.

803. **Tsai, S. S., Ho, L. F., Chang, C. F., and Chu, R. M.**, Cryptosporidiosis in domestic birds, *Chin. J. Microbiol. Immunol.*, 16, 307, 1983.

804. **Tsaihong J. C., Tang, R. B., Wu, K.-K., Wu, T.-C., and Chung, C. H.**, Pediatric cryptosporidiosis: a report of 2 cases, *J. Formosan Med. Assoc.*, 87, 914, 1988 (in Chinese).

805. **Turnwald, G. H., Barta, O., Taylor, H. W., Kreeger, J., Coleman, S. U., and Pourciau, S. S.**, Cryptosporidiosis associated with immunosuppression attributable to distemper in a pup, *J. Am. Vet. Med. Assoc.*, 192, 79, 1988.

806. **Tyzzer, E. E.**, A sporozoan found in the peptic glands of the common mouse, *Proc. Soc. Exp. Biol. Med.*, 5, 12, 1907.

807. **Tyzzer, E. E.**, An extracellular coccidium, *Cryptosporidium muris* (gen. et sp. nov.), of the gastric glands of the common mouse, *J. Med. Res.*, 23, 487, 1910.

808. **Tyzzer, E. E.**, *Cryptosporidium parvum* (sp. nov.), a coccidium found in the small intestine of the common mouse, *Arch. Protistenkd.*, 26, 394, 1912.

809. **Tyzzer, E. E.**, Coccidiosis in gallinaceous birds, *Am. J. Hyg.*, 10, 269, 1929.

810. **Tzipori, S., Campbell, I., Sherwood, D., Snodgrass, D. R., and Whitelaw, A.**, An outbreak of calf diarrhoea attributed to cryptosporidial infection, *Vet. Rec.*, 107, 579, 1980.

811. **Tzipori, S., Angus, K. W., Campbell, I., and Gray, E. W.**, *Cryptosporidium*: evidence for a single-species genus, *Infect. Immun.*, 30, 884, 1980.

812. **Tzipori, S., Angus, K. W., Gray, E. W., and Campbell, I.**, Vomiting and diarrhea associated with cryptosporidial infection, *N. Engl. J. Med.*, 303, 818, 1980.

813. **Tzipori, S., McCartney, E., Lawson, G. H. K., and Rowland, A. C.**, Experimental infection of piglets with cryptosporidium, *Res. Vet. Sci.*, 31, 358, 1981.

814. **Tzipori, S., Angus, K. W., Campbell, I., and Sherwood, D.**, Diarrhea in young red deer associated with infection with *Cryptosporidium*, *J. Infect. Dis.*, 144, 170, 1981.

815. **Tzipori, S., Angus, K. W., Campbell, I., and Clerihew, L. W.**, Diarrhea due to *Cryptosporidium* infection in artificially reared lambs, *J. Clin. Microbiol.*, 14, 100, 1981.

816. **Tzipori, S., Angus, K. W., Gray, E. W., Campbell, I., and Allan, F.**, Diarrhea in lambs experimentally infected with *Cryptosporidium* isolated from calves, *Am. J. Vet. Res.*, 42, 1400, 1981.

817. **Tzipori, S., Sherwood, D., Angus, K. W., Campbell, I., and Gordon, M.**, Diarrhea in lambs: experimental infections with enterotoxigenic *Escherichia coli*, rotavirus, and *Cryptosporidium* sp., *Infect. Immun.*, 33, 401, 1981.

818. **Tzipori, S.**, The aetiology and diagnosis of calf diarrhoea, *Vet. Rec.*, 108, 510, 1981.

819. **Tzipori, S. and Campbell, I.**, Prevalence of *Cryptosporidium* antibodies in 10 animal species, *J. Clin. Microbiol.*, 14, 455, 1981.

820. **Tzipori, S. R., Campbell, I., and Angus, K. W.**, The therapeutic effect of 16 antimicrobial agents on *Cryptosporidium* infection in mice, *Aust. J. Exp. Biol. Med. Sci.*, 60, 187, 1982.

821. **Tzipori, S., Smith, M., Makin, T., and Halpin, C.**, Enterocolitis in piglets caused by *Cryptosporidium* sp. purified from calf faeces, *Vet. Parasitol.*, 11, 121, 1982.

822. **Tzipori, S., Larsen, J., Smith, M., and Leufl, R.**, Diarrhoea in goat kids attributed to cryptosporidium infection, *Vet. Rec.*, 111, 35, 1982.

823. **Tzipori, S., Smith, M., Birch, C., Barnes, G., and Bishop, R.**, Cryptosporidiosis in hospital patients with gastroentertitis, *Am. J. Trop. Med. Hyg.*, 32, 931, 1983.

824. **Tzipori, S.**, Cryptosporidiosis in animals and humans, *Microbiol. Rev.*, 47, 84, 1983.

825. **Tzipori, S., Smith, M., Halpin, C., Angus, K. W., Sherwood, D., and Campbell, I.**, Experimental cryptosporidiosis in calves: clinical manifestations and pathological findings, *Vet. Rec.*, 112, 116, 1983.

827. **Tzipori, S.**, *Cryptosporidium:* notes on epidemiology and pathogenesis, *Parasitol. Today*, 1, 159, 1985.

828. **Tzipori, S.**, The relative importance of enteric pathogens affecting neonates of domestic animals, *Adv. Vet. Sci. Comp. Med.*, 29, 103, 1985.

829. **Tzipori, S.**, A comparative study on important pathogens causing diarrhoea in calves and piglets, in *Infectious Diarrhoea in the Young*, Tzipori, S., Ed. Excerpta Medica, Amsterdam, 1985, 371.

830. **Tzipori, S., Roberton, D., and Chapman, C.**, Remission of diarrheoa due to cryptosporidiosis in an immunodeficient child treated with hyperimmune bovine colostrum, *Br. Med. J.*, 293, 344, 1986.

831. **Tzipori, S.**, Cryptosporidium and routine parasitological diagnosis, *J. Infect. Dis.*, 156, 248, 1987.

832. **Tzipori, S.**, Cryptosporidiosis in childhood, *Aust. Paediatr. J.*, 23, 89, 1987.

833. **Tzipori, S., Roberton, D., Cooper, D. A., and White, L.**, Chronic cryptosporidial diarrhoea and hyperimmune cow colostrum, *Lancet*, 2, 344, 1987.

834. **Tzipori, S.**, Cryptosporidiosis in perspective, *Adv. Parasitol.*, 27, 63, 1988.

836. **Uga, S., Matsmura, T., Yoda, Y., Yatomi, K., and Kotaoka, N.**, The laboratory diagnosis of cryptosporidiosis (in Japanese), *Rinsho Kensa*, 32, 1155, 1988.

837. **Uga, S., Matsumura, T., Ishibashi, K., Yoda, Y., Yatomi, K., and Katoaoka, N.**, Cryptosporidiosis in dogs and cats in Hyogo Prefecture, Japan, *Jpn. J. Parasitol.*, 38, 139, 1989.

838. **Ungar, B. L. P., Soave, R., Fayer, R., and Nash, T. E.**, Enzyme immunoassay detection of immunoglobulin M and G antibodies to *Cryptosporidium* in immunocompetent and immunocompromised persons, *J. Infect. Dis.*, 153, 570, 1986.

839. **Ungar, B. L. P. and Nash, T. E.**, Quantification of specific antibody response to *Cryptosporidium* antigens by laser densitometry, *Infect. Immun.*, 53, 124, 1986.

840. **Ungar, B. L. P., Gilman, R. H., Lanata, C. F., and Perez-Schael, I.**, Seroepidemiology of *Cryptosporidium* infection in two Latin American populations, *J. Infect. Dis.*, 157, 551, 1988.

841. **Ungar, B. L. P., Mulligan, M., and Nutman, T. B.**, Serologic evidence of *Cryptosporidium* Infection in US volunteers before and during Peace Corps Service in Africa, *Arch. Intern. Med.*, 149, 894, 1989.

842. **Ungar, L. L. P., Ward, D. J., Fayer, R., and Quinn, C. A.**, Cessation of *Cryptosporidium*-associated diarrhea in an acquired immune deficiency syndrome patient after treatment with hyperimmune bovine colostrum, *Gastroenterology*, 98, 486, 1990.

842a. **Ungar, B. L. P., Burris, J. A., Quinn, C. A., and Finkelman, F. D.**, New mouse models for chronic *Cryptosporidium* infection in immunodeficient hosts, *Infect. Immun.*, 58, 961, 1990.

843. **Uni, S., Iseki, M., Maekawa, T., Moriya, K., and Takada, S.**, Ultrastructure of *Cryptosporidium muris* (strain RN 66) parasitizing the murine stomach, *Parasitol. Res.*, 74, 123, 1987.

843a. **U.S. Environmental Protection Agency**, national primary drinking water regulations; filtration and disinfection; turbidity; *Giardia lamblia*, viruses, *Legionella*, and heterotrophic bacteria, *Fed. Regist.*, 52, 42178, 1987.

844. **Upton, S. J. and Current, W. L.**, The species of *Cryptosporidium* (Apicomplexa: Cryptosporidiidae) infecting mammals, *J. Parasitol.*, 71, 625, 1985.

845. **Upton, S. J. and Barnard, S. M.**, Two new species of coccidia (Apicomplexa: Eimeriidae) from Madagascar gekkonids, *J. Protozool.*, 34, 452, 1987.

846. **Upton, S. J., Tilley, M. E., Marchin, G. L., and Fina, L. R.**, Efficacy of a pentaiodide resin disinfectant on *Cryptosporidium parvum* (Apicomplexa: Cryptosporidiidae) oocysts in vitro, *J. Parasitol.*, 74, 719, 1988.

847. **Upton, S. J. and McAllister, C. T.**, The coccidia (Apicomplexa: Eimeriidae) of Anura, with descriptions of four new species, *Can. J. Zool.*, 66, 1822, 1988.

848. **Upton, S. J., McAllister, C. T., Freed, P. S., and Barnard, S. M.**, *Cryptosporidium* spp. in wild and captive reptiles, *J. Wildl. Dis.*, 25, 20, 1989.

848a. **Upton, S. J. and Freed, P. S.**, Unpublished data.

849. **Van Winkle, T. J.**, Cryptosporidiosis in young artiodactyls, *J. Am. Vet. Med. Assoc.*, 187, 1170, 1985.

850. **Van Opdenbosch, E. and Wellemans, G.**, Aantonen van Cryptosporidium antistoffen met de indirekte immunofluorescentietest (druppelmethode) voorkomen van antistoffen bij verschillende diersoorten, *Vlaams Diergeneeskd. Tijdschr.*, 54, 49, 1985.

851. **Vandepitte, J., Robrechts, J., and Vanneste, St.**, Cryptosporidium causing severe enteritis in a Belgian immunocompetent patient, *Acta Clin. Belg.*, 40, 43, 1985.

852. **Veldhuyzen van Zanten, S. J. O., Lange, J. M. A., Sauerwein, H. P., Rijpstra, A. C., Laarman, J. J., Rietra, P. J. G. M., and Danner, S. A.**, Amprolium for coccidiosis in AIDS, *Lancet*, 2, 345, 1984.

854. **Vetterling, J. M., Jervis, H. R., Merrill, T. G., and Sprinz, H.**, *Cryptosporidium wrairi* sp. n. from the guinea pig *Cavia porcellus* with an emendation of the genus, *J. Protozool.*, 18, 243, 1971.

855. **Vetterling, J. M., Takeuchi, A., and Madden, P. A.**, Ultrastructure of *Cryptosporidium wrairi* from the guinea pig, *J. Protozool.*, 18, 248, 1971.

856. **Vickers, M.**, Cryptosporidiosis in New Zealand, *N. Z. Vet. J.*, 35, 14, 1987.

856a. **Vila, G., A., Crespo, M. D., Martinez, C., Borras, R., Brines, J., Garcia de Lomas, J., and Codoñer, P.**, Dolor abdominal en la infancia debido a parasitosis por *Cryptosporidium*, *Ann. Esp. Pediatr.*, 25, 111, 1986.

857. **Villacorta, I., Ares, E., Pérez del Molino, L., Bajo, T., Fandiño-Salorio, L.**, Criptosporidiosis experimental en ratones, *Rev. Iber. Parasitol.*, 48, 133, 1988.

857a. **Villalobos, A. N. M. and de Aluja, A. S.**, Evaluación de dos métodos de tinción de frotis fecales para el diagnóstico de la criptosporidiosis en becerros, *Vet. Mex.*, 17, 303, 1986.

857b. **Visconti, S., Canestri-Trotti, G., and Pampiglione, S.**, Infezione accidentale da *Cryptosporidium* sp. nell'uomo in laboratorio. Segnalazione di un nuovo caso, *Ann. 1st. Super Sanita.*, 22, 239, 1986.

857c. **Visconti, S. and Canestri-Trotti, G.**, Indagine sulla fauna protozoaria intestinale in agnelli delle province di Bologna e Ferrara, *Ann. 1st. Super Sanita.*, 22, 529, 1986.

858. **Vítovec, J.**, Enterokolitida při spontánnich kryptosporidiových infecích telat, *Vet. Med. (Prague)*, 57, 411, 1984.
859. **Vítovec, J.**, Variabilita lokalizace kryptosporidií ve střevech spontánně infikovaných telat, *Vet. Med. (Prague)*, 29, 201, 1984.
860. **Vítovec, J. and Koudela, B.**, Natural cryptosporidiosis in a piglet suffering from pseudomembranous typhlitis, *Folia Parasitol. (Prague)*, 34, 378, 1987.
861. **Vítovec, J. and Koudela, B.**, Location and pathogenicity of *Cryptosporidium parvum* in experimentally infected mice, *J. Vet. Med.*, B35, 515, 1988.
862. **Wages, D. P. and Ficken, M. D.**, Cryptosporidiosis and turkey viral hepatitis in turkeys, *Avian Dis.*, 33, 191, 1989.
863. **Waghorn, D. J. and Hinkins, R.**, Screening for cryptosporidial infection, *J. Clin. Pathol.*, 39, 1363, 1986.
864. **Waldman, E., Tzipori, S., and Forsyth, J. R. L.**, Separation of *Cryptosporidium* species oocysts from feces by using a percoll discontinuous density gradient, *J. Clin. Microbiol.*, 23, 199, 1986.
865. **Walters, I. N., Miller, N. M., Van den Ende, Dees, G. C. D., Taylor, L. A., Taynton, L. F., and Bennett, K. J. G.**, Outbreak of cryptosporidiosis among young children attending a day-care centre in Durban, *S. Afr. Med. J.*, 74, 496, 1988.
865a. **Waterman, S., Keene, W., Sterling, C., and Ruiz-Matus, C.**, Cryptosporidiosis in a group of travelers to Puerto Vallarta, Mexico, 1986, *Proc. Int. Conf. on International Travel Medicine*, Zurich, Switzerland, April 5 to 8, 1988, Steffen, R., Lobel, H. O., Haworth, J., and Bradley, D. J., Eds., 1989, 326.
866. **Weber, J. and Philip, S.**, Human cryptosporidiosis, *N. Engl. J. Med.*, 309, 1326, 1983.
867. **Weikel, C., Dixon, B., Conway, M., Guerrant, R., Innes, D. and Sando, J.**, Inhibition by cyclosporin A of *Cryptosporidium* infection in mice, *Clin. Res.*, 34, 537A, 1986.
868. **Weikel, C. A. and Guerrant, R. L.**, Detection of cryptosporidiosis in northeast Brazil, *Clin. Res.*, 32, A517, 1984.
869. **Weikel, C. S., Johnston, L., De Sousa, A., and Guerrant, R. L.**, Cryptosporidiosis in northeastern Brazil: association with sporadic diarrhea, *J. Infect. Dis.*, 151, 963, 1985.
870. **Weinstein, L., Edelstein, S. M., Madara, J. L., Falchuk, K. R., McManus, B. M., and Trier, J. S.**, Intestinal cryptosporidiosis complicated by disseminated cytomegalovirus infection, *Gastroenterology*, 81, 584, 1981.
871. **Weisburger, W. R., Hutcheon, D. F., Yardley, J. H., Roche, J. C., Hillis, W. D., and Charache, P.**, Cryptosporidiosis in an immunosuppressed renal-transplant recipient with IgA deficiency, *Am. J. Clin. Pathol.*, 72, 473, 1979.
872. **Weitz, J. C., Mercado, R., Tassara, R., Espinoza, M., Lorca, P., Herskovic, P., and Atias, A.**, Cryptosporidium en Chile, *Rev. Med. Chil.*, 113, 695, 1985.
873. **Weitz, J. C., Mercado, R., Tassara, R., Espinoza, M., Lorca, P., Herskovic, P., and Atias, A.**, Frecuencia de *Cryptosporidium* sp. en pacientes hospitalizados, con sintomas digestivos, *Rev. Chil. Pediatr.*, 57, 10, 1986.
874. **Weitz, J. C., Tassara, R., Munoz, P., Mercado, R., and Atias, A.**, Cryptosporidiosis del aparato respiratorio, *Rev. Med. Chile.*, 114, 691-692, 1986.
875. **Weitz, J. C., Tassara, R., Mercado, T. M. R., Santa María, F., Maida, M., Lorca, P., and Atías, A.**, Brote de cryptosporidiosis en un centro de recuperací6n nutricional, *Rev. Chil. Pediatr.*, 58, 50, 1987.
876. **Weitz, J. C., Tassara, R., and Mercado, R.**, Cryptosporidiosis in Chilean children, *Trans. R. Soc. Trop. Med. Hyg.*, 82, 355, 1988.
876a. **Weitz, J. C. and Renzo Tassara, V.**, Diagnostico de la cryptosporidiosis: estudio comparativo de las tinciones Ziehl-Neelsen y safranina, *Rev. Med. Chile*, 117, 899, 1989.
877. **Werner, H. and Janitschke, K.**, Bildtafeln für die Laboratoriumsdiagnose von Pneumocystis und Cryptosporidium, *Aerztl. Lab.*, 33, 20, 1987.
878. **Wernicki, A. and Rzedzicki, J.**, Wystpowanie wybranych drobnoustrojow enteropatogenych u cielat zdrowych oraz z objawami biegunki w wieku do trzeciego tygodnia zycia, *Med. Weter.*, 44, 137, 1988.
879. **White, W. L. and Picklo, J.**, Human cryptosporidiosis, *N. Engl. J. Med.*, 309, 1325, 1983.
880. **Wetzel, R.**, Ein neues Coccid (*Cryptosporidium vulpis* sp. nov.) aus dem Rotfuchs, *Arch. Tiereheille*, 74, 39, 1933.
881. **Whiteside, M. E., Barkin, J. S., May, R. G., Weiss, S. D., Fischl, M. A., and MacLeod, C. L.**, Enteric coccidiosis among patients with the acquired immunodeficiency syndrome, *Am. J. Trop. Med. Hyg.*, 33, 1065, 1984.
882. **Whittington, R. J. and Wilson, J. M.**, Cryptosporidiosis of the respiratory tract in a pheasant, *Aust. Vet. J.*, 62, 284, 1985.
883. **Wiata, I. T. and Lennon, D.**, Cryptosporidium infestation in hospitalised urban children, *N. Z. Med. J.*, 349, 1985.
884. **Wickramanayake, G. B., Rubin, A. J., and Sproul, O. J.**, Inactivation of *Naegleria* and *Giardia* cysts in water by ozonation, *J. Water Pollut. Control Fed.*, 56, 983, 1984.
885. **Wickramanayake, G. B., Rubin, A. J., and Sproul, O. J.**, Inactivation of *Giardia lamblia* cysts with ozone, *Appl. Environ. Microbiol.*, 48, 671, 1984.

886. **Wiedermann, B. L., Kaplan, S. L., and Marino, B.**, Prevalence and significance of cryptosporidiosis in children, *Pediatr. Infect. Dis.*, 4, 292, 1985.
887. **Willard, M. D., Sugarman, B., and Walker, R. D.**, Gastrointestinal zoonoses, *Vet. Clin. North Am.*, 17, 145, 1987.
888. **Williams, J. E., Ellis, D. S., Smith, M. D., and Daziel, R.**, Safe method for identifying cryptosporidium cysts in the faeces of patients with suspected AIDS, or those infected with other serious concomitant pathogens, *J. Clin. Pathol.*, 38, 1313, 1985.
889. **Williams, R. O.**, Measurement of class specific antibody against cryptosporidium in serum and faeces from experimentally infected calves, *Res. Vet. Sci.*, 43, 264, 1987.
890. **Wilson, D. W., Day, P. A., and Brummer, M. E. G.**, Diarrhea associated with *Cryptosporidium* spp. in juvenile macaques, *Vet. Pathol.*, 21, 447, 1984.
891. **Willson, P. J. and Acres, S. D.**, A comparison of dichromate solution floatation and fecal smears for diagnosis of cryptosporidiosis in calves, *Can. Vet. J.*, 23, 240, 1982.
892. **Wilson, R. B., Holscher, M. A., and Lyle, S. J.**, Cryptosporidiosis in a pup, *J. Am. Vet. Med. Assoc.*, 183, 1005, 1983.
893. **Wittenberg, D. F., Smith, E. G., Van Den Ende, J., and Becker, P. J.**, Cryptosporidium-associated diarrhoea in children, *Ann. Trop. Paediatr.*, 7, 113, 1987.
894. **Wittenberg, D. F., Miller, N. M., and Van den Ende, J.**, Spiramycin is not effective in treating Cryptosporidium diarrhea in infants: results of a double-blind randomized trial, *J. Infect. Dis.*, 159, 131, 1989.
895. **Wittner, M., Goldfarb, J., Vogl, S., and Tanowitz, H. B.**, Fatal cryptosporidiosis complicating Kaposis sarcoma in an immunocompromised man, *Am. J. Med. Sci.*, 287, 47, 1984.
896. **Wolfson, J. S., Hopkins, C. C., Weber, D. J., Richter, J. M., Waldron, M. A., and McCarthy, D. M.**, An association between Cryptosporidium and Giardia in stool, *N. Engl. J. Med.*, 310, 788, 1984.
897. **Wolfson, J. S., Richter, J. M., Waldron, M. A., Weber, D. J., McCarthy, D. M., and Hopkins, C. C.**, Cryptosporidiosis in immunocompetent patients, *N. Engl. J. Med.*, 312, 1278, 1985.
898. **Woodmansee, D. B.**, An *in vitro* study of sporulation in *Cryptosporidium* species, *J. Parasitol.*, 72, 348, 1986.
899. **Woodmansee, D. B., Powell, E. C., Pohlenz, J. F. L., and Moon, H. W.**, Factors affecting motility and morphology of *Cryptosporidium* sporozoites in vitro, *J. Protozool.*, 34, 295, 1987.
900. **Woodmansee, D. B.**, Studies of in vitro excystation of *Cryptosporidium parvum* from calves, *J. Protozool.*, 34, 398, 1987.
901. **Woodmansee, D. B., Pavlásek, I., Pohlenz, J. F. L., and Moon, H. W.**, Subclinical cryptosporidiosis of turkeys in Iowa, *J. Parasitol.*, 74, 898, 1988.
902. **Woolf, G. M., Townsend, M., and Guyatt, G.**, Treatment of cryptosporidiosis with spiramycin in AIDS, An "N of 1" trial, *J. Clin. Gastroenterol.*, 9, 632, 1987.
903. **World Health Organization Scientific Working Group**, Parasite-related diarrhoeas, *Bull. W.H.O.*, 58, 819, 1980.
904. **Wright, P. A., Harrison, J. M., and Byrom, I.**, Cryptosporidiosis, *Br. Med. J.*, 289, 1148, 1984.
905. **Wu, T.-C., Lo, C.-H., Ting, C.-W., Chang, Y.-R., Tang, R.-B., Wang, H.-C., and Hwang, B.**, Cryptosporidiosis in children (in Chinese), *J. Formosan Med. Assoc.*, 86, 884, 1987.
906. **Wyllie, A. S.**, Cryptosporidiosis, *Br. Med. J.*, 289, 1383, 1984.
907. **Yamini, B. and Raju, N. R.**, Gastroenteritis associated with a *Cryptosporidium* sp in a chinchilla, *J. Am. Vet. Med. Assoc.*, 189, 1158, 1986.
907a. **Yoshinaga, N., Enaga, N., Ichimaru, H., Muraoka, T., Shimohira, H., Uchikoshi, N. Yoshitake, I., and Minamikawa, R.**, Cryptosporidiosis in piglets, *J. Jpn. Vet. Med. Assoc.*, 41, 590, 1988.
908. **Yvore, P., Esnault, A., Naciri, M., Leclerc, C., Bind, J. L., Contrepois, M., Levieux, D., and Laporte, J.**, Enquête épidémiologique sur les diarrhées néonatales des chevreaux dans les élevages de Touraine, in *Les Maladies de la Chévre*, Niort, France, October 9 to 11, 1984, *Les Colloques de l'INRA*, 1984, 437.
908a. **Yvore, P. and Naciri, M.**, Halofuginone lactate in the treatment of cryptosporidiosis in ruminants, *Coccidia and Intestinal Coccidiomorphs*, Proc. 5th Int. Coccidiosis Conf., Tours, France, October 17 to 20, 1989, INRA, 1989, 493.
909. **Zajíček, D. and Dvořáčková, A.**, Diagnostika kryptosporidií při prujmových onemocněních telat, *Veterinarstvi*, 35, 121, 1985.
910. **Zajíček, Dvořáčková, A., and Haisl, K.**, Výskyt kokcidií rodu *Cryptosporidium* (*Cryptosporidiidae*) při průjmových onemocněních telat, *Vet. Med.*, 31, 201, 1986.
911. **Zakowski, P., Fligiel, S., Berlin, G. W., and Johnson, L., Jr.**, Disseminated mycobacterium avium-intracellulare infection in homosexual men dying of acquired immunodeficiency, *J. Am. Med. Assoc.*, 10, 2980, 1982.
912. **Zar, F., Geiseler, P. J., and Brown, V. A.**, Asymptomatic carriage of Cryptosporidium in the stool of a patient with acquired immunodeficiency syndrome, *J. Infect. Dis.*, 151, 195, 1985.
912a. **Zdero, M., Ponce de Leon, P., Boligno, B., and Nocito, I.**, Presencia de *Cryptosporidium* sp. en hecec diarreicas de una población infantil, *Rev. Argent. Microbiol.*, 21, 37, 1989.

913. **Zierdt, W. S.,** Concentration and identification of *Cryptosporidium* sp. by use of a parasite concentrator, *J. Clin. Microbiol.*, 20, 860, 1984.

914. **Zurita, J., Acosta, S., and Soriano, F.,** Prevalencia de ooquistes de *Cryptosporidium* en muestras fecales enviadas para estudio microbiológico de rutina en un hospital general de Madrid, *Rev. Clin. Esp.*, 181, 182, 1987.

INDEX

I

J

K

L

M

Q

R

S